An Introduction to Statistical Genetic Data

MW00528028

An Introduction to Statistical Genetic Data Analysis

Melinda C. Mills, Nicola Barban, and Felix C. Tropf

The MIT Press
Cambridge, Massachusetts
London, England

This book was set in Times Roman by Westchester Publishing Services, Danbury, CT. Printed and bound in the United States of America.

Library of Congress Cataloging-in-Publication Data

Names: Mills, Melinda, author. | Barban, Nicola, author. | Tropf, Felix C., 1984- author.
Title: An introduction to statistical genetic data analysis / Melinda C. Mills, Nicola Barban, and Felix Tropf.
Description: Cambridge, Massachusetts : The MIT Press, [2020] | Includes bibliographical references and index.
Identifiers: LCCN 2019022949 | ISBN 9780262538381 (paperback)
Subjects: MESH: Data Interpretation, Statistical | Genomics | Genome-Wide Association Study | Gene-Environment Interaction | Models, Theoretical
Classification: LCC QH438.4.S73 | NLM QU 460 | DDC 572.8/60727--dc23
LC record available at https://lccn.loc.gov/2019022949

10   9   8   7   6   5   4   3   2   1

Contents

Preface   xiii

I   Foundations   1

1   Introduction: Fundamental Concepts and the Human Genome   3
Objectives   3
1.1   Introduction   3
    1.1.1   Motivation and aim of this book   3
    1.1.2   Overview of topics covered in this book   6
    1.1.3   What are DNA, the genome, a gene, and a chromosome?   8
1.2   Mendel's laws, sexual reproduction, and genetic recombination   9
1.3   Genetic polymorphisms   12
    1.3.1   Alleles, single-nucleotide polymorphisms (SNPs), and minor allele frequency (MAF)   12
    1.3.2   Monogenic, polygenic, and omnigenic effects   13
1.4   From genes to protein and the central dogma of molecular biology   15
    1.4.1   From genes to protein: Genes, amino acids, nucleotides, and proteins   15
    1.4.2   The central dogma of molecular biology: Transcription and translation   18
1.5   Homozygous and heterozygous alleles, dominant and recessive traits   20
1.6   Heritability   22
    1.6.1   Defining heritability: Broad- and narrow-sense heritability   22
    1.6.2   Common misconceptions about heritability   23
    1.6.3   Twin, SNP, and GWAS heritability   24
    1.6.4   Missing and hidden heritability   27
1.7   Conclusion   28
Exercises   28
Further reading and resources   29
References   30

2   A Statistical Primer for Genetic Data Analysis   33
Objectives   33
2.1   Introduction   33
2.2   Basic statistical concepts   34

2.2.1   Mean, standard deviation, and variance   34
2.2.2   Covariance and the variance-covariance matrix   36
2.3   Statistical models   38
2.3.1   Regression models   38
2.3.2   The null and alternative hypothesis and significance thresholds   39
2.4   Correlation, causation, and multivariate causal models   40
2.4.1   Correlation versus causation   40
2.4.2   Multivariate causal models   42
2.5   Fixed-effects models, random-effects models, and mixed models   47
2.6   Replication of results and overfitting   48
2.7   Conclusion   49
Exercises   50
Further reading   52
Software for mixed-model analyses   52
Appendix   52
References   54

3   A Primer in Human Evolution   55
Objectives   55
3.1   Introduction   55
3.2   Human dispersal out of Africa   56
3.3   Population structure and stratification   58
3.3.1   Population structure, genetic admixture, and Principal
         Component Analysis (PCA)   58
3.3.2   Common misnomers of population structure: Ancestry is not race   59
3.3.3   Genetic scores cannot be transferred across ancestry groups   59
3.3.4   How genes mirror geography   61
3.4   Human evolution, selection, and adaptation   63
3.4.1   Evolution, fitness, and natural selection   63
3.4.2   Genetic drift   68
3.5   The Hardy–Weinberg equilibrium   69
3.5.1   Assumptions of the HWE   69
3.5.2   Understanding the notation of the HWE   70
3.6   Linkage disequilibrium and haplotype blocks   71
3.7   Conclusion   73
Exercises   73
Further reading and resources   74
References   74

4   Genome-Wide Association Studies   77
Objectives   77
4.1   Introduction and background   77
4.2   GWAS research design and meta-analysis   79
4.2.1   GWAS research design   79
4.2.2   Data analysis plan   81
4.2.3   Meta-analysis   82

4.3   Statistical inference, methods, and heterogeneity   83
        4.3.1   Nature of the phenotype   83
        4.3.2   *P*-values and Z-scores   83
        4.3.3   Correcting for multiple testing in a GWAS   84
        4.3.4   Manhattan plots   85
        4.3.5   Evaluating dichotomous versus quantitative traits   87
        4.3.6   Fixed-effects versus random-effects models   88
        4.3.7   Weighting, false discovery rate (FDR), and imputation   89
        4.3.8   Sources of heterogeneity   89
4.4   Quality control (QC) of genetic data   90
4.5   The NHGRI-EBI GWAS Catalog   91
        4.5.1   What is the NHGRI-EBI GWAS Catalog?   91
        4.5.2   A brief history of the GWAS   91
        4.5.3   Lack of diversity in GWASs   93
4.6   Conclusion and future directions   97
Exercises   98
Further reading   98
References   99

5   **Introduction to Polygenic Scores and Genetic Architecture   101**
     Objectives   101
5.1   Introduction   101
        5.1.1   What is a polygenic score?   105
        5.1.2   The origins of polygenic scores   105
5.2   Construction of polygenic scores   107
        5.2.1   Large sample sizes required in GWAS discovery   108
        5.2.2   Selection of SNPs to include   108
5.3   Validation and prediction of polygenic scores   108
        5.3.1   Independent target sample   109
        5.3.2   Similar ancestry in target sample   110
        5.3.3   Relatedness, population stratification, and differential bias   110
        5.3.4   Variance explained only by common genetic markers missing
                 rare variants   111
        5.3.5   Missing and hidden heritability in prediction of phenotypes from genetic
                 markers (SNPs)   111
        5.3.6   Trade-off between prediction and understanding biological mechanisms   112
5.4   Shared genetic architecture of phenotypes   113
        5.4.1   Predicting other phenotypes   113
        5.4.2   Phenotypic and genetic correlation   114
        5.4.3   Pleiotropy   115
        5.4.4   Multitrait analysis   119
5.5   Causal modeling with polygenic scores   119
        5.5.1   Genetic confounding   119
        5.5.2   Mendelian Randomization   120
        5.5.3   Controlling for confounders   120
        5.5.4   Gene-environment interaction and heterogeneity   122

5.6   Conclusion   123
       Exercises   124
       Further reading   124
       References   125

6   Gene-Environment Interplay   129
       Objectives   129
       6.1   Introduction: What is gene-environment (G×E) interplay?   129
       6.2   Defining the environment in G×E research   130
              6.2.1   Nature and scope of E: Multilevel, multidomain, and multitemporal   131
              6.2.2   Interdependence of environmental risk factors   132
       6.3   A brief history of G×E research   133
              6.3.1   Classic approaches   133
              6.3.2   Candidate gene cG×E approaches   134
              6.3.3   Genome-wide polygenic score G×E approaches   135
       6.4   Conceptual G×E models   136
              6.4.1   Diathesis-stress, vulnerability, or contextual triggering model   136
              6.4.2   Bioecological or social compensation model   137
              6.4.3   Differential susceptibility model   139
              6.4.4   Social control or social push model   140
              6.4.5   Research designs to study G×E   140
       6.5   Gene-environment correlation (rGE)   143
              6.5.1   Passive gene-environment correlation (rGE)   144
              6.5.2   Evocative (or reactive) rGE   145
              6.5.3   Active rGE   145
              6.5.4   Why are models of rGE important?   145
              6.5.5   Research designs to study rGE   146
       6.6   Conclusion and future directions   146
              6.6.1   Why haven't many G×Es been identified?   146
       Exercises   147
       Further reading   147
       References   147

II   Working with Genetic Data   151

7   Genetic Data and Analytical Challenges   153
       Objectives   153
       7.1   Introduction   153
       7.2   Genotyping and sequencing array   154
              7.2.1   Genotyping and sequencing technologies   154
              7.2.2   Linkage disequilibrium and imputation   155
              7.2.3   Limitations of genotyping arrays and next-generation sequencing   158
              7.2.4   Drop in costs per genome   159
       7.3   Overview of human genetic data for analysis   160
              7.3.1   Prominently used genetic data   161
              7.3.2   Sources that archive and distribute data   163
              7.3.3   Obtaining GWAS summary statistics   164

7.4   Different formats in genomics data   165
          7.4.1   Genomics data is big data   165
          7.4.2   PLINK software and genotype formats   166
          7.4.3   PLINK binary files   170
7.5   Genetic formats for imputed data   171
          7.5.1   PLINK 2.0   171
          7.5.2   Oxford file formats   172
          7.5.3   The variant call format (VCF)   174
7.6   Data used in this book   175
7.7   Data transfer, storage, size, and computing power   176
          7.7.1   Data storage   176
          7.7.2   Data sharing, transfer across borders, and cloud storage   177
          7.7.3   Size of data and computational power   178
7.8   Conclusion   179
Exercises   179
Further reading and resources   179
References   180

8   **Working with Genetic Data, Part I: Data Management, Descriptive Statistics,
      and Quality Control   183**
Objectives   183
8.1   Introduction: Working with genetic data   183
8.2   Getting started with PLINK   184
          8.2.1   The command line   184
          8.2.2   Calling PLINK and the PLINK command line   186
          8.2.3   Running scripts in terminal   188
          8.2.4   Opening PLINK files   189
          8.2.5   Recode binary files to create new readable dataset with .ped
                        and .map files   189
          8.2.6   Import data from other formats   191
8.3   Data management   193
          8.3.1   Select individuals and markers   193
          8.3.2   Merge different genetic files and attaching a phenotype   196
8.4   Descriptive statistics   199
          8.4.1   Allele frequency   199
          8.4.2   Missing values   200
8.5   Quality control of genetic data   202
          8.5.1   Per-individual QC   203
          8.5.2   Per-marker QC   206
          8.5.3   Genome-wide association meta-analysis QC   209
8.6   Conclusion   211
Exercises   214
Further reading and resources   214
References   214

9   Working with Genetic Data, Part II: Association Analysis, Population Stratification, and Genetic Relatedness   217
Objectives   217
9.1   Introduction   217
         9.1.1   Aim of this chapter   217
         9.1.2   Data and computer programs used in this chapter   218
9.2   Association analysis   218
9.3   Linkage disequilibrium   223
9.4   Population stratification   226
9.5   Genetic relatedness   236
9.6   Relatedness matrix and heritability with GCTA   238
9.7   Conclusion   240
Exercises   241
Further reading and resources   241
References   241

10   An Applied Guide to Creating and Validating Polygenic Scores   243
Objectives   243
10.1   Introduction   243
         10.1.1   Creating a polygenic score   243
         10.1.2   Data used in this chapter   244
10.2   How to construct a score with selected variants (monogenic)   245
10.3   Pruning and thresholding method   247
10.4   How to calculate a polygenic score using PRSice 2.0   251
10.5   Validating the PGS   260
10.6   LDpred: Accounting for LD in polygenic score calculations   267
         10.6.1   Introduction and three steps   267
10.7   Conclusion   272
Exercises   273
Further reading and resources   273
References   274

III   Applications and Advanced Topics   275

11   Polygenic Score and Gene-Environment Interaction (G×E) Applications   277
Objectives   277
11.1   Introduction   277
11.2   Polygenic score applications: (Cross-trait) prediction and confounding   278
         11.2.1   Out-of-sample prediction   278
         11.2.2   Cross-trait prediction and genetic covariation   288
         11.2.3   Genetic confounding   295
11.3   Gene-environment interaction   299
         11.3.1   Application: BMI × birth cohort   300
11.4   Challenges in gene-environment interaction research   308
11.5   Conclusion and future directions   310
Exercises   311

Further reading   311
References   311

12   Applying Genome-Wide Association Results   315
Objectives   315
12.1   Introduction   315
12.2   Plotting association results   316
          12.2.1   Manhattan plots   316
          12.1.2   Regional association plots   320
          12.1.3   Quantile-Quantile plots and the λ statistic   320
12.2   Estimating heritability from summary statistics   324
12.3   Estimating genetic correlations from summary statistics   328
12.4   MTAG: Multi-Trait Analysis of Genome-wide association summary statistics   333
12.5   Conclusion   336
Exercises   336
Further reading and resources   336
References   337

13   Mendelian Randomization and Instrumental Variables   339
Objectives   339
13.1   Introduction   339
13.2   Randomized control trials and causality   341
13.3   Mendelian Randomization   341
13.4   Instrumental variables and Mendelian Randomization   343
          13.4.1   The IV model in an MR framework   343
          13.4.2   Violation of statistical assumptions of the IV approach   347
13.5   Extensions of standard MR   349
          13.5.1   Using multiple markers as independent instruments   351
          13.5.2   Using polygenic scores as IVs   351
          13.5.3   Bidirectional MR analyses   352
13.6   Applications of MR   352
          13.6.1   Consequences of alcohol consumption   352
          13.6.2   Body mass index and mortality   353
          13.6.3   Causes of dementia and Alzheimer's disease   354
13.7   Conclusion   355
Exercises   355
Further reading   356
References   356

14   Ethical Issues in Genomics Research   359
Objectives   359
14.1   Introduction   359
14.2   Genetics is not destiny: Genetic determinism   361
          14.2.1   Variation in traits and ability to use individual PGSs as predictors   361
          14.2.2   Heritability and missing heritability   362
14.3   Clinical use of PGSs   363
          14.3.1   Genetics and family history   363
          14.3.2   Genetic scores for screening, intervention, and life planning   364

14.3.3 Pharmacogenetics   365
14.3.4 Public understanding of genetic information and information risks   366
14.4 Lack of diversity in genomics   367
14.4.1 Lack of diversity in GWASs   367
14.4.2 European ancestry bias related to PGS construction   367
14.5 Privacy, consent, legal issues, insurance, and General Data Protection Regulation   367
14.5.1 Privacy in the age of public genetics: Solving crimes and finding people   367
14.5.2 The changing nature of informed consent in genomic research   368
14.5.3 Insurance and genetics   369
14.5.4 GDPR and genetics   370
14.6 Conclusion and future directions   372
Further reading and resources   373
References   373

15   Conclusions and Future Directions   377
15.1 Summary and reflection   377
15.2 Future directions   377
References   380

Appendix 1: Software Used in This Book   381
A1.1 Introduction   381
A1.2 RStudio and R   381
A1.3 PLINK   382
A1.4 GCTA   382
A1.5 PRSice   382
A1.6 Python   383
A1.6.1 How to switch from Python 3 to Python 2   384
A1.6.2 Installing packages in Python   385
A1.7 Git   385
A1.8 LDpred   386
A1.9 LDSC   386
A1.10 MTAG   387
A1.11 Using Windows for this book   388
References   388

Appendix 2: Data Used in This Book   389
A2.1 Introduction   389
A2.2 Description of simulated data   389
A2.3 Health and Retirement Study   391
A2.4 Data used by chapter   395
References   397

Glossary   399
Notes   405
Index   409

Preface

Since the first sequencing of the human genome in 2003, human genetic research has undergone a revolution in the way it is understood and incorporated into research. With advances in computing power, availability of data, and new techniques, this area of research has disrupted many conventions of how we think about disease and behavior. Genetics has now stretched beyond biology, epidemiology, and the medical sciences to psychology, psychiatry, statistics, and social sciences such as demography, sociology, and economics. For the first time in history is it now possible to integrate large-scale molecular genetic information into research across a broad range of topics. Many think that statistical genetic data analysis is relevant only for extensive research teams and highly specialized scientists within the disciplinary boundaries of quantitative genetic or statistical research. The aim of this book is to show applied researchers from various disciplinary backgrounds how to understand, apply, and work with genetic data for your own research topics. The knowledge from this book will allow you to properly and responsibly understand and interpret the data and use it as a blueprint to apply to your own data and research. We also hope that by making this type of data analysis more accessible, we work toward the loftier goal of diversification of both the people that are studied as subjects in human genetics but also the researchers themselves and topics that are covered.

Who should read this book

This book is for current and aspiring students and researchers from any empirically oriented medical, biological, behavioral, or social science discipline who would like to understand the main concepts of human statistical genetic data analysis but also practitioners looking for solutions to enter and undertake this research. Readers are given a blueprint of applied molecular genetic data analysis with hands-on computer exercises and a focus on substantive interpretation. The book operates as a launching pad to enter this seemingly daunting field. It is an introductory book, written for those who do not have a strong background in molecular biology, human genetics, or statistical genetics but would like to integrate genetic data into their research. Elementary knowledge of statistical methods at the first level of a statistics or biostatistics course is optimal. You will get the most out of this book if you first engage in some background tutorial work in R and RStudio and, for some more advanced applications, a basic familiarity with Python (see appendix 1). We also made a concerted effort to focus on the basic terminology and practical aspects of statistical genetic data analysis rather than the math, statistics, and biology behind it. Readers can refer to the further reading section and references in each particular chapter to learn more.

Why we wrote this book

There are many excellent introductory books on human genetics or statistical population genetics, yet most of them are written for advanced graduate and PhD students in biology or genetics. This is in stark contrast with how genetic data are being used in research today, which is increasingly across multiple scientific and research domains. Our journey first started in 2008, now over 10 years ago, when Melinda applied for a grant from the Dutch Science Foundation (NWO) to integrate genetics into demographic research. When granted in 2010, she hired Nicola as a postdoctoral researcher and Felix joined as a PhD student. Funding for Melinda's European Research Council (SOCIOGENOME) grant in 2014 allowed us to continue working together in Oxford. From entering this field we realized that there was an enormous chasm between instructional material and advanced articles and books. We were grateful to receive an additional grant from the National Centre for Research Methods, Economic and Social Research Council, United Kingdom Research and Innovation, to support a summer school and to write this textbook. This textbook reflects the interdisciplinary wave of research and complex skills many of us now require.

Navigating this book

The book is divided into three interdependent parts. Part I provides the foundations including (1) fundamental concepts and the human genome, (2) a statistical primer, (3) a primer in human evolution, (4) genome-wide association studies (GWASs), (5) polygenic scores and genetic architecture, and (6) gene-environment interplay.

Part II delves into the practicalities of how to work with genetic data, including (7) genetic data and analytical challenges; (8) data management, descriptive statistics, and quality control; (9) association analysis, population stratification, and genetic relatedness; and (10) creating and validating polygenic scores.

Part III covers applications and advanced topics, namely (11) polygenic score and gene-environment interaction applications, (12) applying GWAS results, (13) Mendelian Randomization and instrumental variables, (14) ethical issues, and (15) conclusions and future directions.

We also included two appendixes, concerning the software and data used in this book, plus a brief glossary. We all contributed throughout, with Melinda devising the form of the textbook and writing chapters 1–6, 14–15, and parts of chapters 7 and 13. Nicola took the lead on writing chapters 7–10 and 12. Felix led on chapters 11 and 13, with some input in chapters 2, 5, and 14. In the end, we all worked together to read, revise, and test the different chapters.

Online resources

All of the data and code used in this book is on the online companion website that can be found at http://www.intro-statistical-genetics.com. On that website we will also include an updated erratum, additional exercises and solutions and teaching material.

Conventions used in this book

Words in **bold** generally refer to a key concept or term that we define or can be found in the glossary. Text boxes are used to delve into some topics in more detail or to provide examples.

In the applied chapters, commands are shown in nonshaded boxes. For example:

```
./plink --bfile hapmap-ceu --recode --out hapmap-ceu
```

Output is shown in shaded boxes, such as:

```
./plink --bfile hapmap-ceu --recode --out hapmap-ceu
PLINK v1.90b6.7 64-bit (2 Dec 2018) www.cog-genomics.org
/plink/1.9/
(C) 2005-2018 Shaun Purcell, Christopher Chang GNU Gen-
eral Public License v3
Logging to hapmap-ceu.log.
Options in effect:
--bfile hapmap-ceu
--out hapmap-ceu
--recode
```

Acknowledgments

There are countless people who we would like to thank, but at the risk of forgetting anyone we do not dare to produce a long list of names. One standout person is Dr. David Brazel, who tirelessly read virtually all of the chapters, some more than once. Any errors, of course, rest with us and not him. Another person deserving clear thanks is Kayla Schulte, who made many of the beautiful illustrations in the book. Many thanks also to Melinda's postdoctoral researchers and students, Dr. Xueijie Ding, Dr. Riley Taiji, Evelina Akimova, Domante Grendaite, and Dr. Charles Rahal, and others such as Dr. Zachary Van Winkle and Giacomo Vagni, for testing and reading some of the chapters. We also thank our students at our 2017 National Centre for Research Methods (NCRM) Sociogenomics Summer School in Oxford.

Gratitude also to our family, friends, and colleagues who had to endure us while writing this manuscript. We are grateful to our editor at MIT Press, Bob Prior, who urged us to write this book and provided support, and to the very efficient Anne-Marie Bono and John Donohue. As noted earlier, we are deeply grateful for funding to write this book from the UK Research and Innovation/Economic and Social Research Council (UKRI/ ESRC) and NCRM's SOCGEN project (ES/N011856/1) and related funding to support fundamental research from the Netherlands Organisation for Scientific Research (NWO) (M. C. Mills, VIDI Grant, 452-10-012), European Research Council (ERC) (M. C. Mills, Consolidator Grant SOCIOGENOME 61560; Advanced Grant CHRONO 835079), and ESRC (N. Barban, through the Research Center on Micro-Social Change MiSoC at the University of Essex, grant numbers ES/L009153/1 and ES/S012486/1). Melinda is also indebted to the Leverhulme Trust for funding the final stages of this research and supporting interdisciplinary research to build the Leverhulme Centre for Demographic Science.

Due to our condensed and rapid handling of material, we will undoubtedly miss certain topics or inevitability cover others insufficiently. This is also one of the fastest-moving areas in science. Not only are there rapid developments in techniques, approaches, and genotyping technology but also different camps or schools of thought. Even in the final days of finishing the book, new data, techniques, and programs were emerging that we were unable to integrate. Finally, since our aim was to provide the most accessible book possible we hope that it reaches a new and diverse group of researchers. When it does, let us know and we shall be deeply honored.

Melinda Mills, University of Oxford and Nuffield College, United Kingdom
Nicola Barban, University of Essex, United Kingdom
Felix Tropf, École Nationale de la Statistique et de L'administration Économique
(ENSAE), Center for Research in Economics and Statistics (CREST), Paris, France
July 2019

I

Foundations

1

Introduction

Fundamental Concepts and the Human Genome

Objectives

- Understand the *motivation*, aim, target audience, and structure of this book
- Define, recognize, and describe the *fundamental terminology* used in the study of the human genome
- Comprehend the *organization of DNA* in the nucleus of a human cell and the terms genome, gene, and chromosome
- Gain an overview of *Mendel's laws*, *sexual reproduction*, and *genetic recombination*
- Define *genetic polymorphisms* and the terms allele, single-nucleotide polymorphism, minor allele frequency, and unique identifier
- Understand *monogenic*, *polygenic*, and *omnigenic* effects and *polygenic scores*
- Acquire a basic knowledge of the *relationship of genes to proteins*
- Grasp the *central dogma of molecular biology*: transcription and translation
- Understand how polymorphic sites are either *homozygous* or *heterozygous*, and understand the relationship to inheritance of *dominant and recessive traits* using a Punnett Square
- Recognize the meaning of *heritability*, common misnomers, types, and the missing heritability discussion

1.1 Introduction

1.1.1 Motivation and aim of this book

Since the human genome was first sequenced in 2003, human genetics has undergone a veritable revolution. The growth in computing power, the explosion in the availability of datasets with genetic information, and the infusion of bioinformatics into this field have disrupted how we think about disease and behavior. Students and researchers now often possess fundamentally different skills beyond their own disciplinary boundaries or research specializations and increasingly embrace not only diverse disciplinary knowledge but also often have strong computing, statistics, and bioinformatics skills. It feels as if we are in the midst of a scientific renaissance where computing, data, and interdisciplinary knowledge finally frees us to ask and understand basic questions about human origins, health, and behavior that have, until now, evaded empirical study.

The relevance of genetics has also penetrated disciplines far beyond its original homes in biology, epidemiology, and the medical sciences to gain relevance in new areas across the biomedical, social, and psychological sciences. As of 2019, around 4,000 genetic discoveries have been published, linking the genetic basis of thousands of traits ranging from height, type 2 diabetes, and body mass index (BMI) to coffee consumption, depression, neuroticism, and even the age when you have your first child [1]. Researchers with substantive expertise in a myriad of subject areas are now able to integrate genetics across a plethora of topics. For the first time in history we are able to ask fundamentally new types of questions. Researchers in the biomedical sciences can now estimate the genetic component of many major diseases such as type 2 diabetes, breast cancer, or cardiovascular disease. More importantly, new genetic discoveries allow them to understand how genes interact with different lifestyle or environmental factors in a move toward more effective clinical screening and interventions [2]. In the areas of psychiatry and psychology, we can study how genetic markers related to schizophrenia, addiction, neuroticism, and multiple other traits are associated with disease and treatment. Social scientists are now able to empirically study whether it is nature, nurture, or—more likely—a combination or interaction of nature and nurture that shapes our behavior. Genetic scores derived from these many genetic discoveries can be used in data analysis related to reproductive behavior, diabetes, educational attainment, well-being, and countless other outcomes. The genetic component can be used as variables in statistical models to increase the overall predictive power or variance, as control variables, to understand causality or examine how genetics interacts with family background, school, or other environmental factors.

Yet, without a strong background in human genetics, evolution, working with large data, statistical models, computational methods, and computer programming, entering this area of research is not only daunting but for many unattainable. There is, likewise, a growing call for more diversity in both genetic datasets and among researchers, with training and accessibility being one of the main challenges. This book uses the freely available statistical environment R and other useful computer programs such as Python in addition to datasets that can be easily downloaded to carry out the exercises (see appendixes 1 and 2). All computer code, example datasets, and teaching material can be found in the companion website to this book at http://www.intro-statistical-genetics.com.

We were struck, however, by the chasm that existed in textbooks and instruction material in this vital area of research. On the one hand, there are a multitude of excellent introductory human genetics textbooks but these are aimed firmly at genetics or biology students and rarely with applied computer applications. They also generally presuppose a strong background in molecular biology, chemistry, and genetics. On the other hand, there are textbooks that expertly present the advanced statistical and mathematical foundations of population genetics, often in theoretical and formal mathematical terms. There appeared to be no middle ground between the two. Specifically, there was no introductory textbook that integrated the fundamental concepts of human genetics, evolution, theoretical and statistical foundations, with new bioinformatics possibilities, in the form of a very practical and applied computer-based book. Textbooks rarely linked the basic concepts with hands-on computer-based exercises for everyday research.

The aim of this book is to introduce students and researchers to the emerging concepts, data, and methods of statistical genetic data analysis in an accessible, practical, and—we hope—engaging manner. It is written for those who do not have a strong background in molecular biology, human genetics, or cell biology but want to integrate genetic data into their research. This book is written with broad accessibility in mind and will appeal to students and researchers from multiple disciplinary backgrounds who are new to this area of research. An elementary knowledge of statistical methods at the first level of a statistics or biostatistics course is optimal. Our approach is hands-on and applied, focusing on unpacking the basic concepts, the "do's and don'ts," in this research area and how to practically run and interpret analyses. We only provide the basic mathematical and statistical treatments of the material, with references supplied for those wanting to dig deeper. Considering the far reach of genetics, we anticipate interest from students and researchers from the medical and social sciences who increasingly integrate statistical genetic data analysis into their ways of thinking and working.

By virtue of being an introductory text, we do not assert that this is by any means an exhaustive textbook on human genetics and statistical genetic data analysis. We acknowledge that our coverage of such a broad range of fundamental topics makes it at times feel like a crash course. This inevitability means a lack of depth and nuance in certain areas. For this reason, we often include boxes to provide examples or further explanation and a glossary in the backmatter. This glossary is also not all-encompassing, with additional terms sometimes defined within the text or endnotes. We also encourage readers to pursue our "Further reading" sections, the references at the end of each chapter, and the exercises. This book can be used by instructors within a course or by researchers as a self-learning book.

This book falls in the realm of statistical human genetics. Genetics is of course a very old subject about rules of inheritance, and predates molecular biology and DNA. We provide an overview of the origins of this research area but focus largely on current computational approaches using individual-level genetic data. The goal of *statistical genetics* is to explain population variation or, in other words, to ask why we differ in our health outcomes, behavior, or appearance. Here the basic topic of study is the relationship between a genotype and phenotype. A **genotype** is a part of your own unique genome that comprises your complete heritable genetic identity. By part of your genome, we mean that a genotype is actually a particular allele or set of alleles at a locus, all concepts which we define and unpack later in this chapter. A **phenotype**—also often referred to interchangeably as a trait, outcome, or dependent variable—is the outcome we study. It can refer to multiple observable traits of an individual ranging from morphological (i.e., their physical appearance—height, eye color, curly hair) to disease (type 2 diabetes, breast cancer), psychiatric conditions (autism, schizophrenia), personality (neuroticism), or behavioral outcomes (years of education, number of children ever born, well-being, coffee consumption). Phenotypes are called quantitative traits when they are continuous, such as height or blood glucose levels. But they may also be binary or dichotomous, such as whether cilantro has a soapy aftertaste when you eat it (which is the case for around 10% of the population).

The applied techniques in this book focus on a common form of variation across genomes: the **single-nucleotide polymorphism** (**SNP**, pronounced SNIP). As we elaborate upon later in this chapter, SNPs are the way in which we can examine single base differences, or "markers," or flags in DNA that allow us to examine variation in a population. The examination of this variation in SNPs forms the basis of many of the applied analyses we perform in this book. They are the markers that are the focus of the **genome-wide association study** (GWAS, pronounced gee-WAS), a search across the genome, examining each genetic variant (or region) one by one to see if there is a statistical relationship (association) between SNPs and a phenotype. The genetic variants that are isolated from these GWASs are then often used to engage in either further statistical or downstream biological analysis. The reality, however, is that many phenotypes are influenced by both your genotype and the unique circumstances and environment in which you have lived your life. This is often referred to as the "nature" and "nurture" dichotomy or *gene-environment interplay*. This is generally studied in terms of **gene-environment interaction (G×E)**, broadly defined as the interplay between genes (i.e., multiple genetic loci) and one or more environmental factors that in turn affect a phenotype or trait. We unpack this relationship in both theoretical terms and applied examples later in this book.

1.1.2   Overview of topics covered in this book

The book is divided into three parts. Part I, "Foundations," sets the groundwork for this type of research by outlining the basic terminology, theory, and statistical background of the topics we study in this book. The current chapter 1, "Introduction," equips you with some of the basic terminology used in this area of research and serves as a primer on the human genome. Chapter 2, "A Statistical Primer for Genetic Data Analysis," provides an accessible introduction to some of the basic statistical aspects underpinning this research. Chapter 3, "A Primer in Human Evolution," unpacks some of the basic concepts that are required to understand this topic and are essential for you to correctly handle data and interpret results. Chapter 4, "Genome-Wide Association Studies," reviews genetic discovery, which provides the basis of the information that is used for the majority of this book. Chapter 5, "Introduction to Polygenic Scores and Genetic Architecture," zooms in on the topic of polygenic predictive scores and their underlying genetic architecture, which is one of the central ways in which many applied researchers interact with these data. Chapter 6, "Gene-Environment Interplay," elucidates the topics of gene-environment interaction (G×E) and gene-environment correlation (rGE) and the understanding that genetic information in this area of research is rarely examined in isolation.

Part II, "Working with Genetic Data," moves from background topics to hands-on applications within the core computer packages and programs you need to carry out your research. Chapter 7, "Genetic Data and Analytical Challenges," provides an overview of where you can obtain this type of data, the analytical challenges associated with working with it, and different file types. Chapter 8, "Working with Genetic Data, Part I," delves into data management, producing descriptive statistics, and quality control. Chapter 9, "Working with Genetic Data, Part II," provides you with experience to carry out association analysis and the calculation of population stratification, genetic relatedness, and heritability. Finally, chapter 10, "An Applied Guide to Creating and Validating Polygenic Scores," offers a step-by-step guide to create and validate these summary scores.

Part III, "Applications and Advanced Topics," provides more advanced applications and a helicopter view of selected advanced topics. Chapter 11, "Polygenic Score and Gene-Environment Interaction (G×E) Applications," is an applied companion chapter to the theoretical and conceptual introduction of this topic in chapters 5 and 6. This is followed by chapter 12, "Applying Genome-Wide Association Results," where you learn how to visualize genetic findings (Manhattan plots, regional association plots, Quantile-Quantile plots), create a series of estimates from GWAS summary statistics (heritability and genetic correlations), and engage in multi-trait analysis (MTAG). We then turn to chapter 13, "Mendelian Randomization and Instrumental Variables," where we describe these techniques and their extensions, followed by a summary and discussion of limitations. Chapter 14, "Ethical Issues in Genomics Research," delves into ethical issues related to the interpretation of genetic results, the use of polygenic scores, and the legal, social, and policy risks of this research. Chapter 15, "Conclusions and Future Directions," provides a brief summary, followed by insights into current and potential future directions in this area of research. Considering the rapid speed at which this field moves, this area will continue to grow and expand in a multitude of ways, many of which we undoubtably could not include. We also have two appendixes: appendix 1, "Software Used in This Book," and appendix 2, "Data Used in This Book." Since much of the terminology may be new to many readers, we also included a brief glossary.

The current chapter provides a very broad introduction to the area of human genomics. You will encounter many definitions, which we recognize can be daunting. Without a basic of understanding of these main concepts and the processes behind them, however, it is difficult to engage in genetic data analysis. We first define the basics of DNA, followed

by an overview of Mendel's laws including sexual reproduction and genetic recombination, ideas which will become central in later chapters. We then turn to genetic polymorphisms and related terms that form the basis of many topics you will study in this book.
Researchers also need to understand that it is rarely one genetic variant that is related to a
trait, with most complex traits being polygenic (i.e., multiple genetic variants). We then
introduce the relationship between genes to proteins and the central dogma of molecular
biology. Researchers also need to understand that the polymorphic sites we study can be
homozygous or heterozygous and the relationship to inheritance and heritability.

1.1.3   What are DNA, the genome, a gene, and a chromosome?

DNA (deoxyribonucleic acid) is the molecule that makes up the genetic material contained within our bodies' cells. As figure 1.1 illustrates, two long DNA chains, composed
of simpler molecular units (called nucleotides, defined shortly), coil around each other to
form a double helix. DNA contains the genetic instructions that tell each cell which proteins to make. A **genome** is the complete set of genetic material of an organism or, in
other words, the entire set of DNA contained within the nuclei of somatic cells[1] in the
human body. The size of each organisms' genome is the total number of bases in one representative copy of its nuclear DNA. As the figure shows, a **gene** is a section of DNA
found on a chromosome that consists of a particular sequence of nucleotides at a given
position on a given chromosome that in turn codes for a specific protein (or an RNA molecule). As we discuss in a later section, a gene is a segment of DNA that tells the cell how
to make a certain protein. Humans are estimated to have 20,000 to 25,000 genes.

A **chromosome** is a single molecule of DNA that comprises part of the genome. It consists of nucleic acids and protein and is found in the nucleus of somatic cells and carries
genetic information in the form of genes. As figure 1.1 demonstrates, chromosomes are

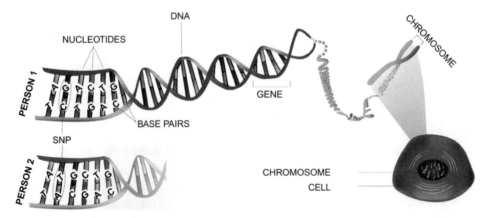

Figure 1.1
Orgaization of DNA (deoxyribonucleic acid) in the cell nucleus.

central to our understanding of genetics. Humans have *23 chromosome pairs* (i.e., 46 chromosomes) in total consisting of 22 autosomal chromosomes and one pair of sex chromosomes, two Xs for females (XX) and an X and a Y for males (XY). *Autosomal chromosomes* are the numbered chromosomes that are not related to sex determination or, in other words, chromosome 1 through 22.

Organisms differ greatly in the size of their genome, with bananas having 22, fruit flies with 8, and goldfish boasting 94 chromosomes. Chromosomes also have different sizes, expressed in terms of megabases. A *megabase* is a measure of the length of a genome segment or, in other words, it represents the physical size of a genomic region. One representative copy of the human genome is estimated to have around 3,200 Mb (megabase pairs) or often described as 3.2×10^9 bp (base pairs). In comparison, fruit flies have around 123 million base pairs and one particular pine tree (the loblolly), for instance, has seven times more than humans with around 23 billion base pairs. In humans, the length ranges between chromosomes from the very largest chromosome 1 (250 Megabases [Mb]) to the smallest chromosome 21 (50 Mb). Sex chromosomes consist of an X chromosome that has two copies in females (155 Mb) and the shorter Y chromosome in males (60 Mb).

1.2 Mendel's laws, sexual reproduction, and genetic recombination

The basis of much of our understanding of genetics stems from *Mendel's laws*. Most readers will be familiar with Gregor Mendel's discovery of the genetic segregation of traits in 1865, which he obtained by crossing garden peas that had different characteristics. Since there are ample descriptions of this discovery and the history of genetics, we do not reiterate it here (see "Further reading and resources" and, for an accessible and engaging read, see, for instance, [3]). Mendel's law consists of two principles. The first law is the *principle of segregation*, which holds that two members of a gene pair (i.e., alleles) segregate from each other in the formation of gametes with half of the gametes carrying one allele and the other half carrying the other allele. *Gametes* are the egg (females) and sperm (males) cells in animals and plants. In other words, each gene has two copies (alleles) and each parent transmits only one copy to a child; the child thereby receives two copies (alleles) in total. The second of Mendel's laws is the *principle of independent assortment*, which suggests that genes for different traits assort independently of one another in gamete production.[2] Put another way, different genes are inherited separately. Using a simple example, this means that the gene(s) that code for height are inherited separately from the gene(s) that code for hair color.

As many might recall from introductory biology, this occurs during *sexual reproduction*, which is when the genetic material of a father and a mother combine to produce offspring that are genetically distinct from either parent. During this process, the mother and father produce gametes (egg cells for females, sperm cells for males). Each gamete has 23 chromosomes, which is half of the number of chromosomes in a typical cell. Since humans are diploids, we have pairs of chromosomes with one set of chromosomes inherited from each parent. During fertilization, the egg of the mother and sperm of the father fuse to form a cell that has the full number of chromosomes consisting of half from the mother and half from the father. Gametes are generated through meiosis and have half of the number of chromosomes of a normal cell. Gametes only have one copy of each chromosome so that when fertilization occurs, the resulting embryo has two copies of each chromosome.

Genetic recombination—also referred to as *genetic reshuffling*—is graphed in figure 1.2 and refers to the sexual reproduction of offspring. This, in turn, produces the many different variations that we observe that are found in either parent. In the human data we examine in this book (i.e., eukaryotes), this novel set of genetic information is generated during meiosis. *Meiosis* is the process by which haploid gametes are generated, during which recombination occurs. As the figure shows, there is mitosis and then two stages of meiosis. In *mitosis*, a cell divides and then replicates its nuclear DNA to form two diploid copies of a cell. These are the primary sex cells. This is the point where somatic mutations

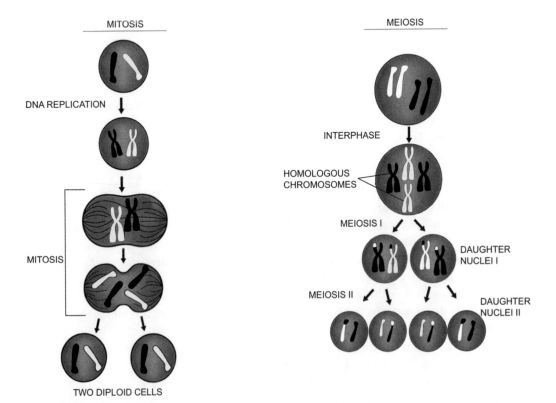

Figure 1.2
Genetic recombination during meiosis.

may occur (see box 1.1 for more on **mutation**). Cells in your body divide to produce new cells in a series of cell cycles. Mitosis is the stage of the cell cycle when the cell divides. Mitosis is used to grow or to replace cells that may have been damaged. Contrary to meiosis, mitosis produces two new daughter cells that are identical to the original parent cell.

The next stage is meiosis. The first division that occurs during meiosis (Meiosis I) is where the chromosome pairs line up in the center of the cell. The pairs are then pulled apart so that each new cell only has one copy of each chromosome. This is when half of the father's chromosomes (black) and half of the mother's chromosomes (white) go into each new cell. The second division consists of the chromosomes lining up again in the center of the cell with the arms of the chromosomes then pulled apart. Finally, four gametes are produced that each contain only a single set of chromosomes. Since chromosomes get randomly "reshuffled" during meiosis, each gamete only acquires half of them. This is the

Box 1.1
What is a mutation and how does it occur?

A mutation is a permanent change in the sequence that makes up a gene. Mutations can affect a single base pair or multiple genes across a large segment of a chromosome. There are two types of gene mutations. The first is the *hereditary mutation*, which is inherited from a parent, is present for an individual's entire life, and is in almost all cells in the body. It is also often referred to as *germ line mutation*, which is a mutation that will be inherited by the offspring of the organism. The second type is *somatic or acquired mutations* that occur during an individual's lifetime and exist only in certain cells. These mutations are generally related to environmental factors. This could include, for example, smoking or exposure to ultraviolet radiation from the sun. They can also occur if there is an error during DNA replication before or during cell division. These acquired mutations in somatic cells are not passed on to the next generation.

Mosaicism can also take place. This is when different populations of cells within an organism have different genotypes. It is now understood that we are all mosaics, to one degree or another. In fact, permanent alterations of the genome are important to the functioning of the immune system and the brain. Whether mosaicism causes disorders depends on the type of mutation and how many cells are affected. What is described as *de novo (i.e., new) mutation* can be hereditary or somatic. These types of mutations explain genetic disorders where an offspring has a mutation in every cell, but the parents do not and there is no family history. They are generally rare alleles that can cause major effects.

What is the difference between gene mutations and genetic polymorphisms? A mutation is an event in an individual whereas a polymorphism is defined across a population. Mutations that cause disease are relatively rare in the general population. Gene mutations thus differ from more common genetic alterations of polymorphisms discussed in this chapter. These are the alterations that are common enough (more than 1% of the population) to be considered as a normal variation in the DNA and are responsible for the normal differences between people that we have described such as hair color, height, or risk of certain diseases.

reason why each of the gametes is genetically different from the others. The importance of these chromosomes being randomly reshuffled will be revisited later in this book in our chapter 13 on Mendelian Randomization. As box 1.1 elaborates, this is also when hereditary gene mutation takes places. If the DNA has a mutation, when an egg and sperm cell unite and the fertilized egg cell receives the DNA from both parents, the offspring will have the mutation across all cells. It is also useful to note that only some of the "breaking down" steps occur in the mitochondria. Meiosis I is similar to mitosis, in that it is a form of cell division that produces two diploid cells, but it is not referred to as mitosis and does differ in some key ways. Genetic recombination during mitosis, for example, occurs between identical copies of a chromosome and does not result in new variability. Although Mendel's experiments were useful to forming the field of genetics, for the vast majority of topics that we study it is not one gene but multiple genetic effects that inform our understanding, which we turn to now.

1.3   Genetic polymorphisms

1.3.1   Alleles, single-nucleotide polymorphisms (SNPs), and minor allele frequency (MAF)

The aim of many contemporary genetic studies has been to identify genetic variants that are associated with phenotypic variation in populations, discussed in detail in chapter 4, on GWASs. The focus of this research is on studying the locations in our DNA where the sequence of *base pairs* differs across individuals. A *genetic polymorphism* refers to the variation in the DNA sequence between individuals. The possible variants of a polymorphism are referred to as alleles, which derives from the Greek prefix "allele," meaning mutual or reciprocal. An **allele** refers to each of the two or more alternative forms of a gene found at the same place on a chromosome that arise by mutation. In figure 1.1 these differences are indicated by single base-pair changes, deletions, or insertions.[3] On the left-hand side of figure 1.1, we see a comparison of two pieces of DNA where Person 1 has CG and Person 2 has AT or, in other words, differences in the base pairs of their SNPs. We discuss the meaning of base pairs and the codes CG and AT in the next section.

Alleles are thus the variations of a locus that codes for a protein.[4] Alleles can come in various forms and one of these are single-nucleotide polymorphisms (SNPs), which are variations in a single nucleotide that occur at a specific position in the genome and are the most common type of genetic variation among humans. SNPs occur around once every 1,000 nucleotides and it is estimated that, as a population, we have at least 10 million SNPs. Humans are around 99% genetically identical in their SNPs or in other words in their genetic makeup. It is the differences in the remaining 0.1% that holds the important information about genetic variation that we study. Alleles have to be in a gene, but SNPs do not always lead to new alleles (e.g., when they occur in noncoding areas as we outline shortly).

Until now we have generally studied *common SNPs* that are *single-nucleotide variants (SNVs)* at a population allele frequency greater than 1%. *Structural variation* is the genomic variation in the structure of a person's chromosome. Around 13% of the human genome is classified as structurally variant in a typical population. It can consist of multiple types of variation such as deletions, copy-number variants, insertions, inversions, and translocations. They seem to be more difficult to detect than SNPs that are part of a particular region of DNA that can be around 1 kb (kilo-base pair, which is a unit of measurement in genetics, equal to 1,000 nucleotides) or larger. *Copy-number variation (CNV)* is a large category of structural variation with CNVs in the human genome thought to affect nucleotides more than SNPs and notably, many CNVs are not in coding regions (see next section for discussion of coding and noncoding regions).

The frequency at which alleles occur in a population is represented as the frequency of the least common or minor allele—called **minor allele frequency (MAF)**—which is one the key statistics used to characterize polymorphisms. The average person has around 4 million SNPs. MAF thus refers to the second most common allele in a particular population. In the literature, polymorphisms are distinguished by their MAF and categorized as *common* (MAF > 0.05), *low-frequency* (0.01 < MAF < 0.05), or *rare* (MAF < 0.01) *variants*. MAF thus provides us with the information to differentiate between common and rare variants in a particular population. In the genome-wide association studies we discuss in chapter 4, the majority of studies up to around early 2018 largely examined common variants (i.e., MAF of less than 5%). There is growing interest and study of *rare variants*, which are alternative forms of a gene that have an MAF of less than 1%.

To date, we have not found polymorphisms at every site in the genome. This is due to the fact that only a selection of people have been genotyped but also that variation at some sites cannot be tolerated, which is explored in further detail in chapter 4. For those interested in understanding sequence variation, see our "Further reading and resources" section for a link to a database that shows the location, frequency, type, and predicted function of each known variant (dbSNP with the Ensembl browser). In humans, variants that have been listed have a *unique identifier* that always begins with "rs," followed by an arbitrary number, such as rs4988235 (linked to lactose intolerance) or rs1799971 (linked to stronger alcohol cravings). As we explore in a later chapter, when you obtain genetic data, each SNP will have an rs ID, a chromosome, and other features such as the position. Researchers should also be aware that there are often many ways of naming genes, or what is known as *nomenclacture*. When in doubt or if you would like to understand the symbols, names, or genes, refer to the HUGO site, which is the HUGO Gene Nomenclature Committee (https://www.genenames.org/).

1.3.2 Monogenic, polygenic, and omnigenic effects

Figure 1.3 provides a broad overview of the spectrum of genetic contributions to particular phenotypes. Here we see the spectrum of allele frequency by effect size. Mendelian

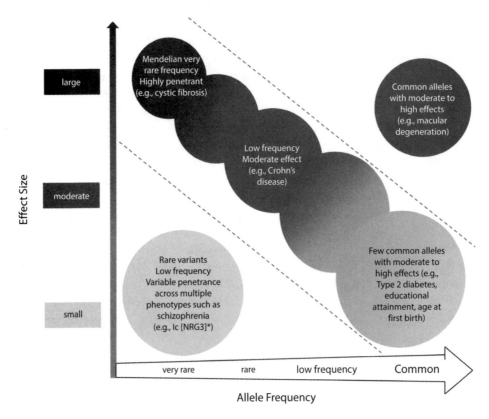

Figure 1.3
The spectrum of genetic contributions to phenotypes.
Source: Produced by authors and adapted from Manolio et al. 2009 [4], figure 1.
*For a list of rare and low-frequency genetic variants in common disease, see Bomba et al. 2017 [5].

traits such as cystic fibrosis or Huntington's disease, although rare, have single variants with large effect sizes that are highly penetrant. *Penetrance* in genetics refers to the proportion of individuals who carry a particular variant or allele of a gene that is associated with a particular phenotype. As we touched upon in the last section, there are rare variants, however, that have very low frequencies but may have small to large effects. There is also a large group with rare to low allele frequency and moderate effects, such as Crohn's disease. Common variants will almost always have small effects when entered into statistical models, which we explore in later chapters (see also box 1.2). This is related to natural selection, which is discussed in detail in chapter 2, and to the fact that they may be distal to biology, be highly behavioral, lifestyle related (e.g., smoking, BMI), or for other reasons.

It is rare that a trait is only associated with one genetic variant, or, in other words, is *monogenic*. The phenomenon of *polygenicity* implies that no single genetic variant determines or is associated with a trait, but rather that it is often hundreds and thousands of genetic variants that each have a small influence on a trait. Such phenotypes are called complex since they have a multifactorial genetic basis. This is often related to the *common disease-common variant (CD-CV)* hypothesis, which holds that common disease associated alleles will be found in all populations that manifest a given common disease. In the most extreme omnigenic model, each variant at each gene is assumed to influence a complex trait and will have a small additive or multiplicative effect on the phenotype. They are often seen to evade evolution in part since it is the small effects of so many genes that influence phenotypes. The most common form of human genome variation is thus these SNPs.

The most important recent developments in statistical genetics surround the discovery of ubiquitous polygenicity in most traits that we study. An intuitive implication of polygenicity is that the effect sizes of individual SNPs are smaller than if only a few SNPs would be associated with an outcome. Small effects are harder to discover given a fixed statistical measure of certainty of the discovery. Polygenicity therefore explains the disappointingly small effects of discovered variants as well as the small number of robustly identified variants. As we explore later in detail in our section on heritability, we often apply *polygenic scores (PGSs)*, which represent numerical summaries for an individual of genetic effects estimated in a GWAS. Sometimes also referred to as polygenic risk scores, a polygenic score is a single quantitative variable that summarizes genetic predisposition to a phenotype by combining multiple genetic loci and their associated weights. PGSs have become a standard tool that we will discuss in detail in chapter 5, provide guidelines on how to create them in chapter 10, and show applied examples in chapter 11. Recently, some researchers have argued that we need to move beyond polygenic models to omnigenic models. In the *omnigenic* model, gene regulatory networks are seen as so interconnected that essentially all genes expressed in trait (disease)-relevant cells are liable to affect the functions of the core trait-related genes. They argue that most associated genetic variants can be explained by effects on genes outside of the core pathways being studied [7]. This is related to the topic of *pleiotropy*, which is said to occur when one gene influences two or more seemingly unrelated phenotypes. Since we explore this in more detail in chapter 5, we do not elaborate upon it here.

1.4   From genes to protein and the central dogma of molecular biology

1.4.1   From genes to protein: Genes, amino acids, nucleotides, and proteins
Recall that a gene is a section of DNA found on a chromosome that consists of a sequence of nucleotides at a given position on a given chromosome that codes for a specific protein

Box 1.2
Common versus rare variants: The case of NBA star Shawn Bradley

Any NBA fan will know that most players are very tall, such as Michael Jordan (6′6″), LeBron James (6′8″), or Wilt Chamberlain (7′1″). But Jordan, for instance, does not have any immediate family members who are over six feet tall. His mother is reported to be 5′5″ and his father 5′9″. A recent study examined the DNA of 7′6″ former NBA star Shawn Bradley, who is in the 99.99999th percentile for height [6]. He agreed to be sequenced, and researchers compared his genome to the polygenic height score produced by a well-known height GWAS conducted by the GIANT (Genetic Investigation of ANthropometric Traits) consortium. The researchers thought that Bradley would have rare, large-effect variants in height-related genes but this was not the case. Instead, when they applied the polygenic score for height it was "off the chart." They found that instead of having a few rare genetic variants he was, in fact, so tall due to the fact that he had many of the common variants positively associated with taller height. He had 198 more height-associated genetic variants than the average member of the general population. As we explore in later chapters that describe polygenic prediction and gene and environment interplay, it is striking that although the genomic analysis predicted his height rank, the individual polygenic score did not accurately predict his actual height.

(or an RNA molecule) (see figure 1.1). Each gene codes for a particular sequence of amino acids[5] that in turn form a protein. In total, 20 amino acids are used to constitute thousands of different proteins. Genes thus indicate to the cell the order in which amino acids should be assembled. A complex system regulates which proteins are produced in each cell such as hemoglobin or keratin, which determines the type of cell that it is (e.g., red blood cell, skin cell).

DNA strands are polymers,[6] which are made up of many repeating units called nucleotides. *Nucleotides* form the structure of DNA and consist of one of four nitrogenous bases—cytosine (C), thymine (T), adenine (A), and guanine (G)—plus a molecule of sugar (deoxyribose) and a phosphate molecule. The sugar and phosphate molecules on the nucleotides alternate but also form the backbone of the DNA strands. One of the four different nitrogenous bases—A, T, C, or G—joins to each sugar. Recall from figure 1.1 that DNA is in the form of a double helix. Each base links to a base on the opposite end of the strand in the double helix. Humans are thus composed of *diploid cells*[7] or in other words, pairs of chromosomes with one set of chromosomes inherited from each parent. We previously described the process of inheritance from parents in the section on genetic recombination. Since we are diploids, DNA's two strands are complementary to each other or in other words they follow complementary base pairing rules. *Complementary base pairing* means that A always pairs with T and C always pairs with G, forming base pairs. The two strands are complementary to each other and therefore contain the same information. As figure 1.1 also illustrates, it is the order of these bases along a single strand that comprises

Shawn Bradley. With permission by Mark A. Philbrik/BYU Photo.

the *genetic code*. In turn, it is the order of bases in a gene that determine the order of amino acids in a protein.[8] Each amino acid is coded for in the *coding regions* of a gene—called *exons*[9]—by a sequence of three bases and the amino acids are joined together to create different proteins, all dependent on the order of the gene's bases. The genetic code is also often referred to as a *triplet code*, which refers to the sequence of three nucleotide codes for one specific amino acid. Only ~1% of the genome is actually translated into protein, which is referred to as the *exome*.[10]

A gene thus consists of *three types of nucleotide sequences*: (1) exons or coding regions, specifying a sequence of amino acids (the exome), but also, beyond the auspices of this introductory chapter; (2) noncoding regions or introns, which do not specify a sequence of amino acids; and, (3) regulatory sequences, determining when, where, and how much of a protein is made. The *noncoding regions* or *introns*[11] do not code for proteins. Once this was referred to as "junk DNA" because it is considered by many biologists as non-functional due to the observation that most of the genome is not conserved.[12] More recently, scientists have come to understand that some of this noncoding DNA regulates the activity of genes and controls the amino acid sequences of the proteins that are synthesized.

A *protein* is a large molecule composed of one or more chains of amino acids that are joined together by peptide bonds[13] in a specific order determined by the base sequence of nucleotides in the messenger RNA for protein (see next section). Proteins are essential for the structure, function, and regulation of our cells, tissues, and organs. Proteins control the shape and structure of cells but also carry out important tasks such as transporting oxygen in blood or digesting food. Proteins have many different functions such as controlling all of the metabolic reactions that take place in the cell and acting as biological catalysts to speed up chemical reactions in the body. Others act as hormones, antibodies, transporters of oxygen (e.g., hemoglobin), or structural proteins (e.g., keratin or collagen).

1.4.2   The central dogma of molecular biology: Transcription and translation

To grasp the processes described in the previous section, it is important to be aware of the central dogma of molecular biology. This describes the two-step process—transcription and translation—through which the information of genes flows into proteins (i.e., DNA to RNA to protein), also visualized in figure 1.4. *RNA (ribonucleic acid)* is a molecule with long chains of nucleotides with a nitrogenous base, a ribose sugar, and a phosphate. RNA on the other hand is single stranded and although it also has adenine, cytosine, and guanine, instead of thymine it has a nucleobase called uracil.

The function of DNA is to hold the original genetic code that is transcribed into RNA, which then codes for protein in a process known as translation. As figure 1.4 displays, *transcription* refers to the synthesis of an RNA copy of a segment of DNA.

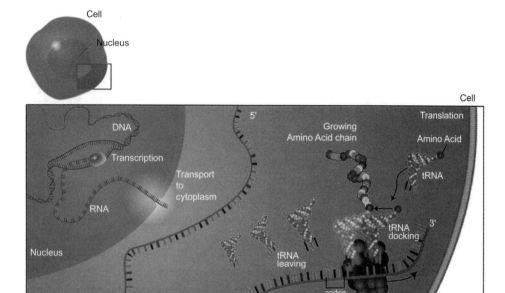

Figure 1.4
The central dogma of molecular biology: Processes of transcription and translation.
Source: Reprinted with permission from NHGRI Digital Media Database, Leja (2010) [8].
Notes: This figure shows the process of transcription from DNA to the messenger ribonucleic acid (mRNA). Outside the nuclear membrane of the cell in the cytoplasm the process of translation occurs where proteins are built from amino acids with the help of transfer RNA (tRNA) molecules, guided by ribosomes.

Translation denotes the RNA being used as the template for a protein. In *eukaryotic* cells,[14] transcription occurs in the nucleus and translation in the cytoplasm.[15] *Messenger RNA (mRNA)* encodes the amino acid sequence of a polypeptide[16] and *transfer RNA (tRNA)* molecules bring amino acids to ribosomes during translation, using the triplet code to determine the amino acid to be integrated into the growing polypeptide chain. In other words, in order to decode the genetic information into proteins, DNA is first transcribed into mRNA, which is then translated into proteins by the aid of tRNA.

For nonbiologists, one way to think of this is that the nucleus is the post office, mRNA is the postal delivery truck, and tRNA is the postal carrier. The DNA cell nucleus has all of the "letters" or code inside. The cell needs to get the code from the DNA's cell nucleus (central post office) to the ribosome. To achieve this, it uses the postal delivery truck of mRNA. During transcription, the DNA information leaves the post office (i.e., the nucleus) as a postal truck full of letters (i.e., coding of genes). Although it is more complex than what this brief introduction affords, the mRNA effectively copies the code from the DNA and acts as a messenger between the DNA and ribosome to carry the code between the two. The correct amino acids are thus brought to the ribosomes in the correct order by these carrier molecules. The tRNAs are the postal carrier, who then reads the address on the letter so that they can deliver it in person. The tRNA reads the mRNA in segments of three bases at a time (called codons) and through an anticodon (three bases complementary to the RNA's codon), binds to the mRNA.

1.5   Homozygous and heterozygous alleles, dominant and recessive traits

Recall that polymorphic refers to the presence of more than one allele at a specific locus. A *locus* (plural loci) is a location on the genome, which could be the location of a gene or marker. When an individual has two of the same allele, regardless of whether it is dominant or recessive, they are called *homozygous*. **Heterozygous** refers to having one of each of the different alleles. A person is heterozygous at a gene locus when their cells contain two different alleles. Heterozygosity thus refers to a specific genotype.

This distinction also explains the difference between *dominant* traits, which is when only one allele of a gene is necessary to express the trait versus *recessive* traits, where both alleles of a gene must be identical to express the trait. For dominant traits we use two capital letters (e.g., AA) and for a recessive trait we use two lowercase letters (e.g., aa). Dominance in melanin deposits results in freckles, for instance. A homozygous freckled person would have the FF genotype while someone without freckles with the homozygous gene would be represented by ff. The Punnett Square is often used to understand these concepts and is shown in figure 1.5. In this example, we have Parent 1 and Parent 2 and walk through the various possible outcomes. Here we refer to a *probability*, which is the chance that the offspring will have a particular genotype such as a 15% chance.

Figure 1.5 provides a stylized example. For example, imagine that Nick and Elizabeth were going to have a baby. Elizabeth has dimples on her cheeks, which is a dominant trait, and Nick does not. We can use a Punnett Square to calculate whether their baby will have dimples. Nick has the recessive trait (no dimples), which means that Nick must have two recessive alleles, or, let us say, aa. Elizabeth has the dominant trait (dimples), but Elizabeth could be homozygous dominant (AA) or heterozygous dominant (Aa).

To find out whether her dimples are homozygous dominant or heterozygous dominant we need to know more about her parents. We discover that Elizabeth's father has dimples on both cheeks and her mother does not. Her mother has the recessive trait and thus has to have the aa genotype. Her father has the dominant trait, but we don't know if he is a

Parent 2

Parent 1		Allele 1	Allele 2
	Allele 1	A1/A1	A1/A1
	Allele 2	A2/A1	A2/A1

Homozygous dominant + Homozygous dominant		
	A	A
A	AA	AA
A	AA	AA

Homozygous dominant + Homozygous recessive		
	a	a
A	Aa	Aa
a	Aa	Aa

Homozygous dominant + Heterozygous		
	A	A
A	AA	AA
a	Aa	Aa

Heterozygous + Heterozygous		
	A	a
A	AA	Aa
a	Aa	aa

Elizabeth

Nick		A	a
	a	Aa	aa
	a	Aa	aa

Figure 1.5
Using the Punnett Square to determine the probability of offspring having a particular genotype.

homozygote (AA) or a heterozygote (aa). We can conclude that she must have the recessive allele (a) since that is all that she could have inherited from her mother and that Elizabeth's father is an Aa. Since Elizabeth has dimples we can conclude that she inherited a dominant allele (A) from her father. Her father passed on the A to her (and it does not matter if he is AA or Aa). To calculate the baby's chance of having dimples we see from the Punnett Square that there is a 50% chance the baby will have genotype Aa and thus have dimples and an equal 50% chance that the baby will have genotype aa and thus not have dimples.

A related topic, which we explore in detail in chapter 3, is **linkage disequilibrium (LD)**, which refers to the fact that alleles are not randomly associated at different loci. Rather they are often in LD, which is when the frequency of a particular association of different alleles is higher or lower than would be expected if they were independent and randomly associated.

1.6   Heritability

1.6.1   Defining heritability: Broad- and narrow-sense heritability

Heritability forms the basis of much of our understanding of genetic and environmental influences on phenotypes. Heritability is defined as the proportion of variation of a trait in a population that is attributable to genetic differences. Or more technically, it is the proportion of the phenotypic variance accounted for by genetic effects. Mathematically it is often abbreviated as h^2, the ratio of the total genetic variance (V_G) to the total phenotypic variance (V_P) where $V_P = V_G + V_E$. This partitioning of the variance assumes that all sources of variance can be reduced to either genetic (V_G) or environmental (V_E) effects.

A distinction is often made between **broad-** and **narrow-sense heritability**. The broad-sense heritability (mathematically abbreviated as H) is represented as:

$$H^2 = \frac{V_G}{V_P}$$

where

$$V_P = V_G + V_E$$

Variance as a measure of individual differences in a phenotype (V_P) can emerge due to genetic differences in a population (V_G) and nongenetic or **environmental variance**, including measurement error (V_E). For details on the calculation of variance and other key statistics, see chapter 2.

The genetic component can be furthermore differentiated into additive (V_A) and nonadditive (V_{NA}, epistatic and dominant) genetic effects:

$$V_G = V_A + V_{NA}$$

Additive genetic effects are when two or more genes contribute to a phenotype or when alleles of a single gene combine such that their combined effects on the phenotype equal the sum of their individual effects. As discussed in the previous section, dominance describes the relationship among alleles of one gene where the effect on the phenotype of one allele masks the contribution of a second allele at the same locus. The first allele is said to be dominant and the second allele is said to be recessive.

Nonadditive genetic effects involve dominance (of alleles at a single locus) or *epistasis* (of alleles at different loci). Dominance differs from epistasis, which is a relationship in which an allele of one gene affects the expression of another allele at a different gene. Dominance variance (V_D) and interaction variance (V_I, epistasis) is thus where the effect of one genotype is influenced by one or more other genotypes. Genotype-by-environment interactions are also possible, which we elaborate upon in chapter 6. If this is the case, we would thus have:

$$V_G = V_A + V_D + V_I + V_G$$

There is considerable debate about how much of the genetic variance in a population is nonadditive and, as we discuss in later chapters, the importance of other factors such as environment. While dominant genetic effects are important for Mendelian traits, for most complex traits, there is little evidence for nonadditive effects and we therefore mostly focus on narrow-sense heritability.

Narrow-sense heritability (h^2) is the proportion of variance in a phenotype within a specific population that is associated with additive genetic variance:

$$h^2 = \frac{V_A}{V_P}$$

Theory suggests selection is more efficient with additive variation but can still occur under other conditions, which we address in more detail in later chapters.

1.6.2   Common misconceptions about heritability

As Visscher and colleagues note, there has been considerable confusion surrounding the term heritability [9].

1. *It is not about individuals.* In other words, a heritability of 25% for obesity does not mean that a quarter of the reason a person is obese is genetic, with the other three-quarters is driven by environmental factors. Rather, it means that 25% of the individual differences in BMI are associated with genetic difference between individuals.

2. *It is only about a single population.* A heritability estimate alone cannot be used to compare genetic differences between different groups or countries. For instance, a heritability estimate of 80% for height does not mean that most of the average difference in height between, for example, the Dutch and Americans is due to genetic differences. Although variation within each particular population may be largely genetic, differences between populations are likely environmental [10]. One caveat is that for some phenotypes, such as a difference in pigmentation or skin color, they are largely genetic (although there is certainly an environmental component).

3. *It is not the same as inheritance.* Inheritance is the relationship between offspring and their biological parents. It measures not only genetic factors but also the environment, cultural, and other factors shared by family members [11].

4. *Very low heritability does not necessarily mean there is very little genetic contribution.* Low heritability could be attributed to the absence of variance in genes that contribute to the phenotype or high environmental variance. A straightforward example is that although the number of cervical vertebrae is highly related to a genetic component, since it is the same in everyone, there is little variance and thus no or low variability attributed to genetic factors.

1.6.3 Twin, SNP, and GWAS heritability

There are multiple ways to estimate heritability (often denoted by h^2), which are twin studies, SNP, and GWAS heritability [10, 12]. Initially and persistently, *twin and family studies* have been conducted to quantify the genetic, common, and unique environmental components in a phenotype. The intuition behind twin (and also family) comparison methods is that if a phenotype is heritable then individuals who are genetically related or identical to one another should be more similar than random members of the population. The problem, however, is that the similarity in environments experienced by twins and family members is likely to be confounded with genetic similarity.

Twins form a perfect natural experiment in which to deal with this confounding. The most common design to study heritability and environmental influences is the classic twin design. As a basis, this design compares *monozygotic (MZ)* or identical twins that are virtually genetically identical with *dizygotic (DZ)* or fraternal twins that share around half of their genetic material, the same amount as any full siblings. If MZ and DZ twins are assumed to share roughly the same environments—known as the equal environment assumption—the difference between their phenotypic correlations should provide an unbiased estimate of heritability. In classic twin models, phenotypic variance is partitioned into three components: A, which is additive genetic effects (heritability); C, referring to the common shared environment of both twins; and, E, the unique environment that is not shared by the twins and also measurement error. The nonshared environment could be peer groups or later in life the impact of partners or coworkers. This is the only component where MZ twins are thought to differ.

Since C is identical for MZ and DZ twins and DZ twins only share half of their alleles, phenotypic correlations can be calculated as follows:

$$r_{MZ} = A + C$$

$$r_{DZ} = \frac{1}{2}A + C$$

The A (i.e., heritability, h^2) component is then calculated by subtracting the correlation between DZ twins (r_{DZ}) from the correlation in MZ twins (r_{MZ}). This is multiplied by the factor 2 since MZs share on average 2 times the genetic alleles compared to DZ twins.

$$A = 2 \times (r_{MZ} - r_{DZ})$$

C (common shared environment) quantifies the extent to which twins are more similar compared to other pairs in the population and is achieved by subtracting the heritability component (A) from the MZ correlation (r_{MZ}):

$$C = r_{MZ} - A$$

Since $A + C + E = 1$ then:

$$E = 1 - r_{MZ}$$

This quantifies the remaining individual differences in families that might be due to non-shared environmental factors or measurement error (E), subtracting the MZ correlation (r_{MZ}) from the standardized total variance of 1. This basic logic is commonly applied in structural equation models (SEM), since these make assumptions more explicit, allow us to calculate uncertainty, compare different model specifications, and are more flexible in modelling joint genetic effects for multiple traits or extended family designs including multiple family members [13]. Heritability estimates from twin studies are the highest and perhaps most well-known estimates of narrow-sense heritability (in the following examples discussed as h^2_{family}).

Second, there is **SNP-heritability** (denoted here by h^2_{SNP})—also sometimes referred to as chip-based heritability, the proportion of phenotypic variance jointly accounted for by all variants on standard GWAS chips [14]. As we discuss in chapter 7, *SNP "chips"* refer to the high-density computer chips that measure SNPs and are of different sizes and vary by population, often referred to in relation to the company that produces them (e.g., Affymetrix, Illumina). Using the computer program *GCTA (Genome-wide Complex Trait Analysis)*, this analysis is based on the statistical method called *GREML (Genome-based Restricted Maximum Likelihood)*, which estimates the variance component to quantify the total narrow-sense (additive) contribution of a trait's heritability [15]. Twin and family studies rely on the estimation of heritability from closely related individuals. This SNP-based GREML heritability technique transcends these designs that relied on comparing twins to estimate heritability across unrelated individuals. Individuals vary in their genetic similarity, and these estimates leverage those differences in the data. If two people within the data are more similar in both their genetic and trait measures, this method indicates the measured genetics that causally influence that trait and how much. It can be extended, for instance, to estimate bivariate correlations between traits, comparing against chromosome length, or changes in heritability by age [16]. These techniques are continuously being evaluated and extended and can be estimated using the summary statistics that are obtained from GWA studies [17]. Typically, h^2_{family} is larger than h^2_{SNP} and h^2_{SNP} is substantially larger than h^2_{GWAS}, which we turn to in the next section. The SNP-based heritability technique provides an "upper level estimate" of the genetic effects that could be identified with a well-powered GWAS. A meta-analysis of virtually all twin studies conducted up to 2012 comes to the conclusion that on average across all traits studied, 50% of the variance is attributed to additive genetic effects and 50% due to non-shared environmental influences or measurement error, respectively—while there is substantial variation across phenotypes [18]. As figure 1.6 illustrates for SNP-based heritability, many phenotypes are partially heritable. This can range from highly heritable

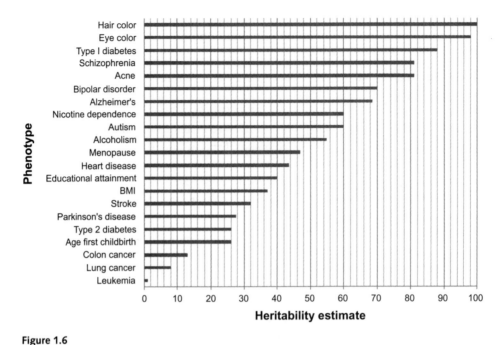

Figure 1.6
Heritability estimates across selected phenotypes.
Sources: Made by authors using https://www.snpedia.com/index.php/Heritability as well as Tropf et al. [19] for age at first childbirth and Branigan et al. [20] for educational attainment.
Notes: Some estimates show a range of heritability estimates from multiple populations or publications. When this occurs, we have taken the average. This is the case for alcoholism (50–60%), Alzheimer's disease (58–79%), autism (30–90%), BMI (23–51%), and Parkinson's disease (25–30%).

traits such as hair or eye color to moderately heritable traits such as autism, alcoholism, or heart disease to some with low heritability such as colon and lung cancer.

Finally, there is **GWAS-based heritability** (denoted by h^2_{GWAS}). This is the proportion of variance accounted for by genetic variants that are known to be robustly associated with the phenotype of interest, derived from a GWAS. This measure of heritability tends to produce the lowest estimates of all three measures discussed here. Although they are lower, with increasing sample sizes and technological developments, more associated variants continue to be discovered resulting in the estimates of the effects of these variants becoming increasingly more accurate. As a result, h^2_{GWAS} increases in tandem with GWAS sample sizes and is expected to approach h^2_{SNP} asymptotically under the assumption that the phenotype of interest is homogeneous in its genetic architecture across different environments. *Genetic architecture* is the mapping of the underlying genetic basis of a genotype of phenotype and determines the variational properties of the phenotype [21].

1.6.4   Missing and hidden heritability

The gap between the h^2_{family} and h^2_{GWAS} is referred to as *missing heritability* [4]. Potential reasons for missing heritability include nonadditive genetic effects (although evidence for this is scarce) [22, 23], large effect rare variants [24], and potentially inflated estimates from twin studies due to shared environmental factors [25]. Missing heritability is commonly defined as the sum of the still-missing and hidden heritability, which we define below [26].

Still-missing heritability. Yang and colleagues [27] argued that most genetic effects are too small to be reliably detected in GWASs of current sample sizes. Studies applying these whole-genome methods typically produce estimates that lie between twin studies and polygenic scores: $h^2_{GWAS} < h^2_{SNP} < h^2_{family}$. The discrepancy $h^2_{SNP} < h^2_{family}$ has been referred to as still-missing heritability [26]. For many traits the still-missing heritability is roughly equal to h^2_{SNP} [28]. It is generally assumed that by genotyping rarer and structural variants, the still-missing heritability will decrease, as the denser arrays will increase h^2_{SNP}.

Hidden heritability. Since we expect to be able to almost fully capture h^2_{SNP} in the long run, the discrepancy between h^2_{SNP} and h^2_{GWAS} is sometimes referred to as "hidden heritability" [26]. One way in which this has been examined is in relation to Fisher's initial formulation of quantitative genetic theory of the *infinitesimal model* of many variants having a small effect [29]. In the GWASs that we discuss in chapter 4, most are very large meta-analyses with in many cases a few hundred loci associated with multiple traits, often explaining a relatively low proportion of the genetic variance. This refers to Fisher's model that heritability may not be missing but rather "hidden" under the significance thresholds that are used to define the alleles with high confidence. Another way to consider hidden heritability is to think in terms of heterogeneity. Since h^2_{GWAS} is usually inferred from GWA study meta-analyses that include multiple populations, heterogeneity in genetic effects on a phenotype between these populations could deflate h^2_{GWAS}. In GWASs we regularly combine multiple data sources across historical time periods and different countries. Combining data in a "mega-analysis" from six countries and several historical periods, we demonstrated that 40% of the genetic effects on education and timing of fertility (age at first birth) was in fact "hidden" or "watered down" when data across populations in different countries and time periods were combined [10]. This increased up to 75% for the number of children. For physical traits such as a height and BMI, however, that have less environmental influence, this was not a serious problem. Missing heritability is thus commonly defined as the sum of the still-missing and hidden heritability [26]. As indicated, the hidden portion will decrease as sample sizes grow and the still-missing portion will decrease with denser forms of genotyping. New techniques are constantly being developed such as the 2018 software to estimate confounding bias, SNP-heritability, enrichments of heritability, and genetic correlations using summary statistics from GWASs [30].

1.7   Conclusion

Entering the subject area of quantitative statistical genetics can be daunting. In this chapter our aim was to provide a helicopter view of the main building blocks and fundamental concepts in this area of research. We recognize that this was a rapid and abridged primer and encourage interested readers to delve further into more detailed literature as you become more comfortable with the concepts. This chapter should arm you with the basic knowledge to distinguish the main concepts such as phenotype, DNA, the genome, genes, and chromosomes and underlying processes. Genetic polymorphisms provide the groundwork for your understanding of this topic, including the terms alleles, single-nucleotide polymorphisms (SNPs), and minor allele frequency (MAF). Throughout this book you will discover that many of the complex traits we often study are highly polygenic. It is likewise important to grasp the processes of how genes relate to proteins and the central dogma of molecular biology. Many of the topics we examine throughout this book are governed by Mendel's laws and our understanding of genetic recombination during sexual reproduction but also transmission of dominant and recessive traits. Finally, we concluded with an overview of heritability, common misnomers, different types, and missing and hidden heritability. In the next chapter we now move from the basics of the human genome to a statistical primer followed by an introduction to human evolution. Here we link the basic concepts of genetics from this introductory chapter to statistical concepts followed by a longer history of human dispersal and evolution.

Exercises

1. Use the Ensembl browser (https://www.ensembl.org/index.html) to explore one of the following SNP variants.

    - rs53576 in the oxytocin receptor associated with social behavior and personality
    - rs6152 in relation to baldness
    - rs1805007, which is associated with red hair and sensitivity to anesthetics
    - rs662799 related to preventing weight gain from high-fat diets
    - s4988235 linked to lactose intolerance
    - rs1333049 associated with coronary heart disease
    - rs1799971 makes alcohol cravings stronger
    - rs7412 and rs429358, which are said to raise the risk of Alzheimer's disease by more than 10 times

2. To examine the association of each SNP with a phenotype, see SNPedia: www.snpedia.com.

3. For a summary of the literature on associations of genotypes and phenotypes, see GeneWiki (https://en.wikipedia.org/wiki/Gene_Wiki) to further investigate. Here you will find various databases to explore further (see additional ones listed in question 1).

Further reading and resources

Excellent online resources for beginners

For a very accessible introduction to many of the terms introduced in this chapter, including films and tutorials, see the website of the National Human Genome Research Institute (NHGRI), https://www.genome.gov.

The American Society for Human Genetics (http://www.ashg.org/education/index .shtml) has various educational resources and teaching material.

The Genetics Society of America (http://www.genetics-gsa.org/education/) has developed teaching materials, including an online resource room, labs, and exercises.

The Human Genome Wikipedia page (https://en.wikipedia.org/wiki/Human_genome) is a nontechnical introduction similar to this textbook.

The GO Consortium for gene ontology (www.geneontology.org) is the largest resource on information about the functions of genes, which serves as a foundation for computational analysis of large-scale molecular biology and genetics experiments.

There are many engaging popular science books on the topic as well. To name only a few, we recommend:

Conely, D., and J. Fletcher. *The genome factor: What the social genomics revolution reveals about ourselves, our history and the future.* Princeton, NJ: Princeton University Press, 2017.

Mukherjee, S. *The gene: An intimate history.* New York: Simon & Schuster, 2016.

Zimmer, C. *She has her mother's laugh: The powers, perversions, and potential of heredity.* New York: Macmillan, 2018.

Online browsers to explore

There are constantly new browsers and online resources available, with this list providing an indication of some of them as of 2019.

To study the structure of individual genes, some good sources are the University of California at Santa Cruz (UCSC) genome browser (genome.ucsu.edu); the US National Center for Biotechnology Information (NCBI) browser (www.ncbi.nlm.nih.gov); and the European Bioinformatics Institute (EBI) browser (https://www.ebi.ac.uk/).

For summaries of genetic, protein, polymorphism, drug, expression, evolutionary, and antibody data, see Israeli Weizmann Institute's Genecards database (www.genecards.org).

To examine the location, frequency, type, and predicted function of each new variant, you can refer to the SNP data called dbSNP (www.ncbi.nlm.nih.gov/SNP). A more user-friendly way to examine this database is to use the Ensembl browser (www.ensembl.org /info/genome/variation/index.html). For a database that catalogs all the known diseases with a genetic component, refer to: OMIM (Mendelian Inheritance in Man) (https://www.omim.org/).

The HUGO site (https://www.genenames.org/) is produced by the HUGO Gene Nomenclacture Committee and provides a useful resource for approved human gene nomenclature (i.e., the body or system of names used in human genetics).

GEO (the Gene Expression Omnibus) (https://www.ncbi.nlm.nih.gov/geo/) and the GO consortium has also produced a Gene Ontology Resource which contains detailed information on the function of genes at: http://geneontology.org/

References

1. M. C. Mills and C. Rahal, A scientometric review of genome-wide association studies. *Commun. Biol.* **2** (2019), doi:10.1038/s42003-018-0261-x.

2. A. Torkamani, N. E. Wineinger, and E. J. Topol, The personal and clinical utility of polygenic risk scores. *Nat. Rev. Genet.* **19**, 581–590 (2018), doi:10.1038/s41576-018-0018-x.

3. S. Mukherjee, *The gene: An intimate history* (London: Penguin Random House, 2016).

4. T. A. Manolio et al., Finding the missing heritability of complex diseases. *Nature* **461**, 747–753 (2009).

5. L. Bomba, K. Walter, and N. Soranzo, The impact of rare and low-frequency genetic variants in common disease. *Genome Biol.* **18** (2017) (available at https://genomebiology.biomedcentral.com/track/pdf/10.1186/s13059-017-1212-4).

6. C. E. Sexton et al., Common DNA variants accurately rank an individual of extreme height. *Int. J. Hum. Genomics* **5121540**, 1–7 (2018).

7. E. A. Boyle, Y. I. Li, and J. K. Pritchard, An expanded view of complex traits: From polygenic to omnigenic. *Cell* **169**, 1177–1186 (2017).

8. D. Leja, NGHRI Digital Media Database (2010) (available at http://www.genome.gov/dmd/img.cfm?node=Photos/Graphics&id=85252).

9. P. M. Visscher, W. G. Hill, and N. R. Wray, Heritability in the genomics era—concepts and misconceptions. *Nat. Rev. Genet.* **9**, 255–266 (2008).

10. F. C. Tropf et al., Hidden heritability due to heterogeneity across seven populations. *Nat. Hum. Behav.* **1**, 757–765 (2017).

11. A. Kong et al., The nature of nurture: Effects of parental genotypes. *Science* **359**, 424–428 (2018).

12. A. Courtiol, F. C. Tropf, and M. C. Mills, When genes and environment disagree: Making sense of trends in recent human evolution. *Proc. Natl. Acad. Sci. USA* **113**, 7693–7695 (2016).

13. M. C. Neale and L. R. Cardon, *Methodology for genetic studies of twins and families* (Dordrecht: Kluwer Academic Publishers, 1992).

14. J. Yang, J. Zeng, M. E. Goddard, N. R. Wray, and P. M. Visscher, Concepts, estimation and interpretation of SNP-based heritability. *Nat. Genet.* **49**, 1304–1310 (2017).

15. J. Yang, S. H. Lee, M. E. Goddard, and P. M. Visscher, GCTA: A tool for genome-wide complex trait analysis. *Am. J. Hum. Genet.* **88**, 76–82 (2011).

16. S. H. Lee, J. Yang, M. E. Goddard, P. M. Visscher, and N. R. Wray, Estimation of pleiotropy between complex diseases using single-nucleotide polymorphism-derived genomic relationships and restricted maximum likelihood. *Bioinformatics* **28**, 2540–2542 (2012).

17. O. Weissbrod, J. Flint, and S. Rosset, Estimating SNP-based heritability and genetic correlation in case-control studies directly and with summary statistics. *Am. J. Hum. Genet.* **103**, 89–99 (2018).

18. T. Polderman et al., Meta-analysis of the heritability of human traits based on fifty years of twin studies. *Nat. Genet.*, **47**, 702–709 (2015).

19. F. C. Tropf, N. Barban, M. C. Mills, H. Snieder, and J. J. Mandemakers, Genetic influence on age at first birth of female twins born in the UK, 1919–68. *Popul. Stud. (NY)* **69**, 129–145 (2015).

20. A. R. Branigan, K. J. McCallum, and J. Freese, Variation in the heritability of educational attainment: An international meta-analysis. *Soc. Forces* **92**, 109–140 (2013).

21. T. F. Hansen, The evolution of genetic architecture. *Annu. Rev. Ecol. Evol. Syst.* **37**, 123–157 (2006).

22. Z. Zhu et al., Dominance genetic variation contributes little to the missing heritability for human complex traits. *Am. J. Hum. Genet.* **96**, 377–385 (2015).

23. N. Barban et al., Genome-wide analysis identifies 12 loci influencing human reproductive behavior. *Nat. Genet.* **48**, 1–7 (2016).

24. J. Yang et al., Genetic variance estimation with imputed variants finds negligible missing heritability for human height and body mass index. *Nat. Genet.* **47**, 1114–1120 (2015).

25. J. Felson, What can we learn from twin studies? A comprehensive evaluation of the equal environments assumption. *Soc. Sci. Res.* **43**, 184–199 (2004).

26. J. S. Witte, P. M. Visscher, and N. R. Wray, The contribution of genetic variants to disease depends on the ruler. *Nat. Rev. Genet.* **15**, 765–776 (2014).

27. J. Yang et al., Common SNPs explain a large proportion of the heritability for human height. *Nat. Genet.* **42**, 565–569 (2010).

28. N. R. Wray and R. Maier, Genetic basis of complex genetic disease: The contribution of disease heterogeneity to missing heritability. *Curr. Epidemiol. Reports* **1**, 220–227 (2014).

29. R. A. Fisher, *The genetical theory of natural selection* (Oxford: Clarendon Press, 1930).

30. D. Speed and D. J. Balding, SumHer better estimates the SNP heritability of complex traits from summary statistics. *Nat. Genet.* **51**, 277–284 (2018).

2

A Statistical Primer for Genetic Data Analysis

Objectives

- Provide a refresher of *basic statistical concepts*, including variance, mean, and standard deviation, as well as covariance and variance-covariance matrices
- Outline the basics and assumptions of *statistical models* used in this research, including the null and alternative hypotheses and significance thresholds
- Distinguish between *correlation versus causation*
- Provide an overview of *causal models*
- Grasp the differences between *fixed-effects*, *random-effects*, and *mixed models*
- Understand the basic issues surrounding *replication of results and overfitting*

2.1 Introduction

Until now, we have concentrated on grasping the fundamental concepts and underpinnings of the human genome. Before moving to more advanced topics and in particular the applied statistical chapters later in this book, it is essential that you also grasp some of the core statistical concepts. As we noted at the onset, this book is written at an introductory level and aims to cater to the diverse group of researchers entering this field for the first time. Readers who already possess some basic statistical knowledge are bound to find this chapter too introductory. For those of whom statistics courses are a distant past or have only followed statistics as part of a larger program, you may find this basic refresher chapter useful. Although many of the concepts introduced in this chapter will be familiar from nongenetic data analysis, we also particularly highlight statistical concepts and issues unique to the analysis of genetic data.

The aim of this chapter is to provide a basic primer to understand the core concepts required for genetic data analysis. We then introduce more advanced topics that you may wish to pursue further. If you are interested in applying these techniques at a higher level of statistical sophistication, we strongly encourage readers to engage in further reading using more advanced or specialized textbooks, some which we refer to in our further reading section at the end of this chapter. We first review some of the basic statistical concepts that are not exclusive to statistical genetic data analysis, such as the central tendency. The bulk of the chapter then unpacks the basics of statistical models and provides a refresher of related concepts such as the null and alternative hypothesis and significance thresholds. We then distinguish between the concepts of correlation versus causation, which is pivotal to distinguish when estimating these models. We also walk the reader through various types of causal models including direct and bidirectional causal relationships, additive and common cause models, and common mediation and moderation (or interaction) models. This is followed by a short introduction to the differences between fixed-effects, random-effects, and mixed models that are often used. We conclude with a short discussion about replication, overfitting and then provide a short summary.

2.2 Basic statistical concepts

Since we know that readers likely come from a variety of disciplines and backgrounds we start with an introductory treatment of the main concepts. As described in chapter 1, you will frequently encounter the terms phenotype, which in these statistical models are the dependent variable, and genotype, which are generally what is known as the covariates, predictors, or independent variables. As we elaborate upon in later chapters, these variables can take various forms depending on their measurement. This measurement in turn impacts the statistical model that we choose. For instance, if a phenotype is measured as a binary variable (1 = disease, 0 = no disease) a logistic or other analogous model would be used. Whereas if the phenotype is measured as a continuous or quantitative outcome (e.g., height), you would need to adopt a model that not only captures the gradual scale of the data but also often the distribution of that measure.

2.2.1 Mean, standard deviation, and variance

A core underlying tenet of statistics is the central tendency or deviation from the central tendency. This deviation is referred to as the distribution or spread around the central tendency, shown in the form of a normal distribution in figure 2.1. Three basic concepts are relevant. First, the *mean* is the average or in other words the sum of the variable values divided by the sample size. Second, the *standard deviation* is the average distance of individual values from that mean across all observations. Third, *variance* (σ^2) is the average of the squared deviation of the observations of an outcome of variable (often **random variable**)[1] from its mean, divided by the number of individuals (sample size).

To calculate the variance, we thus first calculate the mean and then for each number subtract the mean and square the results (i.e., the squared differences or deviation) and then calculate the average of those squared differences. It is the measurement of how numbers are spread or distributed from the mean. It is a critical statistical concept in

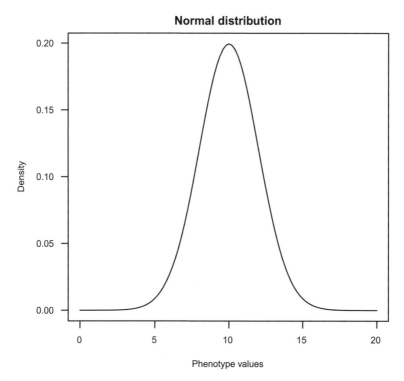

Figure 2.1
Central tendency and the normal distribution.

genetics since it allows us to determine whether different genotypes are associated with a particular trait. We are often interested in knowing how much of the variance of a particular phenotype in the population is explained by a genetic score or variants and how much is related to the environment. It is vital to note that statistical tests rely on sample size, with a difference between the means of the groups and the variance within the groups.

Why do we include the squared term? This is because the difference between each value for our variable such as height, for example, and the overall mean of the height could produce either positive or negative values. If we simply took the average, we could get a mean close to zero regardless of the spread of how tall or short people were in the group. Squaring is therefore used to give a positive value, so the sum is not zero.

$$\sigma^2(X) = \frac{\Sigma(X_i - \bar{X})^2}{N-1}$$

In these types of analyses, we are often studying a sample that is taken from a larger population (N). When we calculate variance for an entire population we use N, but when we calculate variance for a sample (of a larger population) we divide by $n-1$, where X_i is the phenotype (e.g., height) of individual i, $\overline{X}$ is the average height across all individuals, and n is the number of individuals. Recall that the mean $\overline{X}$ is the sum of all of the variable values divided by the total sample size. The *standard deviation* is the square root of the variance σ^2. There are also several methods to fit different *distributions*. The distribution shown in figure 2.1 is the normal distribution. Here you will see that a normal distribution is symmetrical around its center in the shape of a "bell curve," and the mean, median, and mode are all equal (see also box 5.1 in chapter 5 in relation to polygenic scores). There are many different types of distributions depending upon the shape of the variable including Weibull, log-normal, binomial, Poisson, or Gamma distribution to name only a few. For more advanced modeling, we encourage readers to delve into their data and dissect it to understand the distributions of different variables.

Quantiles are the fraction (or, if you prefer, you can think in terms of percentages) of data points below the given value and can be used to compare the expected with the real distribution of variables. The quantile value of 0.3 (or 30%) is, for instance, the point where 30% of the data fall below that value and 70% fall above. See the appendix at the end of this chapter for additional options on how to simulate a random variable with various specifications. The *Quantile-Quantile (Q-Q) plot* is the most frequently used visual graphing technique to evaluate the fit of a variable to a statistical distribution. As shown in figure 2.2, we plot the quantiles of one variable against the theoretical quantiles of a given distribution. The Q-Q plot is often used with genome-wide association studies (GWASs), described in chapter 4, and allows you to visually assess whether the assumption of the distribution that you have specified (e.g., normal) is violated and, if so, which data points are related to that violation. It is essentially a scatterplot of two different sets of quantiles plotted against one another. The vertical axis represents the estimated quantiles from the variable; the horizontal axis shows the theoretical quantiles from a given distribution. We also include a 45-degree reference line. If both of the two sets came from a population with the same distribution (in our example a normal distribution), the points should hover along our reference line as they do in the left panel of figure 2.2. The greater the departure is from the reference line (as in the right panel of figure 2.2), the greater the evidence you would have that the variable has a different distribution. We turn to these Q-Q plots in our chapter on GWASs but also later in chapter 12 in an applied exercise using computer-based output to assess and interpret an applied example. The R code used to create these plots is shown in the appendix.

2.2.2 Covariance and the variance-covariance matrix

Covariance is the measure of joint variance between two (random) variables. The covariance between two variables indicates their similarity. In other words, if the greater values of one variable correspond to the greater values of a second variable, the covariance would be positive. Conversely, we would have a negative covariance if the greater values of one variable are opposite and correspond to the lower values of variable two. For example, if

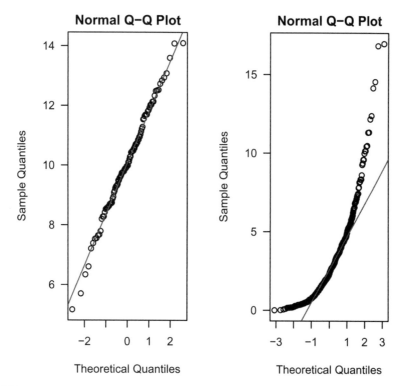

Figure 2.2
Quantile-Quantile (Q-Q) plots.

variable Y (weight) has a high value when variable X (height) is large and low when variable X (height) is small, the variables will have a high covariance. But the covariance would be close to zero if there is little or no connection between the two variables. To estimate the covariance between the two variables of X and Y, we can use the following equation where as previously $\bar{X}$ and $\bar{Y}$ represent the means of the variables.

$$COV(X,Y) = \frac{\sum_{i=1}^{n}(X_i - \bar{X})(Y_i - \bar{Y})}{N-1}$$

Where the covariance of a variable with itself is equal to its variance:

$$COV(X,X) = \frac{\sum_{i=1}^{n}(X_i - \bar{X})(X_i - \bar{X})}{N-1} = \frac{\sum(X_i - \bar{X})^2}{N-1} = \sigma^2(X)$$

A *variance-covariance matrix* is used to describe the variances and covariances between variables. It is sometimes also referred to as the dispersion matrix or simply the covariance matrix. The extension is that the covariance matrix generalizes the variance we discussed earlier to multiple dimensions. With these matrices, you will see that the variances are across the main diagonal (here in bold below) and other values refer to the covariances. This is shown for example as:

$$
\begin{matrix}
\sigma_x^2 & Cov_{x,y} & Cov_{x,z} \\
Cov_{x,y} & \sigma_y^2 & Cov_{y,z} \\
Cov_{x,z} & Cov_{y,z} & \sigma_z^2
\end{matrix}
$$

The actual magnitude of the covariance can be difficult to interpret since it is not standardized and thus depends on the actual magnitude of the variables. When working with data and different types of variables, the size of the variance and covariance thus differs according to the units of the variable. Using the height example once again, if we measured it in meters or feet, the variance would appear smaller than if we measured it in millimeters.

2.3   Statistical models

2.3.1   Regression models

In our applied chapters in the third part of this book (see, e.g., chapter 11), we focus on the estimation of a variety of *statistical models*, which we use to demonstrate how a particular set of variables relate to one another. In these applications, we often use theory to build our thinking about the relationships and which variables to include. This is generally driven by a literature review, which outlines the theory and underlying relationships and mechanisms that isolate and then link key predictors to the outcome. The model is also chosen according to assumptions about the distribution of variables and their covariance. Even with the best theory and literature review, however, it is often impossible to know the distribution or covariance in advance, so initial exploratory models often begin as a springboard to examine relationships and strive for the best fit. Note that there are also a-theoretical or hypothesis-free approaches to study associations and, in fact, a genome-wide association study (chapter 4) represents such a research design.

The equation below describes a simple model of height, genes, and nutrition. It is assumed that height is a normally distributed variable and that both the genetic variants (polygenic score for height, PGS_{height}) and nutrition intake measured in calories have a linear effect on height.

$$height = \mu + \beta_1 PGS_{height} + \beta_2 nutrition + \varepsilon$$

Where *height* is the phenotype of interest, μ is the intercept of the models, or the average value for individuals with the value 0 of the predictors PGS_{height} and *nutrition*, β_1 is the estimated effect of PGS_{height} on *height* and β_2 is the estimated effect of *nutrition* on *height* and

ε is the error term. In a linear model, the regression coefficients directly relate to the covariance as introduced earlier. In a model with only one predictor, we would write:

$$\beta_1 = \frac{Cov\ (Y, X)}{\sigma^2(X)}$$

Due to the standardization of the covariance for the variance of X_i, we can interpret the β_1 coefficient as a one unit change in Y if X changes one unit. This interpretation holds in multivariate models—with multiple predictors—holding everything else in the model constant.

The term ε in the equation represents the stochastic (probabilistic) part of the model. The elements of ε are also called residuals, since they can be thought of as the differences between the outcome variable and the deterministic part of the model ($\mu + \beta_1 PGS_{height} + \beta_{2nutrition}$). In a standard linear model, the elements of ε are assumed to have a normal distribution and to be independent of each other.

Later in this book we will introduce readers to various statistical techniques and types of regression models. *Regression analysis* is the statistical technique used to estimate the relationships among often multiple variables or what is termed multivariate regression. Our goal is to examine the relationship between the genotype and phenotype and thus how the phenotype changes when the genotype and other independent variables (also called covariates, predictors, confounders) are entered into the model. This includes various regression models such as ordinary linear regression (OLS) and maximum likelihood estimation. As we will illustrate, the parameters estimated in these models show the degree that the variables are related, the level of uncertainty (via the standard errors), and statistical significance.

2.3.2   The null and alternative hypothesis and significance thresholds

The goal of the regression approach is generally to test the *null hypothesis*, which is the statistical test to determine that there is no significant difference between specified groups. Recall from your previous introductory statistics courses that this refers to when your parameter of your estimate (Beta, β) is equal to zero. The *alternative hypothesis* is thus when your parameter is not equal to zero. We use data to perform statistical tests and if the null hypothesis is true we then calculate a p-value to determine statistical significance. Simply put, if the p-value is very small, the data are inconsistent with the null hypothesis. If the parameter passes the *significance threshold* (e.g., 0.05, 0.001) then the null hypothesis is rejected in favor of the alternative hypothesis. There has been considerable criticism and heated discussion in the area of statistical significance, primarily surrounding the fact that results are often interpreted only in relation to the null hypothesis. Some have also proposed to change the default p-value for statistical significance from 0.05 to 0.005 [1]. For the moment, however, interpretation in relation to the null hypothesis remains as a dominant approach. As we describe in more detail in chapter 4, *genetic* associations that are considered statistically significant must pass a stringent threshold with a p-value smaller than 5 times 10 to the minus 8 (i.e., 5×10^{-8}). In chapter 4 we also address the multiple-testing problem and why in fact we have this stringent threshold of 5×10^{-8}. In reality, we would not estimate a simple model such as the one shown above, but would estimate a multivariate model with multiple predictors derived from the literature. The goal is often to optimize the accuracy of our prediction or test a particular hypothesis. In our later applied chapters readers will have the opportunity to estimate and interpret statistics from various models.

2.4   Correlation, causation, and multivariate causal models

2.4.1   Correlation versus causation

The terms correlation (r) and causation are used frequently throughout this book, and it is therefore essential to distinguish between the two. Correlation represents the statistical association of two variables with one another. It is a scaled version of the covariance that has a value between minus one and one. A value that is close to zero therefore signals little covariance between the variables. A value close to one represents a strong positive covariance, with a value closer to minus one denoting a strong negative covariance. Figure 2.3 shows three examples of variables with a different correlation structure. The plots on the left and central panels both have a correlation of 0.8, but in the opposite direction. Note that the covariances, correlations, and regression models we discuss here assume a linear relationship between two variables. In other words, we would draw a line through the scatterplots in figure 2.3 and assume constant unit changes of one variable based on another across their distributions. When a correlation is close to 1, variables tend to move in the same direction with relatively little variation around the association line; when it is close to -1, the direction is the reverse (i.e., when X is high, Y is low). The far right panel shows two variables with a very low correlation ($r = 0.1$), meaning more variation in the data. Also some nonlinear associations are established in the literature, such as the U-shaped relationship between age at first birth of parents and the risk to be diagnosed with schizophrenia for the children [2]. Such associations will show low correlations in standard approaches but can be modeled, for example, with transformations of the dependent variables.

A correlation is represented by the equation below where recall that σ represents the standard deviation.

$$r = \frac{\text{Cov}(X, Y)}{\sigma(X)\sigma(Y)}$$

Correlations are the basis for many analyses such as GWASs and gene-environment interaction since they indicate a relationship between two variables. As we explore throughout this book, it is essential to distinguish between correlation and causation. Neophyte researchers can often make some inadvertent mistakes by inferring that variable X causes variable Y,

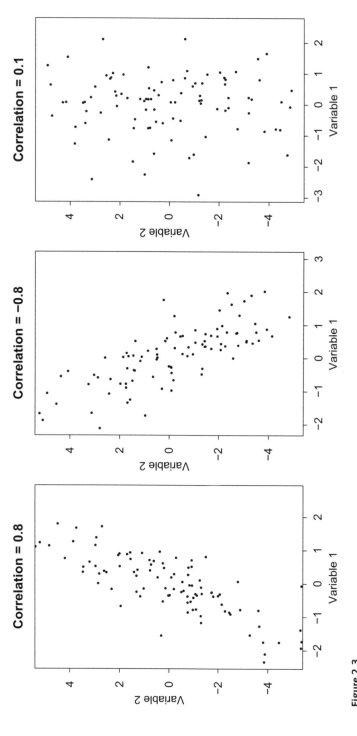

Figure 2.3
Examples of variables with high (0.8) and low (0.1) correlation.

when they are actually only observing a positive or negative correlation. Correlation simply describes the size and direction of the relationship between two or more variables. It does not mean that the change in one variable causes the change in values of the other variable.

Causation, rather, shows that the change in the values in one variable is the result of changes in the other variable. This is referred to as cause and effect. Take smoking, for example. You will find that smoking may be highly correlated with other behaviors such as higher levels of alcohol consumption. You cannot naively infer, however, that smoking causes alcoholism. Causality is often very difficult to expose in virtually all types of data analysis, including statistical genetic data analysis. In medical research, the use of *randomized control trials* is the most effective way to establish causality. In these types of studies, the sample is often split into two groups that are similar in most ways. They then receive different treatments, such as a placebo versus a particular therapy or drug. The outcomes of each group are then assessed. If the outcomes are markedly different, causality can be established. In reality, many of the complex traits that we study in this area of research often evade case-control or randomized control approaches. We are able to look at environmental changes (e.g., policy change, exposure to pollutant), but establishing causation remains a challenge. In later chapters in this book we will show various applied statistical approaches that attempt to establish causality in statistical genetics, including Mendelian Randomization. We will also highlight the challenges and continued areas of debate in this research. The aim of most applied quantitative statistical genetics is to estimate multivariate statistical models, a topic that we turn to now.

2.4.2   Multivariate causal models

In this section, our aim is to describe some basic causal models, which are essential for our theoretical reasoning about associations between variables and crucial for the conceptualization and realization of the applied statistical analyses that follow. To illustrate the different causal models, we provide some diagrams throughout the book with some simple conventions (see box 2.1), showing directional causal links between variables that are represented in boxes with one-headed arrows.

Box 2.1
Conventions for graphical representation of causal relationships

> We use some simple diagrams to illustrate theoretical causal relationships and to illustrate how we will model them in the applied chapters in parts II and III of this book. The key elements are: (1) variables are presented in boxes; (2) directional causal pathways are represented as solid one-headed arrows connecting variables, where the arrow determines the causal direction; and (3) no arrow between variables means no causal relationship between them. Even armed with only these very basic elements, we are able to visually represent more complex causal relationships between multiple variables.

The most basic *direct causal model* can be represented as in figure 2.4, where X has a causal effect on Y. Consider the hypothesis that higher educational attainment (X) causes better health (Y). Some potential underlying causal mechanisms may be that higher educated individuals have better knowledge about healthy lifestyles, are more prone to visit doctors when they are ill, or have better access to health care.

If this is true, we would expect to observe a correlation between educational attainment and health outcomes in our data. However, recall that correlation is not causation and different causal models might lead to the same observation in our data. It is, for example, possible that we observe a *reverse causal relationship* and that poor health (Y) in fact causes lower educational attainment attributed to higher levels of absenteeism or school dropout (see figure 2.5).

Often a more plausible situation is that we observe a *bidirectional causal relationship* and that both relationships are true. In reality, education likely has a causal effect on health and vice versa. In this scenario, we would speak of a bidirectional causal relationship between X and Y (see figure 2.6).

In all three scenarios shown in figures 2.4–2.6, we would observe a correlation between X and Y. Theory can often help us to understand and disentangle the causal direction. It is, for example, implausible to assume that the well-established correlation between height and education is due to the fact that education causes an increase in height. If theory or a literature review does not clarify the relationship, empirical strategies include the collection of longitudinal data in which individuals are observed over time or instrumental variable approaches, which we introduce in the framework of Mendelian Randomization in chapter 13. For example, we could apply Mendelian Randomization to establish whether there is a bidirectional causal relationship between schizophrenia and marijuana use. On the one hand, schizophrenia may increase marijuana use, but on the other hand, marijuana use increases the risk of developing schizophrenia [3, 4]. A particular and unique advantage of genetic analyses derived from the GWA studies we describe in chapter 4 is that genes are fixed at birth and thereby do not suffer reverse causality from phenotype to genotype.

In the majority of cases, however, we are not interested in the simple association between two variables but rather the more complex relationship among multiple variables. Four basic causal models in such a framework are the additive model, model of a common cause, mediation model, and the interaction or moderation model, respectively. In reality again, mostly hybrids of these model types are true.

Figure 2.7 depicts the *additive model* in which X and a second predictor variable Z have independent effects on Y. Think, for example, of a dinner party, where some people drink wine (X) and others drink beer (Z). Some of the guests, however, have both types of drinks. If we are interested in the causal effect of beer and wine consumption on how tipsy our guest will become (Y), we could correlate X with Y or Z with Y. What is important to note in the additive model is that when both X and Z have a causal effect on Y, if we

Figure 2.4
A direct causal effect of X on Y.

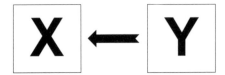

Figure 2.5
Reverse causal relationship between X and Y.

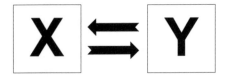

Figure 2.6
Bidirectional causal relationship between X and Y.

would split the group by Z, for example, and differentiate the guests by those who drank some beer and those who did not, within each group we would observe the correlation between drinking wine (X) and getting more tipsy. The same holds if we split the group by X: Z would still be correlated with Y. By splitting the groups, we are simply referring to a multivariate analysis, where we might also use language, such as saying that we *control* for Z when studying the effect of X on Y. Usually, multivariate regression analyses are conducted for this purpose (see next section).

It is also possible that we do not observe a correlation between X and Y among subgroups of Z but rather a causal model with Z as a *common cause* of X and Z. Consider for example the comorbidity (i.e., high correlation) between schizophrenia and bipolar disorder. In this case, the important causal question would be whether we can establish the true causal relationship between these two mental health conditions. One hypothesis is that the correlation is caused by genetic variants (Z) that are affecting both schizophrenia (X) and bipolar disorder (Y; see figure 2.8). If this would be true, we would observe a correlation between X and Y in our data. However, if we split the data in one group with individuals carrying alleles related to schizophrenia and a second group of individuals who do not have these markers, the correlation between X and Y would disappear in both subgroups.

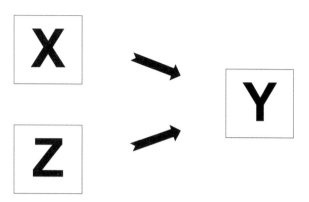

Figure 2.7
Causal model with additive effects of X and Z on Y.

We refer to this model when we say that Z might confound the relationship between X and Y. *Genetic confounding* is a major source of endogeneity, which we discuss in more detail in our chapters on polygenic scores and Mendelian Randomization. There we also provide the computer code to conduct these types of regression analyses in R and how to interpret results.

Another type of causal model is a *mediation model*, where Z mediates the association between X and Y. In genetic research and beyond, we are often interested in the causal mechanisms that connect cause X and outcome Y. Such mechanisms are considered as intermediate variables Z, for example an intermediate phenotype (see figure 2.9). Consider the association between certain genetic variants and lung cancer. If we find a strong association, the aim is then to understand the causal chain in order to, for example, enact targeted intervention strategies. One possibility is that there is an underlying dysfunctional biological pathway. Another example is where a set of a genetic variants such as in a polygenic score (PGS) (X) has a causal effect such as how nicotine dependency causes smoking (Z), which in turn causes lung cancer (Y; see also figure 2.6). Similar to the confounding model, in such a scenario we would expect to observe a correlation between X and Y in the overall sample, but no correlation between the PGS (X) and lung cancer (Y) within the groups of smokers and nonsmokers (at least if all smokers smoke the same amount). Importantly, in contrast to the confounding model, PGSs in this example would represent the X variable and in the confounding model the Z variable. Here, the PGSs help us to theoretically differentiate between models, since in the mediation model genetics and the PGS cannot be the mediator (Z), since it is fixed at birth.

Finally, a variable Z can change the effect of X on Y (figure 2.10). This model is called *moderation model* or *interaction model*, since X and Z interact in a sense that X also changes the effect of Z on Y. The chosen perspective depends on your main research

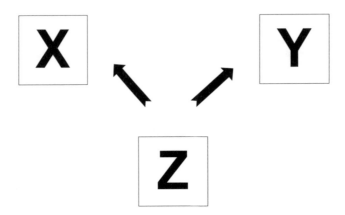

Figure 2.8
Causal model with Z as a common cause of X and Y.

Figure 2.9
Causal model where Z mediates the association between X and Y.

interests, questions and theory. This model is most frequently used in gene-environment interaction (GxE) studies, which is one of the growing topics in statistical genetic data analysis. In this book, we describe gene-environment interplay in detail in chapter 6 and walk through empirical applications later in chapter 11.

A well-established finding in the literature is that the effect of genes (X) on body mass index (BMI) (Y) increases across birth cohorts (Z) [5, 6]. The theoretical reasoning behind this finding is that in more recent birth cohorts, high-fat diets became more easily accessible and were introduced already at young ages. There has been a steady increase in BMI and obesity over the past century in many industrialized countries such as the United States. In terms of interaction, this implies that the effect of birth cohort on BMI depends on genotype (X). Individuals with different genetic predispositions for BMI experience a differential effect of birth cohort on BMI. Empirically, this phenomenon of GxE interaction or heterogeneity in causal effects between groups implies that the effect of X on Y is significantly different if we stratify the sample by Z. It also implies that the overall effect of X on Y is some sort of weighted averaged effect of X on Y across groups of Z. It also implies that there will be smaller or larger effects in some groups or that it may even

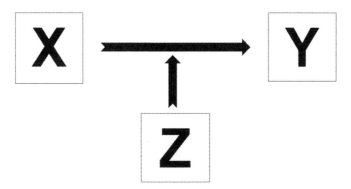

Figure 2.10
Causal model where Z moderates the association between of X and Y; we can also say that X and Z interact.

switch signs across them. Note that although mediation and moderation models are often confused, they are genuinely different in their meaning.

2.5 Fixed-effects models, random-effects models, and mixed models

Until now we have only discussed fixed-effects models, which are when the effect of the covariate on the phenotypic outcome is modeled as fixed or the same per unit increase of the covariate across the sample. Fixed-effect models are thus different from random-effects models or mixed models, where some or all of the model parameters are considered as random variables. Readers should note that the terms are used slightly differently in biostatistics versus econometrics. Andrew Gelman has written an excellent blog describing these differences [7]. In econometrics, fixed-effects models are frequently used to identify a series of individual-specific variables included in hierarchical or panel data. In biostatistics and the genetics texts, "fixed effects" refers to the population-average effects whereas "random effects" denotes the distribution of subject-specific effects. These random subject-specific effects are often considered as unknown latent variables.[2] We often use these models to control for what is termed unobserved heterogeneity. Here, the assumption is often that the heterogeneity is time constant and not correlated with the other covariates. Random-effects models are often very useful since we have subsets of individuals in the data. This includes variation of subsets or clusters of individuals such as by family, school, neighborhood, city, country, or hospital. When examining longitudinal data, the subset could be repeated measurements of the individual. Or if examining recurrent event data, the subset could be repeated disease episodes. We therefore model random effects to account for the subsets in the data that may in turn influence the main effects.

Mixed-models contain both fixed and random effects. They are commonly used to examine repeated measures on the same individuals in longitudinal panel studies or measurements on particular subsets. In the genetics research covered in this book, mixed models are useful for controlling for population structure and estimating heritability. When using these models in relation to population structure, the random effect is the contribution to the genotype-phenotype associations that are due to the relatedness between individuals. The relatedness between individuals is, as we discussed previously, calculated using the *Genomic Relationship Matrix (GRM)*. Mixed models thus account for the genetic distance between individuals in your sample and thus control for the potential confounding due to the association of differences in the genetic profile and differences due to geographic location.

As discussed in our section on heritability in chapter 1, mixed models are also often used to estimate SNP-heritability using *Genome-based Restricted Maximum Likelihood (GREML)*. This is a statistical method of variance component estimation that quantifies the narrow-sense additive contribution to a phenotype's heritability. This is specific to a particular subset of genetic variants, often limited to SNPs with a MAF of $> 1\%$. For this reason it has often been termed "chip" or "SNP" heritability (or h^2_{SNP}, as introduced in chapter 1). As outlined previously, with the arrival of whole-genome data, researchers were able to go beyond the use of twin models to examine the genetic similarity between unrelated individuals. The software that is used to conduct this analysis is *GCTA— Genome-wide Complex Trait Analysis* (see "Software for mixed-model analyses" at the end of this chapter). These estimates produce a lower bound of the genetic contribution of a phenotype or trait without needing to rely upon the often restrictive assumptions from twin or family-family analyses. Simply put, if a particular phenotype is heritable, individuals who are more genetically related should have phenotypic values that are more similar. If the genetic relatedness of individuals is not an indicator of similar phenotypic values, then we can conclude that the particular phenotype is likely not influenced by genetics. Later in this book in chapter 9 we provide an example of how to use GCTA and conduct this type of analysis.

2.6   Replication of results and overfitting

When engaging in an analysis only using one dataset or sample,[3] you may encounter the problem of overfitting, which is tied to the ability to replicate results in a separate sample. *Overfitting* refers to the problem when the predictors in your model predict the outcome better in your particular data or sample than they would in a new independent dataset. Overfitting can firstly, be the result of *multiple testing*. This is when the association between our covariates and phenotype go against the basic premise that this association is the result of both the true population effect and random chance. As we describe in chapter 4, genetic variants from these techniques are the result of testing the association of sometimes millions of genetic variants with a particular phenotype. Those with the largest associations are more likely to have a stronger contribution than we would expect than by chance. When researchers attempt to replicate the results in a smaller sample, they see that the result is often smaller associations. This is attributed to the fact that the top results from the originally overfitted model and their effect estimates (e.g., regression coefficients) are larger or inflated than the true effects. In fact, any joint model with multiple covariates or predictors will be overfitted if it is only built and tested on a single sample. This is due to the fact that we estimate parameters to optimize the fit of the model to that particular data. It is thus logical that the model does not perform well on new independent data.

In this introductory textbook we cannot describe all of the ways in which to deal with overfitting and outline only a few here. One is the use of *training and validation* datasets, which are now more commonly used to counter this problem of replication. One option is to retest the finding in a similar independent dataset to see if the result replicates. Another option, which is becoming increasingly popular due to the release of large datasets such as the UK Biobank (which has around 500,000 individuals), is to split the data in the same sample into a training and validation set. This can then be repeated using different divisions of the data to improve robustness.

Other techniques to deal with this problem are called regularization or shrinkage methods. Shrinkage methods perform variable selection to effectively shrink the parameters so that the predictors are still retained in the model but shrink some portion of the parameter estimates. Lasso regression can be used to perform variable selection and Ridge regression to shrink the parameter estimates. Although they go far beyond this introductory text, they effectively penalize the parameters in the formula for their optimization. There is also the elastic net approach that combines Ridge and Lasso regression. The penalty is formulated as a prior probability in Bayesian shrinkage methods. The penalty can be set in various manners such as penalizing large effects or shrinking small effects to zero or close to zero. The choice depends on the model and analysis. Since the underlying truth is often unknown, multiple analyses should be run to test which leads to the prediction in independent data and cross-validation. These methods are increasingly common in genetics and interested readers should refer to more advanced sources.

2.7 Conclusion

The aim of this chapter was to provide a brief primer of some of the central statistical terms underpinning the core analyses in this book. As noted at the onset of this chapter, our intent was to keep the material as accessible as possible. As you advance in your knowledge, researchers are encouraged to pursue the mathematical and statistical underpinnings of models and methods in the ample advanced books and articles available in this area. Armed with the basic statistical concepts of central tendency and a reminder of the basic premises of statistical models, it is our hope that readers are able to more easily follow the chapters that follow. Here it is likewise vital to be able to distinguish between correlation versus causation and understand the intricacies and possibilities of the causal models you will be able to test. As we note in chapter 4, issues of replication have been key in this scientific enterprise. In chapter 6 we also note problems that have arisen, mostly in the area of early candidate gene studies that did not replicate nor have sufficient sample size, which remain core controversies and areas of discussion.

Exercises

1. Which causal model does each of the following (hypothetical) statements represent: an additive, common cause, mediation, or moderation model?

    a. The rs2777888 SNP reduces male fertility because it reduces sperm quality.

    b. High blood pressure is observed in parents and their children because they share the same dietary habits.

    c. If I do not eat enough before I exercise, my muscle pain will be worse.

    d. Both genes and the environment are important for predicting type 2 diabetes.

2. Think about a causal model that is relevant to your own research and draw it using a diagram.

3. Reproduce figure 2.1 in RStudio (see the appendix) using the following command:

```
curve(dnorm(x,m=10,sd=2),from=0,to=20,main="Normal distri-
bution", xlab="Phenotype values", ylab="Density")
```

4. Simulate a random variable and display the mean, standard deviation, and 30% quantile of this variable.The R code below can be used to simulate a random variable following a Normal distribution with mean 10 and standard deviation equal to 2.

```
X<-rnorm(100, 10, 2)
```

Estimate its empirical mean, standard deviation, and quantiles by typing the following commands in RStudio.

```
mean(X)
sd(X)
quantile(X,.30)
summary(X)
```

5. Repeat the previous exercise and simulate a variable based on a mean of 20, a standard deviation of 5, and on 1,000 individuals.

6. Reproduce figure 2.2 using the following code in R. Note that the variable plotted in the right panel of the figure is obtained by simulating a chi-squared distribution with 3 degrees of freedom.

```
par(mfrow=c(1,2))
qqnorm(X)
qqline(X, col=2)
y <- rchisq(500, df=3)
qqnorm(y)
qqline(y, col=2)
par(mfrow=c(1,2))
```

7. Reproduce figure 2.3 using the following code in R.

```
rho1 <- 0.8
m1 <- 0; s1 <- 1
m2 <- 0; s2 <- 3

m <- c(m1,m2)
sigma1 <- matrix(c(s1^2, s1*s2*rho1, s1*s2*rho1, s2^2), 2)

par(mfrow=c(1,3))

require(MASS)
bivariate.dist1 <- mvrnorm(100, mu=m, Sigma=sigma1)
colnames(bivariate.dist1) <- c("X1"," X2")
plot(bivariate.dist1, pch=20, xlab='Variable 1',
ylab='Variable 2', main='Correlation=0.8')

rho2 <-0.8
sigma2 <- matrix(c(s1^2, s1*s2*rho2, s1*s2*rho2, s2^2), 2)
bivariate.dist2 <- mvrnorm(100, mu = m, Sigma=sigma2)
colnames(bivariate.dist2) <- c("X1"," X2")
plot(bivariate.dist2, pch=20, xlab='Variable 1',
ylab='Variable 2', main='Correlation=-0.8')

rho3 <- 0.1
sigma3 <- matrix(c(s1^2, s1*s2*rho3, s1*s2*rho3, s2^2), 2)
bivariate.dist3 <- mvrnorm(100, mu=m, Sigma=sigma3)
```

```
colnames(bivariate.dist3) <- c("X1"," X2")
plot(bivariate.dist3, pch=20,
xlab='Variable 1', ylab='Variable 2',
main='Correlation=0.1')

par(mfrow=c(1,1))
```

Further reading

Greene, William H. *Econometric analysis*. 4th ed. Upper Saddle River, NJ: Prentice Hall, 2000.

Henderson, C. R., Oscar Kempthorne, S. R. Searle, and C. M. von Krosigk. The estimation of environmental and genetic trends from records subject to culling. *Biometrics, Inter. Bio. Soc.* **15**(2), 192–218 (1959).

Laird, Nan M., and James H. Ware. Random-effects models for longitudinal data. *Biometrics, Inter. Bio. Soc.* **38**(4), 963–974 (1982).

Burgess, S., and S. G. Thompson. *Mendelian randomization: Methods for using genetic variants in causal estimation*. London: Chapman and Hall/CRC, 2015.

Verbeek, Marno. *A guide to modern econometrics*. West Sussex, UK: John Wiley & Sons, 2008.

Yang, Jian, Noah A, Zaitlen, Michael E. Goddard, Peter M. Visscher, and Alkes L. Price. Advantages and pitfalls in the application of mixed-model association methods. *Nat. Gen.* **46**(2), 100–106 (2014).

Software for mixed-model analyses

FastLMM: http://research.microsoft.com/en-us/um/redmond/projects/mscompbio/fastlmm/.

GCTA: https://cnsgenomics.com/software/gcta/#Overview.

GEMMA: http://www.xzlab.org/software.html.

MMM: http://www.helsinki.fi/~mjxpirin/download.html.

Appendix

Figure 2.1 can be reproduced using R by typing the following command:

```
curve(dnorm(x,m=10,sd=2),from=0,to=20,main="Normal distri-
bution", xlab="Phenotype values", ylab="Density")
```

The R code below can be used to simulate a random variable following a Normal distribution with mean 10 and standard deviation equal to 2.

```
X<-rnorm(100, 10, 2)
```

We can then estimate its empirical mean, standard deviation, and quintiles by typing the following commands in RStudio.

```
mean(X)
10
sd(X)
2
quantile(X,.30)
??
summary(X)
??
```

Figure 2.2 is obtained using the following code in R. Note that the variable plotted in the right panel of the figure is obtained by simulating a chi-squared distribution with 3 degrees of freedom.

```
par(mfrow=c(1,2))
qqnorm(X)
qqline(X, col=2)

y <- rchisq(500, df = 3)
qqnorm(y)
qqline(y, col=2)
par(mfrow=c(1,2))
```

Figure 2.3 can be produced using the following code in R.

```
rho1 <- 0.8
m1 <- 0; s1 <- 1
m2 <- 0; s2 <- 3

m <- c(m1,m2)
sigma1 <- matrix(c(s1^2, s1*s2*rho1, s1*s2*rho1, s2^2), 2)

par(mfrow=c(1,3))

require(MASS)
bivariate.dist1 <- mvrnorm(100, mu=m, Sigma=sigma1)
colnames(bivariate.dist1) <- c("X1"," X2")
```

```
plot(bivariate.dist1, pch=20, xlab='Variable 1',
ylab='Variable 2', main='Correlation=0.8')

rho2 <-0.8
sigma2 <- matrix(c(s1^2, s1*s2*rho2, s1*s2*rho2, s2^2), 2)
bivariate.dist2 <- mvrnorm(100, mu=m, Sigma=sigma2)
colnames(bivariate.dist2) <- c("X1"," X2")
plot(bivariate.dist2, pch=20, xlab='Variable 1',
ylab='Variable 2', main='Correlation=-0.8')

rho3 <- 0.1
sigma3 <- matrix(c(s1^2, s1*s2*rho3, s1*s2*rho3, s2^2), 2)
bivariate.dist3 <- mvrnorm(100, mu=m, Sigma=sigma3)
colnames(bivariate.dist3) <- c("X1"," X2")
plot(bivariate.dist3, pch=20,
xlab='Variable 1', ylab='Variable 2',
main='Correlation=0.1')

par(mfrow=c(1,1))
```

References

1. D. J. Benjamin et al., Redefine statistical significance. *Nat. Hum. Behav.* **2**, 6–10 (2018).

2. D. Mehta et al., Evidence for genetic overlap between schizophrenia and age at first birth in women. *JAMA Psychiatry* **73**(5), 497–505 (2016).

3. S. H. Gage et al., Assessing causality in associations between cannabis use and schizophrenia risk: A two-sample Mendelian randomization study. *Psychol. Med.* (2017), doi:10.1017/S0033291716003172.

4. J. Vaucher et al., Cannabis use and risk of schizophrenia: A Mendelian randomization study. *Mol. Psychiatry* (2018), doi:10.1038/mp.2016.252.

5. H. Liu and G. Guo, Lifetime Socioeconomic Status, Historical Context, and Genetic Inheritance in Shaping Body Mass in Middle and Late Adulthood. *Am. Sociol. Rev.* (2015), doi:10.1177/0003122415590627.

6. S. Walter, I. Mejia-Guevara, K. Estrada, S. Y. Liu, and M. M. Glymour, Association of a genetic risk score with body mass index across different birth cohorts. *JAMA—J. Am. Med. Assoc.* (2016), doi:10.1001/jama.2016.8729.

7. A. Gelman, Why I don't use the term "fixed and random effects." *Stat. Model. Causal Inference Soc. Sci.* (2005) (available at https://statmodeling.stat.columbia.edu/2005/01/25/why_i_dont_use/).

3

A Primer in Human Evolution

Objectives

- Gain a basic understanding of *human dispersal out of Africa* and the link with genetic diversity
- Grasp the concept of *population structure* and detecting population stratification with *Principal Component Analysis*
- Understand the common *misnomers of population structure* and that ancestry does not equate to the socially constructed category of race, which is not a biological category
- Realize *how genes mirror geography*
- Identify the fundamentals of *evolution, natural selection, fitness, types of selection,* and related terminology
- Comprehend how evolution can also occur via *genetic drift* in the form of a bottleneck or founder effects
- Understand the assumptions, notation, and implications of the *Hardy–Weinberg equilibrium*
- Grasp the basics about *linkage disequilibrium* and *haplotype blocks*

3.1 Introduction

Human evolution and the history of our species are essential to understanding the human variability that we study in this book. The aim of this chapter is to provide a rudimentary synopsis of this vast area, which in turn forms the foundations of the genetic differences we study throughout this book. The contemporary human genomes we most often examine are the result of genetic reshuffling and recombination, migration patterns, and natural selection. In the next section we document the human dispersal out of Africa that occurred over millions of years. This includes some of the great human migrations and links of

Homo sapiens with *Homo erectus*, Neanderthals, and **Denisovians**. An understanding of where we come from helps us to appreciate why certain populations such as sub-Saharan Africans have more genetic diversity than those of European ancestry, for instance.

We then define the essential concepts of population structure and stratification and how Principal Component Analysis (PCA) is used to detect this. This is followed by debunking common misnomers and misuses of population structure and the clarification that ancestry is not equated with race, nor is race a biological category. We then discuss how genes, in fact, mirror geography and present several examples in this area of research. The broad topics of human evolution, selection, and adaptation are then outlined by first describing evolution and natural selection and linking them to core topics such as beneficial or deleterious alleles. This is followed by further elaboration of fitness and variations of selection, sexual selection, and sexual dimorphism. Evolution can also occur by what is known as genetic drift, and specifically bottleneck and founder effects. You will also need to understand the related notion of the Hardy–Weinberg equilibrium (HWE), including the main assumptions and basic notation. In chapter 1 we discussed how polymorphisms are inherited together through linkage disequilibrium (LD), and in this chapter we now link it to how this results in the haplotype blocks that we observe in populations.

3.2   Human dispersal out of Africa

Archeological and anthropological work has been pivotal in our understanding of human dispersal. (For an overview, also with a visualization using maps, see [1, 2]). *Human dispersal* refers to the early migrations and expansion of modern humans across the continents, which began around 2 million years ago out of Africa. A brief history is mapped in figure 3.1. Humans are primates and the only surviving members of the genus *Homo* and the species *sapiens*. Within the family of **Hominidae**,[1] we are the closest to gorillas and orangutans. Humans split from gorillas and chimpanzees somewhere between 5 to 10 million years ago. Researchers currently estimate that *Homo erectus* migrated out of Africa into Asia and Europe over 1 million years ago. They were then replaced around a half a million years ago by a second lineage of Neanderthals and Denisovans.

Around 50,000 to 100,000 years ago, *Homo sapiens* emerged as a new species in southern Africa and then spread out of Africa. Although most of the populations were isolated, there is evidence that some interbreeding occurred between *Homo sapiens* and Neanderthals, with many individuals of European ancestry still harbouring a small percentage of around 2% Neanderthal. A large migration of *Homo sapiens* formed the ancestors of Australian Aborigines around 40,000 years ago with a second migration colonizing Europe and Asia. Populations in northeast Asia then migrated across the Bering land bridge into northwest North America around 15,000 to 30,000 years ago. Research has suggested that these groups dispersed across the Americas via coastal routes. Two later out-of-Asia

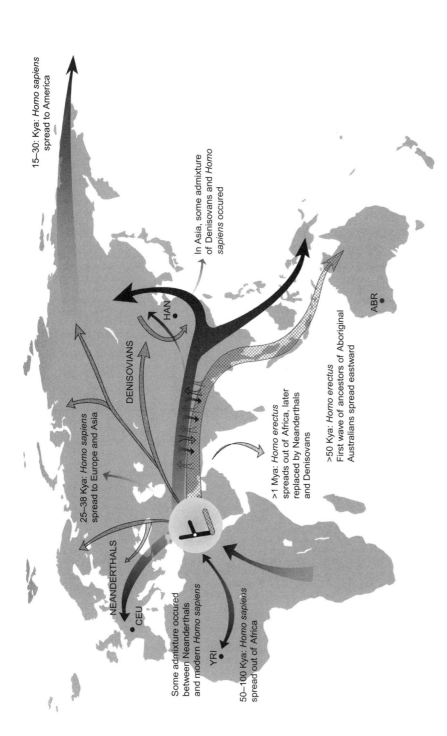

Figure 3.1

A brief history of the great human migrations.

Source: Produced by authors and adapted from the various sources of Stoneking and Krause (2001) [1], Rasmussen et al. (2011) [3], and Gibson (2015) [4]. (For a color version, refer to the online website of the book.) More than 1 million years ago (Mya) *Homo erectus* migrated out of Africa into Asia and Europe and were later replaced by Neanderthals and Denisovans (upper striped lines). The striped upper lines that split refer to an admixture between Denisovan-like populations and *Homo sapiens*. Between 50–100 Kya (i.e., 50,000 to 100,000 years ago), *Homo sapiens* emerged as a new species out of Africa (solid grey lines). Interbreeding occurred between *Homo sapiens* and Neanderthals. Over 50 Kya, a major migration of *Homo sapiens* resulted in the establishment of Australian Aborigines (gridded arrow). From 25–38 Kya, a second migration of *Homo sapiens* colonized Europe and Asia (black arrows). After around 15–30, Kya *Homo sapien* populations in northeastern Asia migrated across the Bering land bridge to northwestern America.

Box 3.1
Why do sub-Saharan African populations have more genetic diversity than other populations?

Most evidence places the origins of *Homo sapiens* in Africa. This is due to the fact that sub-Saharan Africa is where patterns of DNA sequence variation are the greatest. In genetics, the population with the greatest genetic variation is assumed to be the oldest. Current knowledge places the cradle of the human species in what is now modern Namibia and Angola in Southern Africa. This is attributed to the fact that when humans migrated to new regions, they took progressively smaller amounts of genetic variation in the gene pool with them. Each new population is younger than its original source and thus has less time to accumulate new mutations. In fact, sequencing of Khoi-San bushmen showed that even two people from adjacent villages were as different from one another as any two European or non-African ancestry individual [5, 6].

migrations account for around half of the genetic variance in Eskimo-Aleuts and 10% of some Canadian First Nation (NaDene) populations. More recent migrations include Berbers populating North Africa (~10,000 years ago), the Pacific Islands (~3,000 years ago), and New Zealand's Māori (~700 years ago). This was followed by other waves of migration such as the European influx into the Americas in the late 1400s. This human dispersal across the globe in turn impacted genetic diversity between populations (see box 3.1 and also figure 9.1).

3.3   Population structure and stratification

3.3.1   Population structure, genetic admixture, and Principal Component Analysis (PCA)
Although most human variation is the same among all population groups, migration and what is termed "admixture" between different population groups leaves the breadcrumbs of **population structure** patterns. **Genetic admixture** occurs when two or more previously isolated and genetically differentiated populations interbreed. The result is new genetic lineages. The patterns of this population structure allow geneticists to derive ancestry based on genetics. This ability to map or quantify the subdivisions of populations is called population structure. *Population structure* refers to the patterns found in the genetic data that allow us to determine an individual's ancestry. It shows how populations are divided due to genetic admixture.

The most common way to estimate and detect population stratification is by employing the method called Principal Component Analysis, with applications and a more detailed discussion later in this book (chapter 9). **Principal component analysis (PCA)** is a statistical technique used to emphasize variation and bring out strong patterns underlying the data, with the aim of minimal loss of information. It is a way to reduce the dimensionality of data from high to lower dimensional data and allows the principal components (PCs) to be visualized. We need to reduce this dimensionality since one of the largest problems we face when comparing DNA sequences is that each of the 3 billion base pairs represents a possible dimension of similarity. Or, in other words, we would need to compare a pair of individuals nucleotide by nucleotide (i.e., at a single base-pair resolution) along a colossal 3 billion dimensions. To reduce the 3 billion dimensions of the data, we therefore use PCA to identify the axes that have the greatest genetic differentiation among individuals. PCA extracts the principal components from the data in a decreasing order of variance. As we describe shortly in our statistical foundations chapter, some readers will know PCA as multidimensional scaling, which reduces the full matrix of hundreds of thousands of SNPs to "eigenvectors" that capture the central components of the covariance. These are then plotted against each other using a 2-D scatterplot that visualizes the differences between individuals and populations. Each PC represents how much of the variance in genome-wide genotype frequencies are captured by that component. Most human variation is shared among all groups, with only small shifts in some allele frequencies. PCs in fact only explain very little of the overall variance. In many cases, the first two PCs explain the majority of genetic differences.

3.3.2   Common misnomers of population structure: Ancestry is not race

One common misconception is that the population structure or PCAs equate to racial or ethnic differences, a statement that is patently incorrect (see box 3.2). *The terms ancestry and race in human genetics are not interchangeable*, and it is essential when entering this research area to engage in well-informed and careful interpretations of genetic differences between ancestral groups. Genetic variation must be distinguished from the social, cultural, and political meanings ascribed to different human groups. *Race is not a biological category* since as we have described in this chapter, genetic variation is traced to geographical locations and does not map into the perpetually changing and socially and politically defined racial or ethnic groups. Populations are the product of repeated mixtures over tens of thousands of years. The concentration of genetic alleles in some groups is thus related to where they have descended from and has nothing to do with the social category of race.

3.3.3   Genetic scores cannot be transferred across ancestry groups

As we describe shortly in our chapter on genome-wide association studies (GWASs) and elsewhere [7], the majority of discoveries to date have been conducted on European ancestry populations. European ancestry-based polygenic scores derived from GWASs cannot be directly used for prediction in non-European ancestry populations due to differences in linkage disequilibrium (LD), allele frequencies, and genetic architecture. Recall from chapter 1 that LD refers to the fact that alleles are not randomly associated at different loci in a population. For instance, if a T at one SNP locus is almost always observed with a G at another SNP locus, the two SNPs are said to be in LD. This is due to the fact that their co-occurrence is more correlated than you would expect by equilibrium—or in other words random—association conditions.

An excellent study that empirically demonstrates this problem is by Alicia Martin and colleagues [8]. They show that these single-European ancestry GWASs have very limited portability to other populations, emphasizing the need to collect data from more diverse groups. A striking finding, for instance, was that height was predicted to decrease as the genetic distance from Europeans increased. This was contrary to actual observed height, such as those in West Africa. Using simulations, they demonstrated that the PGSs based on European populations were biased by genetic drift in other populations and that biases were unpredictable.

More non-European ancestry-based GWA study discoveries are required to take the different LD population structures into account. In this way, we will be able to discover causal SNPs and those that may differ in their LD with the top hits across populations. Examining the PGSs from type 2 diabetes and coronary heart disease, for instance, Reisberg et al. showed that differences between the distributions of African and European-ancestry populations—and thus also high- and low-risk estimations—can be larger or even the

Box 3.2
Misconceptions of race and ancestry in genetics: Why white supremacists should not be chugging milk

In 2018, Pulitzer Prize–winning *New York Times* author Amy Harmon published the article "Why White Supremacists Are Chugging Milk (and Why Geneticists Are Alarmed)" [10]. The piece addresses the misinterpretation of genetics by right-wing proponents of racial hierarchy. As Harmon notes, the use of the ancestry component of genetics research in the name of white supremacy has been perhaps one of the worst and most incorrect misappropriations of this research. It has been misused by these groups in relation to intelligence, school achievement gaps, immigration, and policing. As we have reviewed elsewhere [11], this links back to the dark history of eugenic policies that emerged in the 1880s and extreme atrocities in recent history. This perspective has been widely, and rightly, condemned. Harmon refers to a lecture of famous geneticist John Novembre, whose research is featured in the next section, where he describes white nationalists chugging milk at a gathering to emphasize their lactose tolerance as adults. There he likewise notes social media discussions of hate speech urging people who can't drink milk to "go back." As Novembre and Harmon both note, beyond this incorrect interpretation, these groups do not realize that cattle breeders in East Africa also have high lactose tolerance. The confusion for the lay public is compounded by the use of ancestry as a main selling point of many commercial direct-to-consumer genetics companies. In these tests a particular ancestral background is often used as Harmon terms as a "racial ID card" or "race realism." Academics increasingly respond to these deep-rooted misunderstandings and misappropriation of research—to those open to listen. In fact in November 2018, the American Society for Human Genetics (ASHG) published a statement denouncing attempts to link genetics and racial supremacy [12]. For a more detailed discussion by Novembre and others see additional reading [13, 14]. In our later chapter 14 on ethics, we return to this topic in more detail.

opposite to each other [9]. The fact that the risk allele frequencies for these diseases of type 2 diabetes and coronary heart disease tend to be higher in African than European populations is the reason for higher-performing PGSs in African ancestry groups. As we describe in later chapters, the frequencies of the SNPs used for PGSs contain a strong population component even without applying any PGS weighting.

3.3.4 How genes mirror geography

PCA methods have been used in multiple studies to stratify populations by continents, countries, or regions within a country without any prior knowledge of their geographical relationships. If a PCA is run including individuals from populations residing in Africa, Asia, and Europe, they are easily differentiated. A now-classic example is the 2008 *Nature* paper by Novembre et al., which mapped how genetic variation mirrors geography in Europe [15]. As shown in figure 3.2, the authors used the Population Reference Sample to show stratification by country of origin. The *Population Reference Sample* is a DNA resource that was assembled from a large number of subjects across the world in order to facilitate exploratory genetics research [16]. It included nearly 6,000 individuals of African American, East Asian, South Asian, Mexican, and European origin. The findings from the Novembre and colleagues study are clear and remarkable given the relatively compact geography and distances within Europe. Another example is the landmark 2014 study published in *Science*, which examined the genomic structure of admixed populations to produce an atlas of worldwide human admixture history [17]. The authors combined this genetic data with over 100 historical events that had occurred over the last 4,000 years. They were able to identify historical events such as the impact of the Mongol empire, Arab slave trade, and European Colonialism to reveal how admixture had shaped human populations.

   This type of research has often been conducted within countries as well. In the Netherlands, for example, a study showed that three principal components were significantly correlated with geography, dividing the country between north and south and east and west, with a distinct middle-band of the country [18]. The strong north–south PC had correlations with genome-wide homozygosity, which reflects a founder effect related to northwards migration. We discuss the founder effect in the next section, which is related to a relatively small number of colonizing ancestors. The authors attributed this divergence between the different geographic subpopulations as signals for diversifying selection (defined in the next section) and as a sign of revealing a particular evolutionary history.

   The broader PCA plot in figure 3.2, however, still shows a broad view of ancestry along different axes. If individuals come from a relatively homogeneous background with limited migration or admixture, these types of metrics are very accurate. They are particularly accurate for European-ancestry individuals where all four grandparents are from the same country. If this is the case, researchers are able to pinpoint ancestry to within a few hundred kilometers [15]. In 2019, PC results are still difficult to interpret for admixed individuals such as Mexican Americans, who have genomic ancestry often from European, Native American, and West African populations.

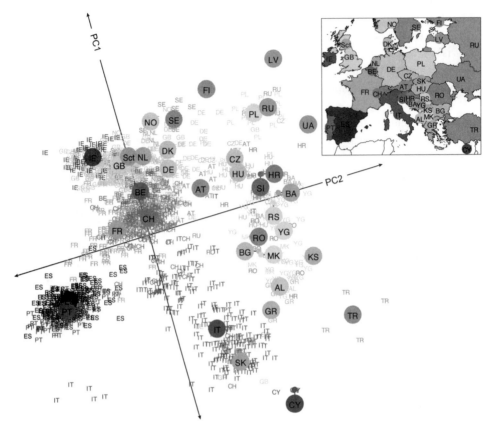

Figure 3.2
Genetic variation mirrors geography in Europe.
Source: Novembre et al. (2008) [15], *Nature*. Reprinted with permission from Macmillian Publishers Ltd.
Notes: This figure shows that performing PCA on 1,387 European individuals results in two principal components (axis 1, PC1; axis 2, PC2) that correlate with geographical axes.

The ability to isolate PCs to very detailed *fine-scaled regional variation* has been further developed by the original authors of the atlas of genetic admixture [17]. One study isolated genetic differentiation and what they termed "footprints of historical migration" in the Iberian Peninsula in Spain [19]. This area in Spain has a very complex demographic history and uniquely a long period of Muslim rule. They were able to measure an admixture event dating from around 860–1120 CE (Common Era, identical to AD, Anno Domini) and the genetic impact of the Muslim conquest, population movements, and the Reconquista. What was remarkable was that they were also able to identify population structure to an unprecedented fine-grained geographic scale of smaller than 10 kilometers in some areas. Here they found the axis of genetic differentiation was east to west of the peninsula and found clear genetic similarity from north to south.

This approach is soon to be extended for other countries such as the UK. Figure 3.3 shows some results of a highly fine-grained geographic genetic mapping of me, the first author. I am Canadian by birth and hold Canadian and Dutch nationality, the latter acquired by residence. If all four of my grandparents were from the United Kingdom, the proportions in the map in figure 3.3 would add up to 1. We see that the figures add up to 0.61, with the majority (0.29) from Anglia (Mills side of the family) and Scotland (Fleming side of the family). This map corroborates the oral family histories of my family migration from U.K.-based predecessors from Norwich (paternal grandfather) and Scotland (paternal grandmother). Also calculated (not shown here) are my apparent Dutch roots (0.16), but also Balkan, Scandinavian, and Russian, which when all added together add up to 1. The Norwich (Anglia) and Dutch link may reflect that this region used to be joined to Continental Europe and it was only around 5000BC at the thawing of the last ice age that it separated. After that trade was frequent across the North Sea. The U.K.-based estimate misses the maternal side that includes a Norwegian grandfather (and two Norwegian great-grandparents) or Greek great-grandfather (and two Greek great-grandparents). Direct-to-consumer companies such as 23andMe, for instance, also continuously update their ancestry data as they calibrate their algorithms and samples diversify. Results as of December 2018, now more accurately capture Greek and Balkan groups, for instance. Different consumer-to-genetics companies often have different ancestry results or the results change over time due to the reference sample in which they compare and methods used. 23andMe for instance previously had around 77% European ancestry clients [20]. It will be fascinating to see how genetic techniques cope with increasing admixture from populations that are migrating and mixing at unprecedented levels.

3.4   Human evolution, selection, and adaptation

3.4.1   Evolution, fitness, and natural selection

We learned in the previous chapter that our genome contains footprints of our ancestral lineage. Any genetic analysis in contemporary human populations therefore carries traces of the human past. **Evolution** refers to the change in heritable characteristics of populations over successive generations. Evolution forms the basis of our understanding not only about the origin of the human species but also the underlying genetic architecture and disease mutations. The forces driving evolution shape the basis for genetic variability within populations but also across different species. Evolution occurs over millions of years, with a spirited debate regarding whether we are able to measure it in contemporary populations (see box 3.3). It is important to note that the environment plays a key role in shaping evolution, which we explore in more detail in our upcoming chapter 5 on gene-environment interplay.

   In biology, evolution is considered to be the study of changes in the gene pool of a population across the generations, governed by processes such as mutation, natural selection,

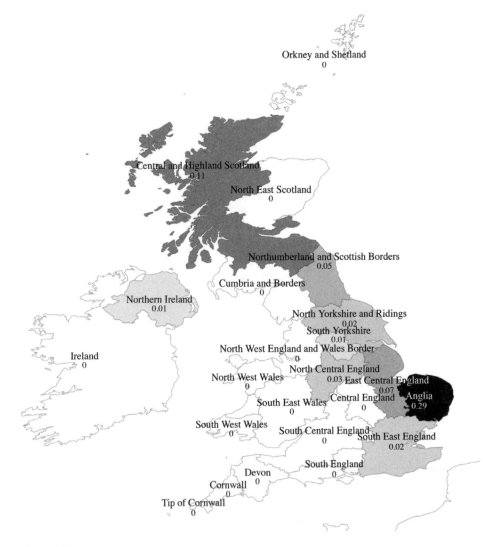

Orkney and Shetland
0

Central and Highland Scotland
0.11

North East Scotland
0

Northumberland and Scottish Borders
0.05

Cumbria and Borders
0

Northern Ireland
0.01

North Yorkshire and Ridings
0.02

South Yorkshire
0.01

Ireland
0

North West England and Wales Border
0

North West Wales
0

North Central England
0.03

East Central England
0.07

Anglia
0.29

South East Wales   Central England
0   0

South West Wales
0

South Central England
0

South East England
0.02

South England
0

Devon
0

Cornwall
0

South West Wales
0

Tip of Cornwall
0

Figure 3.3
Fine-scaled UK regional ancestry of Melinda Mills.
Thanks to Daniel Lawson of University of Bristol and GENSCI Ltd for producing the figure. Copyright University of Bristol and Oxford University. All rights reserved.

Box 3.3
Can we measure evolution and natural selection in contemporary populations? Are we getting dumb and dumber by each generation?

What is the time scale of evolution? Considerable debate exists in this field about whether we are able to measure evolution and selection in contemporary human populations. Some studies, for instance, argued that they have found natural selection on genes implicated in higher educational attainment in contemporary populations in the United States [21] and Iceland [22]. The Icelandic study constructed a score to capture this genetic component and found that individuals with higher polygenic scores (PGSs) for educational attainment had fewer children. They concluded that although small, the rate of decrease was discernable on an evolutionary timescale. A study using data from the United States also showed evidence of negative natural selection on genes implicated in higher educational attainment in a contemporary population the United States.

As we have described elsewhere [23], to claim evidence of natural selection, these studies measure how much the number of children varies to produce some sort of measure of the "magnitude" of natural selection. If the genetic component of a trait is associated with the number of children, the studies conclude that they are evolving as a result of natural selection. In other words, the PGS of educational attainment is applied to assess whether those who have genetic variation related to lower or higher education are predisposed to having more or fewer children.

There are, however, some crucial limitations of these studies. First, selection on education is weak and changes associated with it are very slow. They fail to differentiate between genetic and environmental influence on the number of children individuals have. Natural selection on education may not remain negative over enough generations to shed light on changes, and the period of several decades is very short to examine this question. Second, the "genes for education" are associated with many other cognitive and noncognitive outcomes. The PGS for education reflects associations that might not be causal for education but also other traits. For instance, we found a 0.70 genetic overlap (LD-score regression described later) between the PGS for educational attainment and age at first birth, with similar factors thus driving both [24]. Third, when considering complex behavioral traits such as education, the genetic basis of fertility behavior is also strongly influenced by cultural, economic, and social factors [25]. Educational expansion—particularly of women—has increased by around 2 years of education per generation in addition to other factors such effective contraception, allowing individuals to regulate fertility. This means that the environment has strongly changed and has not been held constant over this period. Fourth, phenotypes such as educational attainment do not perfectly correlate with cognitive skills and education, suggesting we are not getting "dumb and dumber" by the generation as is sometimes implicitly suggested. Finally, most of the data used to examine contemporary natural selection suffer from a healthy volunteer effect and mortality bias of including only people that are healthier and more likely to survive [7, 26, 27].

and genetic drift. *Mutation* refers to a change to the actual sequence of a genome (see chapter 1, box 1.1). The relevant mutations to consider are those that occur in the *germ line mutation*[2] of the organism that can be passed on to offspring across the generations. **Natural selection** is the increase or decrease of particular genetic traits as a function of the differential fitness and the reproductive success of individuals. In other words, natural selection operates when particular genetic variants render the individuals who bear them more likely to survive. As a consequence, those genetic variants increase in frequency in the next generation. Natural selection is said to drive adaptive evolution to select for traits that are beneficial to a particular population within an environment. One way to think about selection is that it is a filter that removes suboptimal alleles from a population so that it is better adapted to its environment.

The process of *adaptive evolution* refers to the selection of *beneficial alleles*, or those that are useful in particular environments and thus increases their frequencies in a population. This is in contrast to decreasing the frequency of deleterious alleles. Someone who carries a single recessive *deleterious allele*—sometimes called a "harmful" allele—will not experience the impact of this allele but can easily pass it on to the next generation. If we have a large population, this is not usually a problem since the population carries many deleterious alleles that are rarely expressed.

Fitness—also sometimes referred to as evolutionary fitness—is how well a species adapts to its environment. If a species is no longer reproducing, they are considered as no longer evolutionary fit. This was first coined by Herbert Spencer but more famously to the extensions and work of Charles Darwin. *Relative fitness* compares an individual's fitness to others in the population or, in other words, which individuals contribute to offspring in the next generation. This in turn allows us to establish how a population might evolve. For example, since height is highly heritable, if taller individuals have more children, genes important for being tall become more frequent in future generations [28]. The term *fecundity* is also often used in this research to refer to the number, rate, or capacity to produce offspring. This can be confusing for interdisciplinary or medical researchers, since this term is often used in the medical and demographic sciences to represent the biological ability to conceive within one year and related to infertility [29]. *Darwinian fitness* is also a frequently used expression, which refers to the average contribution to the gene pool to the next generation by an average individual's genotype or phenotype. Natural selection only operates on traits that are heritable.

Different *variations of selection* contribute to the way in which natural selection can affect variation within a population, which is visualized in figure 3.4. Panel *a* in this figure shows the classic normal distribution of the trait. Without any selection pressure on height, for instance, the height of people within this population would vary, with most being average and very few being extremely short or tall. *Stabilizing selection*, shown in panel *b*, is when an average (non-linear) trait is favored. This happens when selective pressures choose

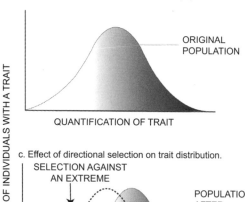

a. Standard distribution of trait across a population.

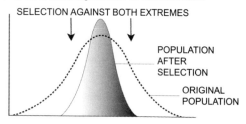

b. Effect of stabilizing selection on trait distribution.

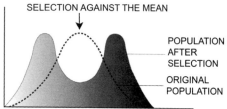

Figure 3.4
Types of natural selection.

against two extremes of a trait. Using the height example it would mean that the very short and very tall have difficulty competing, with these two selection pressures resulting in selecting only average height people. *Directional selection*, in panel *c*, is when an extreme trait is favored over others causing the allele frequency to shift over time toward the direction of that phenotype. Or in other words, one extreme of the trait distribution experiences selection against it. As the figure illustrates, the entire population's trait distribution shifts to the other extreme. The majority of examples come from other nonhuman species. Black bears in Europe, for example, decreased during interglacial periods and increased during each glacial period. *Diversifying or disruptive selection* (panel *d*) refers to genetic changes in which extreme values for a trait are favoured over more moderate or intermediate values. The variance of the trait thus increases, dividing the population into a bimodal curve. As we touched upon previously, the strong North–South principal component divide between the North and South of the Netherlands has been argued as an example of this type of diversifying selection [18].

There are other types of selection in the literature that we only briefly touch upon here. *Frequency-dependent selection* is when traits that are either common (positive frequency-dependent selection) or rare (negative frequency-dependent selection) and are favoured through natural selection. *Sexual selection* refers to natural selection that emerges due to the preference by one sex for particular characteristics in another. *Intersexual selection* is when members of a competitive sex show off attractive traits to gain the attention of a mate and thus increase their chances to be selected and have better reproductive success.[3] This can be physical traits such as height [28, 30], but within human populations there is also a rich literature on assortative mating by traits that represent a more successful partner such as educational level [31–33]. Sexual selection thus results in the development of secondary sexual characteristics that help to maximize reproductive success. *Sexual dimorphism* describes the differences (e.g., physical, cognitive, behavioural) between parental investments of the same species. Genes can thus be expressed differently between the sexes, or in other words, different genes are involved with traits in men and women (i.e., dimorphism). Elsewhere, for instance, we explored sexual dimorphism in different genetic loci operating in male and female fertility and that these genes are potentially passed on to the next generation [34]. There is also *genetic hitchhiking*, known as *genetic draft* (not to be confused with genetic drift). This is when an allele changes frequency not due to the fact that it is under natural selection but due to the fact that it is near another gene undergoing a selective sweep (an allele is on the same DNA chain). When one gene goes through a selective sweep, other nearby polymorphisms that are in linkage disequilibrium (LD), which we turn to now, generally change their allele frequencies.

3.4.2   Genetic drift

Although natural selection is an important aspect of evolution, it is not the only mechanism. Evolution can also occur by chance or what is referred to as genetic drift. **Genetic drift** is a mechanism where allele frequencies of a population change over generations due to chance, often quantified by sampling error. It is measured as change due to sampling error in selecting the alleles for the next generation from the gene pool of the current generation. The effects of genetic drift are the strongest in smaller populations and can result in the loss of some alleles. Both beneficial and deleterious alleles are subject to selection and drift but a very small population with strong drift may cause the loss of a beneficial allele and what is known as the *fixation* or carrying on of a deleterious allele. Two main types of genetic drift have been identified in the literature: the bottleneck effect and the founder effect.

A **bottleneck effect** is an extreme example of genetic drift that has large effects due to a drastic reduction in population size by an exogenous factor such as a natural disaster. This is often attributed to natural disasters such as earthquakes, floods, and fire that have the potential to decimate entire populations and leave behind a limited number of survivors. Since the allele frequency of the survivors may be very different from the population composition prior to the natural disaster, some alleles may not be present at all. This means the smaller population is more susceptible to the impact of genetic drift for multiple generations and the loss of more alleles. It is called the bottleneck effect since we can use the analogy that a bottle is filled with marbles that represent a population. After a natural disaster, which is represented by the small opening of the bottle, a small random group of individuals (marbles) pass through the bottleneck. These then form the new population, with the large majority of individuals (marbles) in all of their variety remaining in the bottle. In humans, the example often used to describe population bottlenecks is of the Greenlandic Inuit. A study in 2017 by Pedersen and colleagues [35] showed that the Inuit experienced a severe and prolonged (~20,000-year long) bottleneck. The result was the most extreme allele frequency distribution tested to date. This Inuit population carries fewer deleterious variants than other human populations but those that are present are at higher frequencies than other populations. They argue that this population is thus ideal to study the effect of bottleneck patterns of deleterious variation.

The **founder effect** is another type of genetic drift, which is when a small group splits from the main population to found a colony. This newly formed colony is isolated from the original larger population, and the founders of the colony may be "selective" and thus not represent the full genetic diversity of the original group. The allele frequencies may thus differ between the original and founder (colony) population and the founders may even miss a selection of alleles. A common example is the Amish in North America. In Eastern Pennsylvania, a small group of around 200 German immigrants moved to form a small, closed colony. The group carries a usual concentration of gene mutations that results in various inherited disorders that are otherwise rare. One is Ellis-van Creveld syndrome, which causes a form of dwarfism [36]. Since the colony of these founders was relatively small, and they tended to marry within the same group, there was a greater likelihood that the recessive genes of these founders combined and showed up more frequently (see chapter 1). Another study of British Pakistani adults with high parental relatedness discovered rare-variant homozygous genotypes that predicted "knockouts" (loss of gene function) in hundreds of genes [37].

Population size is an important component in understanding genetic drift. We know that larger populations are less likely to change very quickly due to genetic drift. Since it is due to chance it is somewhat akin to the example of flipping a coin in a small versus large population. If you were to flip the coin only a few times (i.e., the small population), it is possible to get a heads-to-tails ratio that is quite different from the 50:50 ratio that you would achieve if you flipped the coin many times (i.e., a large population). Genetic drift differs from natural selection. Whereas natural selection takes into account whether an allele is beneficial or deleterious, in genetic drift a deleterious allele could be fixed by chance and a beneficial allele might even be lost.

3.5   The Hardy–Weinberg equilibrium

3.5.1   Assumptions of the HWE

Another central concept within this area of research is the **Hardy–Weinberg equilibrium (HWE),** which is a theoretical mathematical model describing the probability and distribution of genotype frequencies in a population. The main purpose of the HWE is to express the principle that the amount of genetic variation (allele and genotype frequencies) in a population will remain constant from one generation to the next in the absence of evolutionary influences. The HWE is used to model and predict genotype frequencies in large, stable populations. Put another way, when a population is in HWE for a gene, it is not evolving and allele frequencies will remain the same across generations.

The HWE dictates that the frequencies and relative proportions of genotypes remain stable—or in other words in equilibrium—over time if all *assumptions of the HWE* are met. The proportions will remain constant at this equilibrium if these five assumptions hold:

1. There is no natural selection (i.e., all genotypes have equal fitness)
2. There is no genetic drift (i.e., stable population size)
3. A closed population (there is no significant migration in or out of the population)
4. Mutation does not occur
5. There is no assortative mating

These assumptions thus entail that the population structure is not from two or more sub-populations, there is no **inbreeding** (i.e., mating without one or more **common ancestors**), males and females have similar allele frequencies (i.e., more likely on the autosomal locus), all members of the population have equal reproductive success and the population is infinitely large.

If the basic assumptions are not met for a particular gene, the population may evolve. Or in other words, genotype frequencies might change. Most will immediately realize that these are assumptions that are likely to be violated in many instances. The HWE is thus a useful test since any deviation suggests that the locus has been influenced by non-equilibrium forces. This includes factors such as mutation or natural selection. In practice, violation of the HWE may also point to measurement error in genetic data. Testing the HWE is therefore a crucial part of the quality control process in handling genetic data that we describe in the second part of this book. Perhaps it is important to also clarify that evolution does not mean that all populations are all moving toward one state of similar perfection. Rather evolution means that a population changes genetic makeup over different generations. The changes are often subtle and take multiple generations over a long period of time (see box 3.3).

3.5.2 Understanding the notation of the HWE

The notation often used to describe the HWE is p^2, $2pq$, and q^2 to represent the three genotype probabilities. Let us first use an example where the population is not evolving. Here note that:

p = the frequency for the major allele (A)

q = the frequency for the minor allele (a)

Let us assume that the allele A has a frequency of $p = 0.3$ and allele a has a frequency of $q = 0.7$.

The Hardy–Weinberg equation is thus:

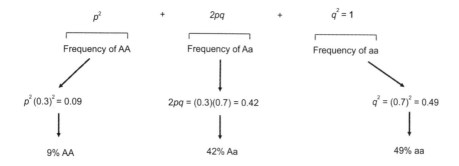

We expect the genotype frequencies of 9% AA, 42% Aa, and 49% aa.

To predict the genotype frequencies of the next generation several assumptions need to be made. If we first make the assumption that none of the genotypes is superior to the other in terms of fitness or finding mates we can set the frequencies of A and a alleles in the pool of gametes (i.e., sperm and eggs) that will in turn produce the next generation. Second, if we assume that there is no **assortative mating** (i.e., individuals mate randomly), reproduction is assumed to be the result of two random events—selection of a sperm and egg from the same gene pool. Recall the Punnett Square from the previous chapter. In practice the HWE can be used to determine the frequency of individuals that may be affected by a diseases caused by what is known as recessive deleterious mutations. One example is Tay-Sachs disease, which results in mental and physical deterioration and early childhood death. This mutation has a higher frequency of up to 2% in Ashkenazi Jews. If two people of that ancestry had a child and are both homozygous for the disease mutation, we can use the HWE to determine that $0.02^2 = 0.0004$, which is 0.04%.

3.6 Linkage disequilibrium and haplotype blocks

Recall from chapter 1 that during recombination, variants that are located near one another on the same chromosome have a high probability of being transmitted together. This is due to the fact that there are only one or two recombination events per chromosome. Polymorphisms are inherited together through what is called linkage disequilibrium (LD), which is the nonrandom occurrence in members of a population of the combinations of 2 or more linked genomic loci. For instance, if a T at one SNP locus is generally observed with a G at another SNP locus, these two SNPs are said to be in linkage disequilibrium. Their co-occurrence is more correlated than we would expect by random (equilibrium) conditions. Two alleles (i.e., that are variants of polymorphisms) which are located at different positions at the same chromosome are in LD if they are not inherited independently from one another. In general, alleles which are located close together at the same chromosome will have stronger LD.

Why is this the case? Recall from our discussion of genetic recombination in chapter 1 that when a chromosome is transmitted to a sperm or egg cell during the process of meiosis, the two neighbouring SNPs are transmitted together or there is a recombination in between. The probability of recombination is low for a very small section with on average around 35 recombinations per meiosis. This is around one recombination every 100 Mb. This in turn leads to the correlated inheritance of linked SNPs over time, or LD. Conversely, when two SNPs are inherited randomly (i.e., unlinked), they are said to be in equilibrium. High LD thus means that two SNPs are linked, which is measured by R^2. This is the level of correlation between SNPs where a perfect correlation of 1 means fully linked SNPs and two random SNPs would be $R^2 = 0$. One way to think of it is like the shuffling of a deck of cards where we often take "chunks" of cards instead of only individual cards.

As a result of LD, we often measure one *tag SNP* in order to predict the genotype of another. This is referred to as imputation, which we explore later in the applied chapters. In practice, it is more difficult as we cannot always isolate a SNP in perfect LD where $R^2 = 1$. As noted in the previous chapters, these SNPs are the core markers or flags that we examine [38]. Even if there is no direct association between a SNP marker and a trait, it may be the case that it is indirectly associated with the trait if it has been transmitted together with a causal genetic variant—or in LD. The markers we often observe thus are often more flags on the genome to mark the area where a causal genetic variant might be present but may not be the causal variant itself. If LD is 0.9 for instance, it is hard to prioritize the causal SNP. In many cases SNPs are in LD with nearby genetic variants that are not captured in the particular GWAS array. (We discuss different arrays in more detail in chapter 7.) The GWA studies we discuss in an upcoming chapter thus also detect these unobserved variants or also the low-frequency and common genetic variants across the genome.

As a result of LD, we observe haplotype blocks in populations. When variant alleles are transmitted together during recombination, they are more likely to be found in the larger population and form what is called haplotype blocks. **Haplotype** blocks are a sort of mutation or recombination of a set of closely linked alleles on a chromosome that evolutionarily over time tend to be inherited together in LD [38, 39]. These are the blocks of SNPs that are inherited as a group. Although LD is roughly similar across populations, haplotype blocks vary. They are considerably shorter in African populations where there is more diversity that has been present for a longer period of time. Interested readers should refer to the International HapMap project, which has detailed reviews, tutorials and multiple resources on this topic (see "Further reading").

3.7 Conclusion

The goal of this chapter was to provide readers with a very brief chronicle of the human dispersal out of Africa over millions of years in order to understand patterns of interbreeding, migration, and admixture. This was then linked to contemporary population structure and stratification found in contemporary data, which is detected using Principal Component Analysis. We also elaborated on common misnomers of population structure and mistakes when population structure and ancestry are incorrectly equated or race is misinterpreted as a biological rather than a socially and politically constructed term. Although only fleetingly, we show how these methods can be used to illustrate how genes mirror geography by continents, countries, or regions. We then turn to the fundamental topics of evolution, natural selection, and fitness (and variations of fitness). These help us to comprehend the increase or decrease of particular genetic traits in relation to reproductive success. We then differentiate this from bottleneck and founder effects in the form of genetic drift. Theoretical mathematical models have also been developed to test for deviations in the probability and distribution of genotype frequencies in a population in the form of the Hardy–Weinberg equilibrium, which follow five key assumptions (no natural selection, no genetic drift, no assortative mating, closed population, no mutation). Finally, we explain haplotype blocks, which is when variant alleles are transmitted together during recombination and thus more likely to be found in the larger population. We now turn to the basis in which the majority of genetic data we use in this book is derived from, which are genome-wide association studies.

Exercises

1. Go to the website that accompanies the Hellenthal et al. (2014) *Science* article: http://admixturemap.paintmychromosomes.com/. Read the two tutorials under the "Historical Event" menu. Click on a labeled population on the map or selection from the "Target Population" drop-down menu. Examine some of the details of the past admixture events that they use to infer how that population has been formed. Pick one historical event and read the historical interpretation of the admixture signals that we see.

2. Go to the site of the International HapMap project and explore: https://www.genome.gov/10001688/international-hapmap-project/.

3. Explore ANCESTRYMAP 2.0 from https://reich.hms.harvard.edu/software.

ANCESTRYMAP 2.0 from https://reich.hms.harvard.edu/software

ANCESTRYMAP (Patterson et al., 2004) finds skews in ancestry that are potentially associated with disease genes in recently mixed populations like African Americans. It can be downloaded on various operating systems and see the tutorial: https://reich.hms.harvard.edu/software/tutorial. The ANCESTRYMAP Software Documentation is available at https://reich.hms.harvard.edu/sites/reich.hms.harvard.edu/files/inline-files/ANCESTRYMAP_documentation.pdf.

Further reading and resources

Basic books on evolution and migration

Ayala, F. J., and C. J. Cela-Conde. *Processes in human evolution: The journey from early hominins to neanderthals and modern humans.* Oxford: Oxford University Press, 2017.

Harari, Y. N. *Sapiens: A brief history of humankind.* London: Harvill Secker, 2014.

Knight, J. C. *Human genetic diversity.* Oxford: Oxford University Press, 2009.

Rutherford, A. *A brief history of everyone who ever lived.* London: Weidenfeld & Nicolson, 2016.

Wood, B. *Human evolution: A very short introduction.* Oxford: Oxford University Press, 2019.

Readings on methods, structure, and Principal Component Analysis

Conomos, M. P., M. B. Miller, and T. A. Thornton. Robust inference of population structure for ancestry prediction and correction of stratification in the presence of relatedness. *Genet. Epidemiol.* **39**(4), 276–293 (2015).

Gopalan, P. et al. Scaling probalistic models of genetic variation to millions of humans. *Nat. Gen.* **48**(12), 1587–1590 (2016).

Liu, Y. Software and methods for estimating genetic ancestry in human populations. *Hum. Gen.* **7**(1) (20130), doi: 10.1186/1479-7364-7-1.

Price, A. et al. Principal components analysis corrects for stratification in genome-wide association studies. *Nat. Gen.* **38**, 904–909 (2006).

Pritchard, J. J. et al. Association mapping in structured populations. *Am. J. Hum. Gen.* **67**(1), 170–181 (2000).

References

1. M. Stoneking and J. Krause, Learning about human population history from ancient and modern genomes. *Nat. Rev. Genet.* **12**, 603–614 (2011).

2. S. López, L. Van Dorp, and G. Hellenthal, Human dispersal out of Africa: A lasting debate. *Evol. Bioinforma.* **21**(11), 57–68 (2016).

3. M. Rasmussen et al., An aboriginal Australian genome reveals separate human dispersals into Asia. *Science* **334**, 94–98 (2011).

4. G. Gibson, *A primer of human genetics* (Sunderland, MA: Sinauer Associates, 2015).

5. S. C. Schuster et al., Complete Khoisan and Bantu genomes from southern Africa. *Nature* **463**, 943–947 (2010).

6. J. Lachance et al., Evolutionary history and adaptation from high-coverage whole-genome sequences of diverse African hunter-gatherers. *Cell* **150**, 457–469 (2012).

7. M. C. Mills and C. Rahal, A scientometric review of genome-wide association studies. *Commun. Biol.* **2** (2019), doi:10.1038/s42003-018-0261-x.

8.   A. R. Martin et al., Human demographic history impacts genetic risk prediction across diverse popula-tions. *Am. J. Hum. Genet.* **100**, 635–649 (2017).

9.   S. Reisberg et al., Comparing distributions of polygenic risk scores of type 2 diabetes and coronary heart disease within different populations. *PLoS One* **12**, e0179238.

10. A. Harmon, Why white supremacists are chugging milk (and why geneticists are alarmed). *New York Times*, October 17, 2018 (available at https://www.nytimes.com/2018/10/17/us/white-supremacists-science -dna.html).

11. M. C. Mills and F. C. Tropf, The biodemography of fertility: A review and future research frontiers. *Kolner Z. Soz. Sozpsychol.* **55**, 397–424 (2016).

12. ASHG denounces attempts to link genetics and racial supremacy. *Am. J. Hum. Genet.* **103**, 636 (2018).

13. C. D. Royal et al., Inferring genetic ancestry: Opportunities, challenges, and implications. *Am. J. Hum. Genet.* **86**, 661–673 (2010).

14. J. L. Baker, C. N. Rotimi, and D. Shriner, Human ancestry correlates with language and reveals that race is not an objective genomic classifier. *Sci. Rep.* **7**, 1572 (2017).

15. J. Novembre et al., Genes mirror geography within Europe. *Nature* **456**, 98–101 (2008).

16. M. R. Nelson et al., The population reference sample, POPRES: A resource for population, disease, and pharmacological genetics research. *Am. J. Hum. Genet.* **83**, 347–358 (2008).

17. G. Hellenthal et al., A genetic atlas of human admixture history. *Science* **343**, 747–751 (2014).

18. A. Abdellaoui et al., Population structure, migration and diversifying selection in the Netherlands. *Eur. J. Hum. Genet.* **21**, 1277–1285 (2013).

19. Clare Bycroft et al., Patterns of genetic differentiation and the footprints of historical migrations in the Iberian Peninsula. *bioRxiv* (2018), doi:https://doi.org/10.1101/250191.

20. K. Servick, Can 23andMe have it all? *Science* **349**, 1472–1477 (2015).

21. J. P. Beauchamp, Genetic evidence for natural selection in humans in the contemporary United States. *Proc. Natl. Acad. Sci. USA* **113**, 7774–7779 (2016).

22. A. Kong et al., Selection against variants in the genome associated with educational attainment. *Proc. Natl. Acad. Sci. USA* **114**, E727–E732 (2017).

23. A. Courtiol, F. C. Tropf, and M. C. Mills, When genes and environment disagree: Making sense of trends in recent human evolution. *Proc Natl Acad Sci USA* **113**, 7693–7695 (2016).

24. N. Barban, M. C. Mills et al., Genome-wide analysis identifies 12 loci influencing human reproductive behavior. *Nat. Genet.* **48**, 1–7 (2016).

25. F. C. Tropf et al., Human fertility, molecular genetics, and natural selection in modern societies. *PLoS One* **10**, e0126821 (2015).

26. B. W. Domingue et al., Mortality selection in a genetic sample and implications for association studies. *Int. J. Epidemiol.* **46**, 1285–1294 (2017).

27. A. Fry et al., Comparison of sociodemographic and health-related characteristics of UK biobank partici-pants with those of the general population. *Am. J. Epidemiol.* **186**, 1026–1034 (2017).

28. G. Stulp, L. Barrett, F. C. Tropf, and M. Mills, Does natural selection favour taller stature among the tall-est people on earth? *Proc. R. Soc. B Biol. Sci.* **282**, 20150211 (2015).

29. M. C. Mills, R. R. Rindfuss, P. McDonald, and E. te Velde, Why do people postpone parenthood? Rea-sons and social policy incentives. *Hum. Reprod. Update* **17**, 848–860 (2011).

30. G. Stulp, M. Mills, T. V. Pollet, and L. Barrett, Non-linear associations between stature and mate choice characteristics for American men and their spouses. *Am. J. Hum. Biol.* **26**, 530–537 (2014).

31. M. R. Robinson et al., Genetic evidence of assortative mating in humans. *Nat. Hum. Behav.* **1**, 1–13 (2017).

32. C. Quintana-Domeque, N. Barban, E. De Cao, and S. Oreffice, Assortative mating on education: A gene-tic assessment. *Econ. Ser. Work. Pap.* (2016) (available at http://ideas.repec.org/p/oxf/wpaper/791.html).

33. B. W. Domingue, J. Fletcher, D. Conley, and J. D. Boardman, Genetic and educational assortative mating among US adults. *Proc. Natl. Acad. Sci. USA* **111**, 7996–8000 (2014).

34. R. M. Verweij et al., Sexual dimorphism in the genetic influence on human childlessness. *Eur. J. Hum. Genet.* **25**, 1067–1074 (2017).

35. C.-E. T. Pedersen et al., The effect of an extreme and prolonged population bottleneck on patterns of deleterious variation: Insights from the Greenlandic Inuit. *Genetics* **205**, 787–801 (2017).

36. V. A. McKusick, Ellis-van Creveld syndrome and the Amish. *Nat. Genet.* **24**, 203–204 (2000).

37. V. M. Narasimhan et al., Health and population effects of rare gene knockouts in adult humans with related parents. *Science* **352**, 474–477 (2016).

38. B. M. Neale, M. A. R. Ferreira, S. E. Medland, and D. Posthuma, *Statistical genetics: Gene mapping through linkage and association* (New York: Taylor & Francis Group, 2008).

39. J. D. Wall and J. K. Pritchard, Haplotype blocks and linkage disequilibrium in the human genome. *Nat. Rev. Genet.* **4**, 587–597 (2003).

40. L. Aldén, L. Edlund, M. Hammarstedt, and M. Mueller-Smith, Effect of registered partnership on labor earnings and fertility for same-sex couples: Evidence from Swedish register data. *Demography* **52**, 1243–1268 (2015).

4

Genome-Wide Association Studies

Objectives

- Understand a *genome-wide association study*
- Grasp the basics and limitations of *genotyping and sequencing arrays* and their relationship to linkage disequlibrium and imputation
- Understand *genome-wide association study research design*, *meta-analysis*, and a *data analysis plan*
- Know the fundamental aspects of *statistical inference*, *methods*, and *heterogeneity* for genome-wide association studies
- Grasp the types of *quality control*
- Gain knowledge of the *NHGRI-EBI GWAS Catalog* for an overview of genome-wide association studies
- Recognize the *lack of diversity* in terms of ancestral, geographical, temporal, and demographic diversity of genome-wide association studies to date and implications for research
- Become aware of *future directions* in this area of research

4.1 Introduction and background

The design of genetic association studies has dramatically changed over the past decades in tandem with developments in genotyping technology, a reduction in costs, and the development of advanced data analysis methods. Although high-throughput, whole-genome profiling is now standard, early research focused only on a limited number of "candidate" loci. The term **candidate gene studies** refers to early work in this field that focussed on predefined loci of interest thought a priori to be relevant to the trait under examination. As we discuss in detail in our chapter 6 on gene-environment interplay, many of the early candidate gene studies were problematic for multiple reasons, primarily due to a lack of replication [1]. Although our aim is to inoculate researchers new to this field against making similar mistakes, we should note that some candidate gene studies are still performed quite successfully for a variety of nonbehavioral medical phenotypes. The extreme polygenicity of many traits and failure of candidate genes as drug targets (e.g., depression) came as a genuine surprise to many at the time. An alternative, the genome-wide association study (GWAS), emerged, in which millions of genetic loci are measured simultaneously.

GWASs are currently the primary method used to identify associations between single-nucleotide polymorphisms (SNPs) and phenotypes. As we discuss in more detail shortly—GWASs test millions of separate regression models for associations between genetic variants and a phenotype. Recall from chapter 1 that phenotypes can be monogenic traits, strongly influenced by variation within a single gene. But many are polygenic complex traits, which are the result of variation within multiple genes and their interaction with behavioral and environmental factors. The results of a GWAS show the association of each individual SNP with a particular trait or phenotype. In contrast to candidate gene studies, GWASs are hypothesis-free and search for associations across all genotyped regions. As discussed previously in chapter 1, GWAS examines the polymorphisms that distinguish us from each other. With the exception of monozygotic (i.e., identical) twins, it is this 0.1% by which we differ that makes us all unique.

Since many traits are complex and linked to multiple genetic loci (i.e., polygenic), a GWAS often identifies many genetic variants that each have a small influence on a phenotype. Due to small effect sizes, very large data sources are required and the GWAS discovery typically culminates in many GWAS analyses conducted on multiple data sources and then combined into one meta-analysis. The majority of variants that are identified in GWASs are not assumed to be biologically causal, but rather, due to linkage disequilibrium (LD), may identify a region that contains one or more of the biologically functional variants. Almost 4,000 GWASs have been conducted by early 2019, which have agnostically identified thousands of genetic variants [2, 3]. Traits that have been studied include many common human diseases such as breast cancer, Alzheimer's disease, and type 2 diabetes, but also anthropometric (height, weight) and behavioral traits such as age at first childbirth or educational attainment.

The current chapter provides an introduction to GWAS research and fundamental concepts. Since the results from GWASs often serve as a basis for many practical applications, this chapter is essential for later applied chapters in part II, including how to conduct quality control (QC) on genetic data (chapter 8). In this chapter, we unpack the basics of GWAS methodology including the nuts and bolts in terms of genetic data collection, research design and approach, and the need to correct for multiple testing. This is followed by an introduction to the types of individual-level and genetic-marker level QC that we later conduct in chapter 8. Section 4 briefly describes GWAS meta-analysis and further extensions. Finally, we provide a comprehensive overview of the NHGRI-EBI (National

Human Genome Research Institute–European Bioinformatics Institute) GWAS Catalog, followed by a brief history of GWA discoveries from 2005 until late 2018. We note the lack of various types of diversity in GWAS samples, such as lack of ancestral and demographic diversity and concentration of subjects in a particular set of countries. We conclude with a brief summary and provide an indication of future directions for research.

4.2 GWAS research design and meta-analysis

4.2.1 GWAS research design

Genetic discovery is not only an intellectual but also an organizational and logistic challenge. Since the quality and success of a GWAS traditionally depended upon gathering a large sample size, large consortia have been formed who conduct independent GWASs, which are subsequently meta-analyzed by the core group leading the project. Figure 4.1 depicts GWAS stages in perhaps one of the largest types of collaborative efforts in modern science. Considering the broad expertise required, consortiums that need to be formed, and long-term and time-consuming investment, it is rare that researchers new to this area would initiate their own independent GWAS. It is useful, however, to understand the process of how a GWAS is conceived.

It first starts with a general *feasibility analysis*, where researchers need to understand the phenotype, what has been researched to date, measurement, and previous heritability estimates or other GWAS results, if available. This area of research continues to flourish in terms of online tools and packages that summarize existing results. You can, for instance, refer to a comprehensive analysis of the heritability of many human traits over 50 years of twin studies (see [4]). It is also accompanied by a web application called MaTCH (Meta-Analysis of Twin Correlations and Heritability) that is accessible via http://match.ctglab .nl/. There are also other websites, such as SNPedia (https://www.snpedia.com/index.php /Heritability), that catalog heritability estimates linked with particular studies. Ben Neale's lab also has an incredible website examining heritability of many traits in the UK Biobank (http://www.nealelab.is/uk-biobank/). You can also produce a visualization of results including Manhattan plots and many others from the Complex-Traits Genetics Virtual Lab (CTG-VL) for post-GWAS analyses [5], available at https://genoma.io and http://atlas .ctglab.nl/.

The next stage is to isolate which data sources may have the phenotype you are interested in and, if applicable, form or approach a consortium or gain access to existing or publicly available data (e.g., UK Biobank). It takes considerable time and effort to form a consortium, including often waiting for ethics and access clearance and in some cases processing the payments to use the data. Although large datasets such as the UK Biobank (~500,000) have more recently become available, it has been common to form large consortiums where multiple datasets are combined to produce the largest sample possible. In many cases, individual analysts for each data source are responsible for conducting the

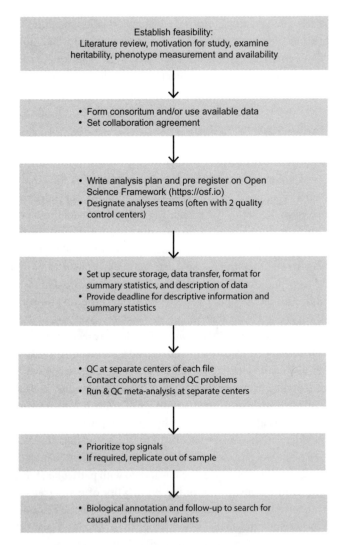

Figure 4.1
GWAS research design and meta-analysis.

GWAS internally and sending the results back to the consortium leaders. This is often related to privacy and consent issues to the data, described in the final part of this book in chapter 14. Meta-analysis of GWAS summary statistics is thus the most popular method to find genetic variants related to a phenotype. Since genetic effects due to common alleles are small, we know from the previous discussion in chapter 1 that the detection of signals requires ever larger sample sizes. Since single GWASs are underpowered, researchers need to engage in a meta-analysis and combine multiple data sources.

4.2.2 Data analysis plan

The first step in conducting a GWAS is to produce an analysis plan, which most now post on the Open Science Framework (https://osf.io/). The full analysis plan for our study of human reproductive behavior [6] is, for example, available at: https://osf.io/53tea/. Here we examined the two phenotypes of age at first birth (AFB) and number of children ever born (NEB). Analysis plans may differ depending on the consortium or trait studied but generally involve the following aspects.

1. Background and motivation for the study.

2. Clear definitions of the traits and examples of potential questions that capture that trait.

For age at first birth (AFB) shown above, for example, we stated: AFB can be treated as a continuous measure, which has generally been asked directly or can be imputed from several survey questions (such as date of birth respondent and date of birth first child). The common question is:
How old were you when you had your first child?
Another variant is:
What is the date of birth of your first child?
In the case of the latter, you can simply impute this variable to get the AFB by subtracting the date of birth of the first child from the date of birth of the subject. In many surveys, this question has been adapted for male and female subjects and in several cases only women have been asked.

3. Instruction on how to opt-in to the consortium and key deadlines if your aim is to gather a large sample.

4. Detailed sample inclusion criteria are then often listed. For instance, in our study of human reproduction, we also examined the number of children ever born (NEB) and only included those who had reached the end of their reproductive period (at least age 45 for women, 55 for men) and clarified that we also wanted analysts to include individuals who had never given birth. This is also where you specify any ancestry requirements, relevant covariates, genotyping rate (> 95%), and additional quality controls (see also chapter 8).

5. Information on genotypes and imputation, including any recommended marker filters that need to be applied before imputation, which we will discuss shortly. In the example analysis plan referred to previously, it was SNP call > 95%, HWE p > 10^{-6}, MAF > 5%. The logic behind these values is discussed in more detail in chapter 8.

6. Clear specification of the models to be used to test for association. In our study, for example, we required that regression models for two phenotypes (AFB, NEB) were estimated for men and women and then pooled. For example, an equation is $Y = m + SNP_i \beta_i + Z\gamma + e$. Many studies also often include family-based data in which

case clear instructions should be provided to consider the fraility structure in the data or selection household members. We specified linear regression models, which include several covariates (e.g., controls for population stratification, birth cohort to control for nonlinear effects or any study-specific covariates).

7. Specification of file formats for the results. Many, for instance, often opted for the CHARGE consortium sharing format.[1] The file-naming scheme is likewise essential because you will be receiving of hundreds of different files.

8. Data exchange and security procedures are also important and more recently for many working in Europe need to be GDPR (General Data Protection Regulation) compliant (see chapter 14, on ethics).

9. Description of the meta-analysis is then also often included. This includes marker exclusion filters, genomic control, significance thresholds, and the way in which top SNPs are reported.

Each participating data source (often called a cohort in this area of research) runs the analysis separately or may grant access to the data. The summary statistic results for each study are then generally uploaded together with some descriptive information about the data for that particular data source. These results are then combined and a *meta-analysis* is conducted.

4.2.3   Meta-analysis
Meta-analysis is the statistical synthesis of information from multiple independent studies that increases power and subsequently reduces the risk of false-positive findings [7]. It is also suggested that all researchers in the consortium should sign a collaboration agreement that includes, for instance, not publishing a GWAS on that phenotype before the current consortium does.

GWAS meta-analyses use what is called *summary data*, which provide regression coefficients, standard errors, and so on for each genetic marker in a population following a prespecified analysis plan. It is thus not individual-level data but the aggregated summary results. Our 2016 study on reproductive behavior [6], for instance, involved a meta-analysis that used the summary statistics from over 60 different data sources. In chapter 8 we describe how to engage in QC at the individual level, before conducting, for example, a GWAS (such as removing variants with low allele frequency, low imputation quality, allele frequency that diverges substantially from a reference sample, or results driven by a specific study that are not replicated elsewhere). An important and time-consuming step in the GWAS meta-analysis is a second set of quality control, which is basically harmonizing the results across studies. Despite providing a unified analysis plan, this cleaning process might take the longest time in an initial project, since analysts might use different software or there are other inconsistencies in the results. An excellent protocol for the meta-QC process based on the work of the GIANT consortium is provided by Winkler et al. [8].

4.3   Statistical inference, methods, and heterogeneity

4.3.1   Nature of the phenotype

The core premise of GWA studies is to perform what are now millions of hypothesis tests, or in other words one hypothesis test per variant, simultaneously on a large sample of individuals in a particular population. Each genetic association study employs statistical inference to establish and quantify the strength of the association between a genetic locus and a phenotype. The choice of the association method generally depends on the *nature of the phenotype* and whether it is dichotomous (i.e., binary) or quantitative (i.e., continuous), but also it is common to take potential confounders into account (e.g., sex, age, birth cohort).

For *quantitative* or *continuous traits* (e.g., age at first birth or body mass index), the analysis compares individuals over a range of the continuous distribution of the phenotype, most often using *linear regression*. Here we compare the distribution of test statistics based relative to the null hypothesis of no association at any marker and take into account the standard error. Additional extensions of survival models for censored data are also increasingly possible. For *binary or dichotomous traits*, it compares those categorized into, for instance, high (cases) versus low (control) values, generally using *logistic regression*. Just as with typical logistic models, it is assumed that the logit transformation of the trait under study has a linear relationship with the alleles, but it is often interpreted in terms of an odds ratio.

4.3.2   *P*-values and Z-scores

We elaborated upon the statistical underpinnings of this type of research in more detail in chapter 2. Briefly, the goal is to produce an estimate of *statistical significance* for each true association between a genetic locus and the phenotype that is being studied. As most readers will know and as discussed previously in chapter 2, statistical significance is generally determined by a *p*-value. A *p*-value estimates the probability of obtaining a test statistic value as extreme as the one estimated (i.e., under the null) for a potential association by the chosen statistical method you use. It is not the probability of a locus being associated with a trait. When we perform such a regression, we use a **test statistic** such as a *t*-test to test whether the beta parameter of the particular genetic variant is significantly different from zero. A **test statistic** is a numerical summary of the data used to measure support for the null hypothesis. A test statistic may have a known probability distribution (e.g., such as x^2) under the null hypothesis or its null distribution is estimated. Recall that a *null hypothesis* is a statistical test of the hypothesis that there is no significant difference between the specified populations, which in the case of GWAS is between the cases and controls. Any observed difference is attributed to sampling or experimental error. If the value of the test statistic generated from the genetic locus deviates significantly from what we would expect from the null hypothesis, there is evidence of the alternative hypothesis that there is a significant difference between the groups (case versus controls) or significant relationship with a quantitative trait.

The disadvantage with p-values in meta-analyses, which has been widely discussed, is that it cannot provide an overall estimate of effect size. Also, between-dataset heterogeneity cannot be assessed. A related statistic that is also used is the *Z-score*, which is based on the average of Z_i values where it is the Z-score of the ith study. Although the p-value and Z-scores are highly correlated, an advantage of using Z-scores is that they take the direction of effect into account and you are able to introduce weights (for example, if you want a particular study have a higher or lower weight). SNPs are flagged or considered as "hits" by the measure of a p-value.

As noted earlier, the agreed upon *genome-wide significant threshold* is $p < 5 \times 10^{-8}$. This corresponds to a Bonferroni correction, addressed in the next section. The genome-wide significance threshold may vary across populations due to variation in SNPs, MAFs, LD patterns, or arrays. In populations that have a lower LD, such as African ancestry groups, stricter thresholds should be used [9].

4.3.3 Correcting for multiple testing in a GWAS

DNA microarrays and next-generation sequencing allow us to test for associations for a large number of genomic loci in tandem. The magnitude of comparisons that are made in GWAS results is known as the *multiple testing problem*. This is the potential for both false positives (Type 1 errors) and if the correction for multiple comparisons is overly conservative or power are inadequate, false negative (Type 2 errors) results. We test for the associations of millions of genetic variants across the genome, but only a very small fraction will actually be associated at the genome-wide significance level with the phenotype. The issue is that when we conduct so many tests, we are also in danger of finding many strong associations merely by chance.

In a GWAS, a statistical test is performed for each genetic locus and the phenotype to produce a test statistic and associated p-value. If we took the standard p-value of 0.05 (i.e., 5%), even if a given genetic variant was not associated with our phenotype, there is a 1 in 20 chance that we would find a significant association. This is what is referred to as a **type 1 error** or a false positive. Since in a GWAS we perform literally millions of tests in parallel, we are highly likely to reap many false positives if we were to adopt the standard 0.05 significance threshold. To solve this multiple testing problem the most commonly used and straightforward correction is the **Bonferroni correction**. Simply put, we divide the chosen significance threshold (p-value) by the number of tests that are performed. If 10 tests were performed, we would only state that results are significant if they have a p-value smaller than 0.005. In the case of the genome, we are testing 1 million independent genetic variants for common sequence variation and for this reason the Bonferroni-corrected p-value of significance is $p < 5 \times 10^{-8}$. This relates to the basic *assumption of independence* in statistics or that you should get results from your sample that reflect what you would find in a population. If there is even the smallest of dependence in your data and you violate this assumption, biased results are produced. One statistical issue with GWASs is that there is often a strong correlation between the genotypes at nearby genetic variants. Or in other words, actually testing 1 million genetic variants is in reality more like testing 700,000 to 800,000 genetic variants that are not correlated. In a GWAS, statistical thresholds are thus adopted with $p < 5 \times 10^{-8}$ (i.e., $p < .00000005$) the standard for genome-wide statistical significance and $p < 5 \times 10^{-6}$ often used to represent "suggestive hits."

Some argue that Bonferroni correction is too conservative and leads to an increase in the proportion of false negative findings and the assumption of independence that every genetic variant is tested independent of the rest [10]. Although a detailed explanation of alternative methods goes beyond the auspices of this introductory book, there are other ways to correct for multiple testing. *Permutation-based testing* engages in a permutation of the phenotype a large number of times and then recalculation of the statistical tests each time to produce an empirical null distribution that can be used by hypothesis testing. It may be more intuitive to think of this as a shuffling of labels. To calculate permutation-based p-values, the outcome measure labels are randomly permuted or shuffled multiple (e.g., 1,000–1,000,000) times, which effectively removes any true association between the genotype and phenotype. For all permuted datasets, statistical tests are then performed. This provides the empirical distribution of the test-statistic and the p-values under the null hypothesis of no association. The original test statistic or p-value obtained from the observed data is then compared to the empirical distribution of p-values to determine an empirically adjusted p-value. Permutation-based testing is computationally intensive, especially if many permutations are required, which is necessary to calculate very small p-values accurately [11].

Another technique is the *Benjamini–Hochberg false discovery rate (FDR)*, which is less conservative than Bonferroni correction. It controls for the expected proportion of false positives among all signals, with an FDR value below a fixed threshold and assumes that SNPs are independent. The method minimizes the expected proportion of false positives but does not imply statistical significance. A limitation is that the FDR method still has the assumption that SNPs and p-values are independent.

4.3.4 Manhattan plots

The main results of a GWAS are generally presented in what is called a **Manhattan plot**, shown for the trait of age at first childbirth in figure 4.2. This figure is a scatterplot that plots the negative logarithm (base 10) of the p-value (y axis) and the significance of the association of the SNP ordered by position along the chromosomes (x axis). The genome-wide significant threshold of $p < 5 \times 10^{-8}$ is represented by the top line in the figure. The threshold for suggestive hits of $p < 5 \times 10^{-6}$ is shown by the bottom red line in the figure. The SNPs shown in the figure are markers, and many will not be the actual causal variant

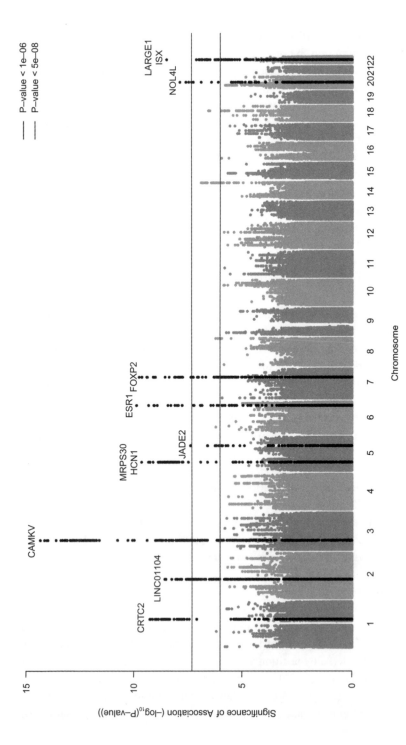

Figure 4.2
Example of a Manhattan plot, GWAS age at first birth.
Source: Barban et al. (2016) [6]; figure made by the authors.

but rather a "tag." In other words, they are tags since nearby variants might actually be driving the association. Remember that this is a study of correlation and not causation, so further biological and downstream work is then required to understand the biological function of the marker or those in its vicinity. We provide a more detailed case study of how this might work using FTO, often referred to as the "fat gene," in chapter 10, section 10.2. Chapter 8 describes various additional diagnostic checks that we also undertake during a GWAS, including examining the heterogeneity of results by sex or data source using the forest plot and Quantile-Quantile (Q-Q) plots. Chapter 9 also goes into detail on the mechanics of controlling for population stratification, a concept that was introduced earlier in chapter 3.

4.3.5 Evaluating dichotomous versus quantitative traits

To *evaluate dichotomous traits*, a *chi-squared test* is often used to test for differences in the frequencies of distributions between the cases and controls. It calculates the expected allele frequencies of cases and controls as if the SNP was not associated with the phenotype. It then measures the deviations from this expectation in the form of the chi-square statistic (χ^2). The test is reported by the p-value of the probability that these deviations occur by chance given that the SNP and the trait are not associated. If the p-value is below the defined significance threshold (after controlling for multiple testing, discussed shortly), the finding is significant.

We often then also *estimate an effect size*, which is important in understanding the magnitude or strength of the association. To calculate the effect size of dichotomous traits, different methods can be used such as the odds ratio (OR). This is the odds of having the phenotype given the phenotype-associated allele, divided by the odds of having the phenotype given the non-associated allele. Note that this should not be interpreted at the individual level as a "personal risk" but rather it is the calculation of a risk compared to another genome. The p-value indicates whether a genetic association conforms to our chosen statistically significant threshold but cannot be used to compare genetic associations. This is due to the fact that p-values are strongly influenced by the sample size, power of the statistical test, and other factors outside of the relationship being studied. It is for this reason that we use an effect size to compare two SNPs; you need to know both the p-value and effect size estimate for genetic associations for a proper assessment of the strength and interpretation of the associations.

To evaluate quantitative traits, such as height, we often use *linear regression*, where we aim to correlate the trait and each SNP of interest. As with the previous test, the regression model produces a measure of significance in the form of a p-value and effect size defined by a beta coefficient. This regression is then run for each SNP to search for significant associations by the deemed genome-wide significance threshold ($p \leq 5 \times 10^{-8}$). To interpret effect size of quantitative traits, we use the beta coefficient, where the occurrence of each risk allele corresponds to an increase in the quantitative trait equal to the

Box 4.1
Genetic power calculations

The Genetic Power Calculator (http://zzz.bwh.harvard.edu/gpc/) by Purcell and Sham is a useful online tool for genetic power calculations based on specific study characteristics [12]. The Genetic Association Study (GAS) Power Calculator, which is used to compute statistical power for large one-stage genetic association studies, can be found at http://csg.sph .umich.edu/abecasis/cats/gas_power_calculator/index.html. Other power calculators include the Effect Size Calculator for T-Test at http://www.socscistatistics.com/effectsize/Default3 .aspx, and for a very nice interactive visualization see "Interpreting Cohen's *d* Effect Size: An Interactive Visualization" at https://rpsychologist.com/d3/cohend/. To calculate Bonferroni correction, see http://www.quantitativeskills.com/sisa/calculations/bonfer.htm. For a tutorial on how to conduct a power analysis using dichotomous (i.e., discrete) and quantitative (i.e., continuous) see Purcell and Sham's aforementioned website and the supplementary material from Stringer et al. (2015) [13]. You are required to specify parameters such as the allele frequency of the high-risk allele, prevalence of the phenotype in the population, effect size of the relative risk of genotype (i.e, Aa and AA relative to aa), D-Prime (correspondence between the two genetic variants), total sample size, alpha, and power.

beta coefficient. For example, assume that we correlate a SNP with genotypes AA, AG, and GG with height in centimeters. If we find that A is the "tall" allele with a beta coefficient of 0.5, each A allele is predicted to contribute 0.5 cm to an individual's height.

Effect size, sample size, and statistical power are important interlinked aspects in this analysis. Although we do not explore this in detail here, power also depends on other factors such as the MAF of a genetic variant. Rare causal variants are much more difficult to detect than common causal variants since the statistical power to significant associations is low and demands a very large sample size. Or, in the context of a case-control study, it is not only the sample size that is important but also the relative number of cases and controls. An equal number of cases and controls is the most optimal for power. To explore genetic power calculations further, see box 4.1.

4.3.6   Fixed-effects versus random-effects models

As we discussed in chapter 2, fixed-effects models rely on the assumption that the true effect of each risk allele is the same in each dataset. Although the assumption can be tenuous, these models are able to maximize discovery in comparison to random effect models [14]. There are various fixed-effects models that we do not describe in detail but include inverse variance weighting and Cochran-Mantel-Haenszel. Random-effects models do not assume that all studies are functionally equivalent and are less often used for discovery since they have limited power. These models are more often applied when the aim is to attempt to generalize the observed association outside of the population and estimate the average effect size of the associated variant and across different populations for predictive purposes.

4.3.7 Weighting, false discovery rate (FDR), and imputation

When multiple data sources are combined, some studies will have more data and thus should count more or have a larger *weight* in meta-analysis results than smaller ones. The optimal weight that is most often used is inverse variance weighting (each study is weighted according to the inverse of its squared standard error). *False discovery rate (FDR)* refers to the estimate of the proportion of associations that are discovered but deemed to be false positives. Here we calculate what is called a Q value, which is the minimum FDR that is possible in order to claim an association. As shown shortly in our applied chapters, we also test for the reliability of *imputation*. It can be a problem when there are polymorphisms with low MAFs, since imputed variants with MAFs $< 5\%$ are excluded from the analysis.

4.3.8 Sources of heterogeneity

Some phenotypes may be difficult to measure or have high measurement variability. In large GWA studies there is often the need to harmonize different data sources and construct one comparable phenotype. Since most phenotypes have already been collected, it is often difficult to engage in a perfectly harmonized analysis. A 2018 study examining the genetic underpinnings of years of education, for instance, engaged in a detailed examination of how variation in the categorization of the phenotype impacted results [15]. They concluded that when possible, the most detailed measure was the best. Yet when harmonizing multiple datasets, many GWASs often harmonize to the highest common— and thus generally least detailed—categorizations.

Beyond ancestry-based heterogeneity, which was discussed at length in chapter 3, there may be inconsistencies such as birth cohort, country, or sex. In chapter 3 we demonstrated how even in relatively small countries such as the Netherlands or in the United Kingdom there are different population stratification patterns. GWAS often combines data from multiple countries and historical time periods to gain a large enough sample size. The implicit assumption is that the influence of genetics on individuals is universal across time and place. In a previous study published in *Nature Human Behavior*, we demonstrated that this is not the case and that combining these disparate datasets has the potential to mask differences, particularly for behavioral phenotypes [16]. In what is called a "mega-analysis," we demonstrated that around 40% of the genetic effects on education and timing of first child is hidden or watered down when data is combined, which increased to 75% for the number of children ever born. In contrast, we found that the genetic variants associated with height seemed to be the same across populations. Sex differences may also induce heterogeneity, which is why some analyses such as those related to reproduction or reproductive behavior examine females, males, and pooled results separately [6, 17]. Obviously, this could be extended to think about other types of heterogeneity such as age or life course effects or socioeconomic status.

4.4 Quality control (QC) of genetic data

The analysis of genetic data to conduct a GWAS entails an understanding of statistical inference in this setting but also numerous quality checks—referred to as quality control (QC). QC is one of the central aspects of working with genetic data. We discuss QC related to GWASs in chapter 8 (see section 8.5). QC is necessary for reliable GWAS results because raw genotyped data are inherently problematic (see box 4.2). For instance, you might have missing data in a large proportion of individuals or high rates of missing genotypes within individuals or other issues related to low sample quality. As we outline in more detail in chapter 8, QC can be divided into individual-level QC and marker-level QC.

Individual-level QC often checks for (1) poor DNA data quality, (2) high or low heterozygosity across autosomal chromosomes, (3) discordant sex information, (4) duplicated or related individuals, and, (5) divergent ancestry. A second set of quality control analyses focuses on the data quality of genotypes or what we discuss in chapter 8 as per-marker QC. Here we take several steps to remove variants that may introduce bias in the study, namely: (1) exclusion of low call rate SNPs; (2) removal of SNPs with very low allele frequency (rare variants); (3) identification and exclusion of variants with extreme deviation from the Hardy–Weinberg equilibrium; (4) in case-control studies, exclusion of SNPs

Box 4.2
Retraction of *Science* longevity paper

Failure to properly check for errors in the data can have serious consequences. Perhaps one of the most well-known cases is the retraction of the exceptional longevity study by Sebastiani et al. (2010) [18] in the journal *Science*. The paper originally identified 19 genes associated with extreme longevity in centenarians. However, days after it was published experts and critics started to ask whether the strong correlation that the authors found was actually due to a technical error in how the different sequencing chips were assigned to samples (Illumina 610-Quad array) that the team used that could in turn produce false positive associations. The problem was that the chip was only used for the centenarians, with a different one used for the control group. The issue is that if you have different call rates or other systematic biases that are different between the chips, it introduces an artifactual association (see chapter 7, section 7.2, where we discuss these aspects in more detail). Since the new analyses deviated quite seriously from the original article and the authors could not replicate the original finding, the article was retracted. Interested readers can refer to the discussion in particular around this error and subsequent retraction.

with extreme differential call rates between groups; and (5) in the case of dealing with imputed SNPs, exclusion from the study of variants with low imputation quality.

4.5 The NHGRI-EBI GWAS Catalog

4.5.1 What is the NHGRI-EBI GWAS Catalog?

Researchers new to the field often want to know which phenotypes have already been studied and the various SNPs that have been identified. The primary resource is the NHGRI-EBI GWAS Catalog (hereafter referred to as the Catalog) and includes data from all published GWASs, located at https://www.ebi.ac.uk/gwas/. It is produced by the U.S. National Human Genome Research Institute (NHGRI) [19] in conjunction with the European Bioinformatics Institute (EBI) [20]. To be included in the Catalog, studies must meet very strict criteria (see www.ebi.ac.uk/gwas/docs/methods), include an array-based GWAS and an analysis of more than 100,000 SNPs with genome-wide coverage. SNP-trait associations that are reported in the Catalog are those with at least a p-value of $< 1 \times 10^{-5}$. The Catalog researchers locate studies via an automated PubMed search and then manually curate them for assessment and inclusion. All GWAS traits are mapped to terms from the *Experimental Factor Ontology* (EFO) [21], which is an ontology of variables used in molecular biology including aspects of disease, anatomy, cell type, cell lines, chemical compounds, and assay information. If you search for "cardiovascular disease," for instance, the Catalog provides the results and visualizations of all studies and associations for this specific trait and its subtraits. In this example, subtraits might be "myocardial infarction" or "coronary heart disease."

Figure 4.3 provides a visualization of the NHGRI-EBI GWAS Catalog, illustrating reported genetic associations based on their genomic locations across all (human) chromosomes. Each line links to a locus that has been associated with a trait that has a p-value threshold of $p \leq 5 \times 10^{-8}$, and each circle is color-coded to represent a distinct trait. They are grouped according to 17 main trait categories such as digestive system disease, hemotological measurement, cancer, or response to drug. It is possible to search the Catalog by publications, variants, traits, or genes, which is continuously updated with new publications.

4.5.2 A brief history of the GWAS

There are several excellent narrative reviews of GWASs describing the underlying rationale and scientific conclusions and highlighting key milestones [2, 22, 23]. Although the first GWAS was published in 2005, the major breakthrough was a paper published by the Wellcome Trust Case Control Consortium in 2007 [24], which was heralded as a masterwork in diplomacy due the need to collaborate to combine multiple sources of data [23]. As noted earlier, to conduct a successful GWAS, a large sample size is required to afford sufficient statistical power [25]. This means that a majority of the GWASs published to

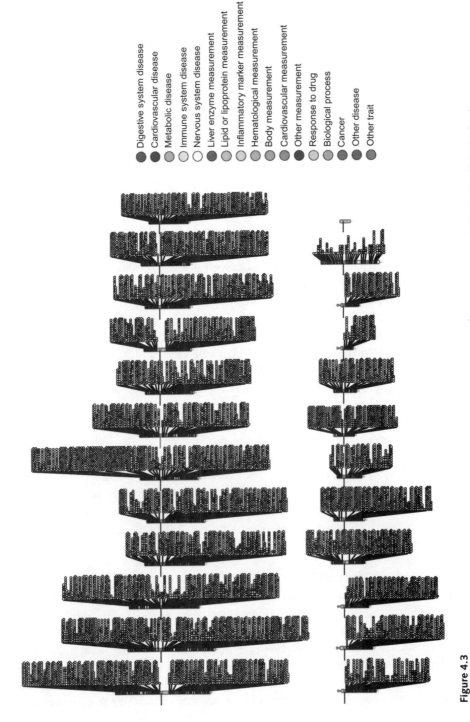

Digestive system disease
Cardiovascular disease
Metabolic disease
Immune system disease
Nervous system disease
Liver enzyme measurement
Lipid or lipoprotein measurement
Inflammatory marker measurement
Hematological measurement
Body measurement
Cardiovascular measurement
Other measurement
Response to drug
Biological process
Cancer
Other disease
Other trait

Figure 4.3

Chromosomal map of published genome-wide associations as of May 2018, $p \leq 5 \times 10^{-8}$ for 17 trait categories by the NHGRI-EBI Catalog.

Source: NHGRI-EBI Catalog and diagram are freely available from https://www.ebi.ac.uk/gwas/docs/diagram-downloads for use under the general EMBL-EBI terms of use (http://www.ebi.ac.uk/about/terms-of-use) [19, 20]. The image is generated by the NHGRI-EBI on a quarterly basis.

date often pool the summary results of separate analyses from multiple data sources in a meta-analysis in order to obtain the largest sample size possible. Advances in technology, methods, theory, computational power, and funding have drastically changed the GWAS landscape over the past decades.

In our previous work, Mills and Rahal (2019) [3] carried out a systematic and computational review of all GWASs in the 13 years from 2005 until October 2018. We used the NHGRI-EBI GWAS Catalog and linked it to external databases such as PubMed. It is important to note that we included all code that we used on a publicly available GitHub site in addition to making this a living database (https://github.com/crahal/GWASReview). In other words, with the update of each Catalog, our database and the figures and numbers described here will automatically update over time. As figure 4.4 shows, there has been a remarkable growth in the number of GWASs published, sample size, number of associations, and diseases studied over time.

In the upper panel we see the large jump in the number of studies published over time (divided according to sample size). Here we see the incredible gains in sample sizes over time, with those published in the late 2018s and early 2019s sometimes containing over 1 million individuals. These larger studies are mostly attributed to the UK Biobank (around 500,000 individuals) [26, 27] and large direct-to-consumer companies such as 23andMe that participate in this research [28]. The lower-left panel shows the strong positive correlation between the number of associations found and the number of participants used in GWASs over time. The lower-right panel shows the growth in the number of unique traits and in journals publishing GWASs. As of October 2018, we found that 3,639 studies were published covering 5,849 unique study accessions (identifiers ascribed to traits within a paper) across 3,508 unique traits mapped to 2,532 EFO traits. These traits can include anything from height to male-pattern baldness, Alzheimer's disease, breast cancer, coffee consumption, or neuroticism. The average number of hits per study is 15.3, with an average p-value for the strongest risk allele of 1.3729×10^{-6}. Around 55% of the reported associations met the standard threshold of $p \leq 5 \times 10^{-8}$.

4.5.3   Lack of diversity in GWASs

For researchers new to the field, it is essential to note the current lack of diversity in genetic samples. As we discussed in previous chapters, disparities in *ancestral diversity* of subjects has been related to technical issues such as population stratification [29], reduced linkage disequilibria [30], genetic diversity, and admixture [31] but also refusal to participate in studies due to cultural distrust and social misuses of data [32, 33]. Figure 4.5 shows that although there was a veritable explosion in the number of GWASs and traits over time, it still remained largely within European ancestry populations, with non-European populations more often examined in the replication phase. What this means is that these non-European populations were often used to test whether the results found in the

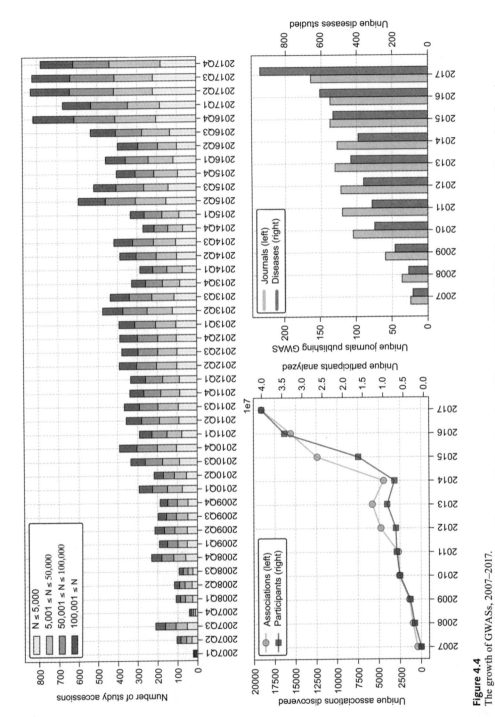

Figure 4.4

The growth of GWASs, 2007–2017.

Sources: NHGRI-EBI GWAS Catalog and figures made by authors from Mills and Rahal (2019) [3]. Creative Commons 4.0 license: http://creativecommons.org/licenses/by/4.0/.

Notes: The upper panel shows the number of study accessions published per quarter over time colored according to sample size to show the growth of larger (100,001 ≤N) GWASs. The lower-left panel shows the strong positive correlation between the number of associations found and the number of participants used in GWASs over time. The lower-right panel shows the growth in the number of unique traits examined as well as the number of unique journals publishing GWASs over time. The period from 2007 to 2017 is selected because only 10 entries occurred before 2007. Each panel contains full calendar years only.

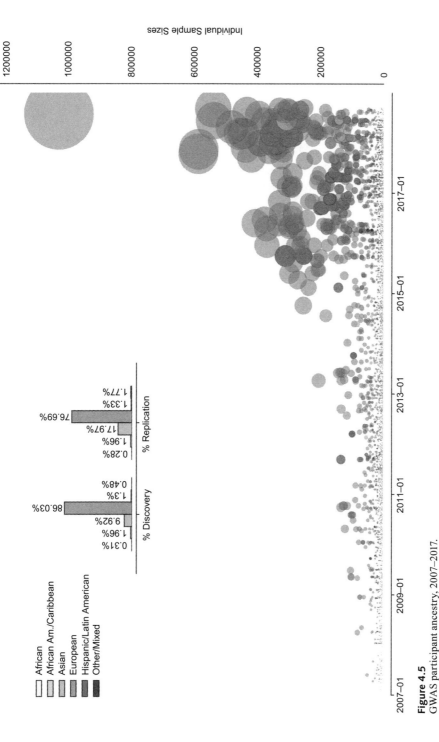

Figure 4.5
GWAS participant ancestry, 2007–2017.

Sources: NHGRI-EBI GWAS Catalog and author mapping; figure produced by authors and reproduced from Mills and Rahal (2019) [3]. Creative Commons 4.0 license: http://creativecommons.org/licenses/by/4.0/.

Notes: The main panel shows a disaggregation of our broader ancestral categories field, which is a direct mapping from the 17 broad ancestral categories identified in the Catalog. We drop all rows where any proportion of the ancestry is not recorded, and for combinations of ancestries (e.g., European and African) we create a new field: Other/Mixed. The inset aggregates this across the entire sample but partitions the data across discovery and replication phases. The period from 2007 to 2017 is selected since only 10 entries occurred before 2007, and we have complete information for the year 2017.

European ancestry population would replicate in other ancestry groups and therefore not often as a basis for fundamental genetic discovery in those populations.

Figure 4.5 shows ancestry groups by the commonly used six broad ancestral categories. Those of European ancestry have been examined the most, ranging from as high as 95% of subjects in 2007–2008 to 88% in 2017. Particularly since 2011, there has been a strong and steady rise of research in Asian populations (see box 4.3). As described in Mills and Rahal (2019, table 2) [3], this is primarily the Japanese, Chinese, and South Korean populations. African populations have been the least studied over time, with the hope that projects such as the African Genome Variation Project [34] and others promoting diversity will continue to increase and alter these trends.

Diversity in relation to GWA studies is almost exclusively discussed in relation to ancestry, yet we also found a striking lack of geographical, environmental, temporal and demographic (e.g., age, sex) diversity in our GWAS review [3]. As we note, although around 76.2% of the current world population resides in Asia or Africa, 72% of genetic discoveries emanate from participants residing in only three countries (the United States, the United Kingdom, and Iceland). As we elaborate upon in this chapter and elsewhere [16], more work needs to be done to understand how environmental exposure and geographical concentration influence results. For example, those with a predisposition for obesity face radically different environmental stimuli in the the United States, Mexico, and the United Kingdom compared to some other nations that have markedly lower obesity

Box 4.3
The rise of Asian ancestry genetic research

As shown in figure 4.5, there has been marked rise in GWA studies in Asian populations, particularly since 2011. The most frequent studies have emanated from Japan, China, and South Korea. As of the end of 2018, 7.7% of recorded studies involved Japanese participants, representing 14.3% of all participants contributing to GWAS research; Chinese, 3.7% of recorded studies representing 8% of all participants, and South Korean, 1.5% of recorded studies representing 4% of all GWAS participants. For the full tables, see Mills and Rahal (2019) [3] and our regularly updated GitHub site, https://github.com/crahal/GWASReview /blob/master/tables/CountryOfRecruitment.csv. As noted, these Asian countries rank close to the countries with the largest number of recorded studies and percentage of participants in the United Kingdom (40.5%; 10.5%), the United States (19.8%; 41%), and Iceland (11.5%; 1.1%), respectively. When looking at the most frequently used data sources in the largest 1,250 GWASs as of August 2018, we see that these numbers are related to the inclusion of studies such as the Biobank Japan Project (BBJ), Korean Association Resource Project (KARE), Korean Genome Epidemiology Study (KOGES), Korean National Cancer Center Study (KNCC), Japanese Millenium Genome Project (JMGP), Guangxi Fancchenggang Area Male Health and Examination Survey (FAMHES), the Japan Multi-Institutional Collaborative Cohort Study (J-MICC), and others such as the Chinese Kadoorie Biobank.

levels, such as Japan, Korea, Italy, and the Netherlands. We also found a lack of temporal and demographic diversity of birth cohorts, historical periods, and life course stages. The most frequently used data in GWASs are often disproportionately older, higher socioeconomic status, frequently more women, and often compounded by a "healthy volunteer" selection such as in the UK Biobank [35].

4.6 Conclusion and future directions

There have been considerable changes in this area of research since the first GWAS in 2005. We introduced readers to the NHGRI-EBI GWAS Catalog that contains a summary of all published GWASs to date. We also chronicled how this field has exploded not only due to the sheer number of studies, diseases, and associations studied but also burgeoning sample sizes. As of 2019, many large studies have a combined sample of over 1 million cases. We note, however, that this growth has not been even across different ancestral or geographical groups, with the majority of research still within European ancestry populations. Asian studies in particular have grown, with new investments across the world, such as in Africa, to enhance further diversity. An emerging and exciting line of research will be discoveries of the genetic diversity of non-European ancestry populations. We should also note that forming these large consortiums may also be something of the past. With a growing number of large data sets such as UK Biobank and direct-to-consumer companies such as 23andMe, gathering many small data cohorts to produce a large sample appears to be increasingly less common.

Readers will have also gained a basic idea about the methodology underlying GWA studies. Although this remains an introductory book, our hope is that you have gained a rudimentary understanding of how this type of research is conducted, the meaning of statistical inference in GWASs, and why and how we need to correct for multiple testing. The importance of quality control (QC) at the individual and genetic marker level was also described, with hands-on applications in chapter 8 of this book.

As our brief history of the GWAS demonstrated, this is a rapidly moving area of study. As we elaborate upon in chapter 14 and 15 on ethical issues and future directions, GWASs have also not been entirely free from controversy. There has been some concern that the long lists of prioritized "hits" have not brought the heralded personalized medicine and new therapies and risk prediction tools that some had promised. Although beyond the auspices of this book, the biological follow-up of many GWAS hits has located variants linked to known biological pathways but also others that had not been clinically targeted. A growing number of studies are moving to examine not only common but also rare variants. Further developments in sequencing data will also likely uncover exciting new findings, areas of research, and new methods. New ways of analyzing and synthesizing GWAS data have also emerged, such as the work by the Complex-Traits Genetics Virtual Lab for post-GWAS analyses (https://genoma.io/updates).

Exercises

Choose a phenotype you are interested in and explore what has already been done.

1. Go the GWAS Catalog and examine the diagram that contains the various traits that have been studied to date at each chromosome and see if the trait you are interested in has been studied: https://www.ebi.ac.uk/gwas/diagram.

2. Examine twin-heritability using the application MaTCH (Meta-Analysis of Twin Correlations and Heritability) accessible via http://match.ctglab.nl/; Gephi, http://gephi.github.io/.

3. Go to SNPedia (https://www.snpedia.com/index.php/Heritability) to see if SNP-based heritability has been examined.

4. Go to Ben Neale's site using the UK Biobank and see what you uncover: (http://www.nealelab.is/uk-biobank/).

5. Produce a visualization of results including Manhattan plots and many others from the Complex-Traits Genetics Virtual Lab (CTG-VL) for post-GWAS analyses [5]: https://genoma.io.

Further reading

For introductory articles on GWASs, see the following:

Attia, J. et al. How to use an article about genetic association. A: Background Concepts. *JAMA* **301**(1), 74–81 (2009).

Attia, J. et al. How to use an article about genetic association. B: Are the results of the study valid? *JAMA* **301**(2), 191–197 (2009).

Lunetta, K. L. et al. Genetic association studies. *Circulation* **111**(1), 96–101 (2008).

For an early introduction to gene mapping and association studies, see the following:

Neale, Benjamin et al. (2007). *Statistical genetics: Gene mapping through linkage and association*. London: Taylor and Francis. For recent reviews, see references 2, 3, and 29.

For further reading in the area of power analyses, see the following:

Hong, E. P., and J. W. Park. Sample size and statistical power calculation in genetic association studies. *Gen. & Inform.* **10**(2), 117–122 (2012).

Sham, P. C., and S. M. Purcell. Statistical power and significance testing in large-scale genetic studies. *Nat. Rev. Gen.* **15**(5), 335 (2014).

Online resources

dbSNP: https://www.ncbi.nlm.nih.gov/snp/.

NHGRI-EBI GWAS Catalog: https://www.ebi.ac.uk/gwas/.

UCSC Genome Browser: https://genome.ucsc.edu/.

For a longer list of genetic datasets as reported in Mills and Rahal (2019), see https://github.com/crahal/GWASReview/blob/master/tables/Manually_Curated_Cohorts.csv.

References

1. L. E. Duncan and M. C. Keller, A critical review of the first 10 years of candidate gene-by-environment interaction research in psychiatry. *Am. J. Psychiatry* **168**, 1041–1049 (2011).

2. P. M. Visscher et al., 10 years of GWAS discovery: Biology, function, and translation. *Am. J. Hum. Genet.* **101**, 5–22 (2017).

3. M. C. Mills and C. Rahal, A scientometric review of genome-wide association studies. *Commun. Biol.* **2** (2019), doi:10.1038/s42003-018-0261-x.

4. T. Polderman et al., *Nat. Genet.*, in press.

5. Gabriel Cuellar-Partida et al., Complex-Traits Genetics Virtual Lab: A community-driven web platform for post-GWAS analyses. *bioRxiv Prepr.* (2019), doi:10.1101/518027.

6. N. Barban et al., Genome-wide analysis identifies 12 loci influencing human reproductive behavior. *Nat. Genet.* **48**, 1–7 (2016).

7. E. Evangelou and J. P. A. Ioannidis, Meta-analysis methods for genome-wide association studies and beyond. *Nat. Rev. Genet.* **14**, 379–389 (2013).

8. T. W. Winkler et al., Quality control and conduct of genome-wide association meta-analyses. *Nat. Protoc.* (2014), doi:10.1038/nprot.2014.071.

9. N. D. Palmer et al., A genome-wide association search for type 2 diabetes genes in African Americans. *PLoS One* **7**, e29202 (2012).

10. J. Fadista, A. K. Manning, J. C. Florez, and L. Groop, The (in)famous GWAS P-value threshold revisited and updated for low-frequency variants. *Eur. J. Hum. Genet.* **24**, 1202–1205 (2016).

11. B. V. North, D. Curtis, and P. C. Sham, A note on the calculation of empirical P values from Monte Carlo procedures. *Am. J. Hum. Genet.* **72**, 498–499 (2003).

12. P. C. Sham and S. M. Purcell, Statistical power and significance testing in large-scale genetic studies. *Nat. Rev. Genet.* **15**, 335–346 (2014).

13. S. Stringer et al., A guide on gene prioritization in studies of psychiatric disorders. *Int. J. Methods Psychiatr. Res.* **24**, 245–256 (2015).

14. T. V. Pereira, N. A. Patsopoulos, G. Salanti, and J. P. A. Ioannidis, Discovery properties of genome-wide association signals from cumulatively combined data sets. *Am. J. Epidemiol.* **170**, 1197–1206 (2009).

15. J. J. Lee et al., Gene discovery and polygenic prediction from a genome-wide association study of educational attainment in 1.1 million individuals. *Nat. Genet.* **50**, 1112–1121 (2018).

16. F. C. Tropf et al., Hidden heritability due to heterogeneity across seven populations. *Nat. Hum. Behav.* **1**, 757–765 (2017).

17. R. M. Verweij et al., Sexual dimorphism in the genetic influence on human childlessness. *Eur. J. Hum. Genet.* **25**, 1067–1074 (2017).

18. P. Sebastiani et al., Genetic signatures of exceptional longevity in humans. *Science* (2010), doi:10.1126/science.1190532.

19. D. Welter et al., The NHGRI GWAS Catalog, a curated resource of SNP-trait associations. *Nucleic Acids Res.* **42**, D1001–D1006 (2014).

20. J. MacArthur et al., The new NHGRI-EBI Catalog of published genome-wide association studies (GWAS Catalog). *Nucleic Acids Res.* **45**, D896–D901 (2017).

21. J. Malone et al., Modeling sample variables with an experimental factor ontology. *Bioinformatics* **26**, 1112–1118 (2010).

22. P. M. Visscher, M. A. Brown, M. I. McCarthy, and J. Yang, Five years of GWAS discovery. *Am. J. Hum. Genet.* **90**, 7–24 (2012).

23. T. A. Manolio, In retrospect: A decade of shared genomic associations. *Nature* **546**, 360–361 (2017).

24. P. R. Burton et al., Genome-wide association study of 14,000 cases of seven common diseases and 3,000 shared controls. *Nature* **447**, 661–678 (2007), doi:10.1038/nature05911.

25. F. Dudbridge, Power and predictive accuracy of polygenic risk scores. *PLoS Genet.* (2013) (available at https://doi.org/10.1371/journal.pgen.1003348).

26. C. Sudlow et al., UK Biobank: An open access resource for identifying the causes of a wide range of complex diseases of middle and old age. *PLOS Med.* **12**, e1001779 (2015).

27. C. Bycroft et al., The UK Biobank resource with deep phenotyping and genomic data. *Nature* **562**, 203–209 (2018).

28. K. Servick, Can 23andMe have it all? *Science* **349**, 1472–1477 (2015).

29. D. Hamer and L. Sirota, Beware the chopsticks gene. *Mol. Psychiatry* **5**, 11–13 (2000).

30. A. C. Need and D. B. Goldstein, Next generation disparities in human genomics: Concerns and remedies. *Trends Genet.* **25**, 489–494 (2009).

31. M. P. Conomos et al., Genetic diversity and association studies in US Hispanic/Latino populations: Applications in the Hispanic community health study/study of Latinos. *Am. J. Hum. Genet.* **98**, 165–184 (2016).

32. After Havasupai litigation, Native Americans wary of genetic research. *Am. J. Med. Genet. A.* **152**, 33592 (2010).

33. V. L. Shavers-Hornaday, C. F. Lynch, L. F. Burmeister, and J. C. Torner, Why are African Americans under-represented in medical research studies? Impediments to participation. *Ethn. Health* **2**, 31–45 (1997).

34. D. Gurdasani et al., The African Genome Variation Project shapes medical genetics in Africa. *Nature* **517**, 327–332 (2015).

35. A. Fry et al., Comparison of sociodemographic and health-related characteristics of UK Biobank participants with those of the general population. *Am. J. Epidemiol.* **186**, 1026–1034 (2017).

5

Introduction to Polygenic Scores and Genetic Architecture

Objectives

- *Define* and understand the *origins* of the polygenic score
- Understand the process and *flowchart of working* with polygenic scores
- Comprehend the main principles of *constructing* a polygenic score
- Know the basics of *validation and prediction* of polygenic scores
- Grasp concepts surrounding the *shared genetic architecture of phenotypes* and potential ways to examine this (correlation, pleiotropy, multitrait analysis)
- Be introduced to *applications* of causal modeling with polygenic scores (genetic confounding, Mendelian Randomization, gene-environment interaction)
- Recognize the *central challenges*, why they are problematic, and potential solutions in working with polygenic scores

5.1 Introduction

The genetic architecture of most phenotypes and health conditions are polygenic in nature. *Polygenic* refers to the fact that it is not a single or handful of variants but instead hundreds or thousands of variants that each has very small effects on the phenotype. Although there are monogenic diseases such as Huntington's disease that has a mono-causal effect, most of the traits that we study are polygenic. With the growth of genome-wide association studies (GWASs) and larger samples, PGSs have increasingly emerged as a major tool in several areas of quantitative genetic research.

The aim of this chapter is to provide you first and foremost with an understanding of polygenic scores, how they emerged, and the central challenges and potential solutions to effectively apply them. A secondary aim is to provide you with a blueprint of how to carry out your own research in this area. Our flowchart in figure 5.1 presents an overview of the steps but also possibilities of working with PGSs for those entering the field for the first

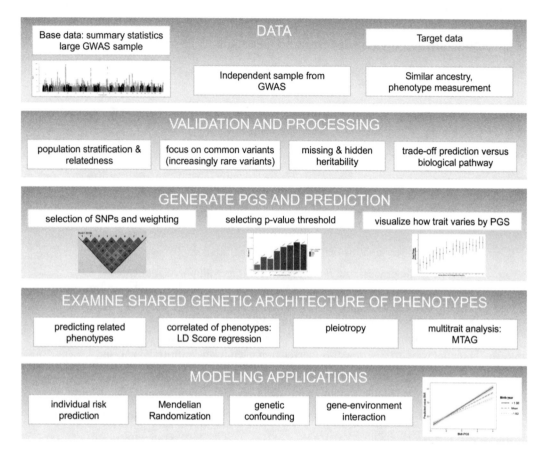

Figure 5.1
Flowchart of development and working with polygenic scores.

time. This includes the initial phases of data, validation and processing, generation, and prediction of PGS. Some readers may also desire to dig deeper and examine the shared genetic architecture of phenotypes. This is then followed by various modeling applications that are discussed in this chapter and then applied in parts II and III of this book. In table 5.1, we provide an additional summary of the main challenges of working with PGSs, explain why these challenges are problematic, and offer potential solutions and further reading on the topic. The current chapter provides the necessary background that you will need to create and validate PGSs in chapter 10 and then properly apply them in statistical models across various situations in chapters 11–13.

Table 5.1

Summary of challenges and solutions in polygenic score application

Challenge	Why is this problematic?	Potential solutions	Example of study, further explanation
Requires large sample sizes	Large sampling error due to large number unknown SNPs, true effects being very small	Create PGSs from sufficiently large GWAS sample sizes for accuracy	Daetwyler et al. (2008) [1] Dudbridge (2013) [2]
Selection of SNPs to include in PGS	Trade-off: if you include more (or all) variants you will increase the prediction (i.e., R^2) but introduce noise and noncausal SNPs	Choice depends on phenotype and research question.	Wray et al. (2013) [3]
Weighting of effects in PGS		Highly polygenic traits, more lenient *p*-value thresholds (i.e., include all SNPs), LD pruning (isolate independent SNPs), and thresholding methods	
Overlap in discovery and target sample	Overestimate the accuracy of prediction via over-fitting	Use an independent sample Remove target sample from the GWAS Use summary statistics of GWASs from similar large study	Wray et al. (2013) [3] https://github.com/Nealelab/UK _Biobank_GWAS
Prediction possible only within similar populations	Different allele frequencies and LD across ancestry populations	Collect more genetic data from more diverse populations	Martin et al. (2017) [4]—limited portability of scores derived from European ancestry GWAS Lee et al. (2018) [5]—nontransferability of PGS on African Americans De La Vega and Bustamante (2018) [6]
Population stratification in samples and relatedness	Will produce potential inflation in R^2 of PGS on your target sample	Fit ancestry principal components Use conventionally unrelated individuals in discovery and validation stages	Belgard et al. (2014) [7]—critique genetic autism study confounded by population structure Makowsky et al. (2011) [8]—for height Wray et al. (2013) [3]—for height in Framingham Heart Study

(continued)

Table 5.1 (continued)

Challenge	Why is this problematic?	Potential solutions	Example of study, further explanation
Differential bias in population stratification between cases and controls	Leads to spurious prediction of R^2 if discovery and target sample have same differential bias	Perform stringent QC and/or validate in a different independent sample	Lee et al. (2011) [9] QC step—use genotyped SNPs in PGS and quantify estimated relatedness between target and discovery sample in a PCA If target sample is an outlier on PCA, prediction accuracy in target is less than expected
Variants only explained by common markers, missing rare variants	Some SNPs have not been identified in GWASs Results in discrepancy between family and SNP-based heritability estimates	Incorporate rare variants into predictions and GWASs	Kemper et al. (2002) [10]
Missing and hidden heritability	Lower than expected prediction of PGS (i.e., low R^2) Bias due to noisy estimates, heterogeneity in sample	Attempt to eliminate error, including accurate or harmonized phenotype, sufficient sample size, consideration of interaction with environment	Tropf et al. (2017) [11] Courtiol et al. (2016) [12]
Trade-off of prediction and understanding biological mechanisms	Obtain more predictive PGSs for highly polygenic traits by including more (all) SNPs in PGS, yet lose biological specificity	Understand trade-off for phenotype and research question If interested in interventions, be aware of interpretations Examine downstream biological analyses on prioritized genes	Goodarzi (2018) [13]—discussion of biological functional information required for better insights into biology of obesity
Shared genetic architecture with other phenotypes	Overlap and co-occurrence important when studying or designing potential treatments	Examine whether PGS predicts other phenotypes	Purcell et al. (2009) [14]—example schizophrenia and bipolar disorder
Causal modeling using PGS with Mendelian Randomization	Minimize risk of including noise due to directional pleiotropy in PGSs	Do not use high p-value thresholds, which may violate the assumptions of MR	Hemani et al. (2013) [15]
Using PGS in gene-environment interaction studies	Summary in table 6.2	Summary in table 6.2	Duncan and Keller (2011) [16]; Keller (2014) [17]
Utility for personalized medicine	Discussion in chapter 14	Discussion in chapter 14	Torkamini et al. (2018) [18]

5.1.1 What is a polygenic score?

A *polygenic score* (PGS) is a numeric summary of the relationship between multiple genetic loci and a phenotype. A PGS is sometimes referred to as a polygenic profile score, genetic profile score, genotype score, or, when discussing disease, as a polygenic risk score. We adopt the more neutral term—polygenic score—since it is less intuitive to speak in terms of "risk" when we discuss nondisease-related behavioral phenotypes. Polygenic scores are derived directly from the genome-wide associations in GWASs that we outlined in chapter 4. We use the summary statistics from these to construct an estimate of how *single-nucleotide polymorphisms* (SNPs) combine to explain the trait of interest.

In practice, PGSs are linear combinations of the phenotype-associated alleles across the genome, typically weighted by GWAS effect sizes. It is thus a single quantitative measure that can be interpreted as a measure of an individual's genetic propensity toward a phenotype relative to a population. Individual SNPs (i.e., monogenic, as discussed in chapter 1) are weak predictors for most of the traits that we are interested in. Complex traits are associated with many genetic variants, each of which account for a small percentage of variance. PGSs are a solution to aggregate this information across the genome. In general, we can define a polygenic score for an individual as the *weighted sum* of a person's genotypes at M loci. A PGS for individual i can be calculated as the sum of the allele counts a_{ij} (0, 1, or 2) for each SNP $j = 1, \ldots M$, multiplied by a weight w_j

$$PGS_i = \sum_{j=1}^{M} a_{ij} w_j$$

where the weights w_j are transformations of GWAS coefficients. This equation shows that it is a linear combination of the effects of multiple SNPs on phenotype. The underlying model in a PGS is also usually additive, since we count the number of "risk alleles" for each SNP included in the score. We note, however, that recessive or dominant models can be used in the construction of a PGS. Due to the large number of SNPs that are included in their construction, they also follow a normal distribution (see box 5.1). An additional assumption is the absence of gene-gene interactions (or epistasis) since SNP effects are assumed to be independent.

5.1.2 The origins of polygenic scores

Many studies and lines of thinking culminated in the production of PGSs. One of the early studies that introduced the concept of PGSs was a study on the genetic architecture of schizophrenia in 2009 [14]. Schizophrenia is a severe mental disorder with a high level of heritability (h^2) of up to 0.80 [20, 21]. The disease is characterized by hallucination, delusion, and cognitive deficits and has a prevalence of around 7 in 1,000 people. Researchers had observed that it was passed along in families and assumed to be polygenic even as early as the 1970s [22, 23]. The field then developed further such as a seminal paper by

Box 5.1
Why do polygenic scores have a normal distribution? An application of the central limit theorem

Polygenic scores can be thought of as a sum of many independent genetic signals. A central premise of probability theory in statistics, known as the *central limit theorem*, establishes that when many independent random variables are added, their sum tends toward a normal distribution regardless of the original distribution of the single variables. This is what is often informally referred to as the "bell curve." As our simulation below demonstrates, the larger the number of alleles, the better the approximation will be to a normal distribution. Polygenic scores therefore tend to have a normal distribution since the number of SNPs that are included in the score is sufficiently large [19].

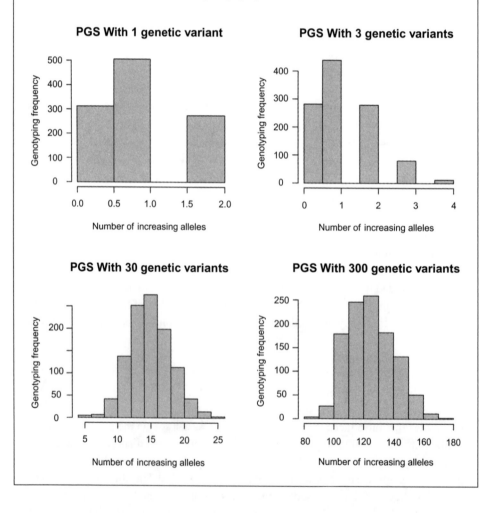

Risch, Merikangas and colleagues in 1996 in *Science* demonstrating that for complex phenotypes, GWASs had superior power over the genome-wide linkage studies used at the time [24]. The first GWAS for schizophrenia was published in 2008 [25]. This was then followed by a larger 2009 study published in *Nature* (~13,000 cases; 35,000 controls) [26].

One of the key shifts toward the creation of PGSs was in 2009, when the International Schizophrenia Consortium "failed" to identify any specific SNPs predicting this highly heritable mental disorder. The research team decided to dig deeper and investigate the role of all SNPs, revisiting one of the most classic theories of polygenic inheritance in the form of Fisher's 1918 infinitesimal model [27]. Recall that the *infinitesimal model* hypothesizes that a quantitative (continuous) phenotype is controlled by an infinite number of loci and that each locus has an infinitely small effect. Rather than searching for a small number of genes with larger predictive power, the group claimed that there could be potentially thousands of very small individual effects that collectively accounted for a substantial part of the heritability. Those variants that were derived from a GWAS of a smaller sample size, however, would not show up in a GWAS since they did not reach genome-wide significance.

Consider, for example, a SNP, for which a risk allele increases the relative risk to develop schizophrenia by only 5%. Such a small effect would need to be estimated with an extremely small standard error in order to fall below the significance threshold of 5×10^{-8}, the standard criterion for genome-wide significance in a GWAS (see chapter 4). Therefore, it would be highly likely for it to remain undetected even in a relatively large sample. The team therefore first calculated the score only including highly significant SNPs and then recalculated it by continuously relaxing the *p*-value threshold up to 0.5, basically including 50% of all SNPs. They used this battery of scores and generated a sample that was not part of the original GWAS to predict schizophrenia. They found that the explanation of variance increased as the *p*-value threshold was relaxed. This implied that even supposedly "nonsignificant" genetic variants explained variation in the phenotype, although their individual effects and mechanisms remain unspecified. Although this original study already suggested that schizophrenia is highly polygenic, later studies quantified expectations more precisely, finding that around 8,300 independent SNPs contribute to the phenotype [28]. Since then multiple GWASs from different groups have been published, with larger studies leading to more precise PGS estimates.

5.2 Construction of polygenic scores

In chapter 10 we demonstrate the practicalities of how to construct PGSs, followed by how to validate and apply them across multiple applications for trait prediction, as confounders and to examine gene-enviornment interaction in chapter 11. We discussed the discovery of

SNPs in considerable detail in chapter 4. In this section we highlight pitfalls and dangers of construction of PGSs, but note that some of the solutions involve detailed statistical techniques that remain beyond the scope of this introductory textbook.

5.2.1 Large sample sizes required in GWAS discovery

It is no coincidence that there has been a rapid growth in the sample size of GWASs over time (see figure 4.5). In order to estimate the effects of SNPs on a phenotype, it is important to reduce sampling error, which can be achieved by including a large sample size in the discovery of genetic markers. We have repeatedly noted that complex phenotypes are influenced by a large number of unknown SNPs that have very small effects, thereby necessitating large discovery samples. As noted in chapter 4, for many common traits, discovery sample sizes are now reaching around 1 million individuals. Multiple authors have demonstrated how the accuracy of SNP effects, and by extension PGSs, increase with sample size [1, 2, 29]. Others are now increasingly questioning whether we have hit a point of diminished returns and should now shift the focus from discovery of more loci to a deeper understanding of the biological function of loci.

5.2.2 Selection of SNPs to include

As we explore in chapter 10, two key decisions are required to construct a PGS: the number of genetic variants to include and how to weight their effects. The most commonly used method is a straightforward least squares prediction [30]. Since we discuss pruning and threshold methods and weighting in chapter 10 (section 10.3), we do not reiterate it here. It is possible to select only GWAS-significant SNPs (p-value $< 5 \times 10^{-8}$), something in between, or all SNPs (p-value $<=1$). The choice depends on the phenotype and the type of application you will conduct. Stricter p-value thresholds are generally considered to be more suitable for traits that are not polygenic, while more lenient thresholds perform the best for polygenic traits. In the case where traits are not polygenic, which researchers are now realizing is actually quite rare, only genome-wide significant variants are included to increase the accuracy of the predictive score. You can expect to have more predictive results when all SNPs are included in the calculation of a PGSs for highly polygenic traits. A challenge we discuss shortly, however, is the trade-off of including more variants in the analysis to increase prediction, which in turn adds potential "noise" of noncausal variants but also causal variants that are proxy SNPs (see box 10.2).

5.3 Validation and prediction of polygenic scores

Validation of the PGS underpins its usefulness. If incorrect decisions or conclusions are drawn at this initial stage, the PGS may lack precision and accuracy. Validation is also inherently intertwined with prediction. In this section, we focus on the basic and common errors that can either lead to an overestimation of the PGS or misinterpretation of results, sometimes using examples from the literature. *Prediction* is the estimation of R^2, which is the proportion of variance explained by the regression model. In that sense, we note that prediction is somewhat of a misleading term since we are generally interested in understanding the amount of variability that can be explained by including a particular PGS in a model. Most applied researchers are often interested in understanding the incremental increase in the R^2 when you enter your PGS into a model compared to the baseline model. The *baseline model* is the simplest possible prediction, which you use as a starting point in which to benchmark against when additional variables are added. Here we then also generally include the population stratification variables (e.g., the first 10 or 20 PCAs) and other relevant covariates. In chapters 10 and 11, we will demonstrate how to engage in prediction and how to deal with some of the issues discussed below and summarized in table 5.1.

5.3.1 Independent target sample

When you engage in prediction, it is essential that the data that you are using is an *independent sample* or, that is to say, that there is no overlap in the discovery and target samples. In other words, the target sample that you are using should either not be one of the datasets included in the original GWAS or you need to remove it from the GWAS summary results. We discuss how and where to obtain GWAS summary statistics in chapter 7 (section 7.3.3).

If you attempt to validate or predict the performance of the score using the same data that was used in the original GWAS to also estimate the effect of the SNPs on the phenotype, you overestimate the accuracy of the prediction via *over-fitting* [3]. To ensure that the association results do not overlap with your genotyped data, it is a good practice to first examine which cohorts were included in the discovery analysis. This information is usually reported in the initial tables of the Supplementary Material in the published GWAS article. Many authors increasingly have a pipeline that has results ready and prepared to apply for each of the cohorts in the study. Increasingly, studies also add the PGS as part of their data (e.g., the Health and Retirement Study). If this is not the case, it is good practice to inquire directly with the researchers who conducted the study and ask if they are willing to share the results of the meta-analysis excluding the cohort that you want to analyze. Note that this does require a certain amount of effort, also on the side of the original authors. Alternatively, it is possible to use another sufficiently large dataset and the summary statistics from a GWAS calculated on a single very large study. One solution is to use information from Ben Neale's lab, which openly produced results from more than 4,000 phenotypes using the UK Biobank, also with 20 principal components and covariates (e.g., age, age^2, sex, age*sex) (http://www.nealelab.is/uk-biobank/). They also generated sex-specific results and included all of the code used to run their analyses on GitHub (https://github.com/Nealelab/UK_Biobank_GWAS). The degree of bias also depends on various factors, including heritability of trait, genetic heterogeneity across studies, and sample size, which we discuss shortly in relation to missing and hidden heritability. If the sample size of the genotyped data that you plan to use is much smaller than the sample size of the entire GWAS, the bias is probably limited. However, this aspect still needs to be taken into account.

5.3.2 Similar ancestry in target sample

When selecting your target sample, the ancestry composition should not markedly differ from your initial baseline sample. Recall from chapter 4 that most GWASs have been performed on people of European ancestry, and that these results cannot be directly transferred to other populations due to differences in allele frequencies, LD, and genetic architecture. Using the 1000 Genomes reference panel, Martin and colleagues [4] used European-ancestry GWAS summary statistics and calculated PGSs for eight phenotypes. They concluded that these findings from large-scale GWASs have limited portability to other groups, which we discussed earlier in relation to population stratification (chapter 3). Since allele frequencies differ between ancestry groups (see box 3.2), for instance, using a PGS that is derived from one ancestral population to a very different one would result in a very imprecise and biased score in the target population, even if the phenotype was highly heritable. Later in chapter 9 (section 9.4) we visualize how we can distinguish different ancestry groups in the population by virtue of their clustering across different principal components.

5.3.3 Relatedness, population stratification, and differential bias

When you are choosing the data that you would like to analyze for your research, it is essential to be aware of the potential of inflation in your PGS on your target sample due to population stratification. A study that used the PGS for height in the Framingham Heart Study, for instance, showed that when related individuals were included in the analysis, the R^2 was inflated from 0.15 to 0.25 [8]. Wray et al. [3] also examine differences when related individuals are removed from a sample and controlling by various population stratification principal components in relation to the inflation of the R^2. As we also outline in table 5.1, they suggest using conventionally unrelated individuals in the discovery and validation stages. In later chapters where we describe quality control (QC), we demonstrate how to remove related individuals. This mistake has occurred in published research. For instance, Belgard and colleagues [7] argue that the 2014 genetic study of autism in *Molecular Psychiatry* [31] suffered from a lack of controlling for population stratification. Another issue that researchers may encounter is differential bias of population stratification between cases and controls. This can lead to a spurious prediction of the R^2 but can be countered by performing stringent QC or validating results in a separate sample [3].

5.3.4   Variance explained only by common genetic markers missing rare variants

There have been a variety of genome-wide "SNP-chips" used to identify SNPs, which we discuss in more detail in chapter 7. The data that have been collected largely until around 2018 gathers information on common genetic variation. We discuss the limitations of this genotyping and future directions elsewhere (see section 7.2.3 and chapter 15). Due to LD, many of the SNPs that we identify in a GWAS and use in our PGSs may not be the actual causal SNP but could be in LD with one or more of the causal variants. The SNPs measured on most chips (until more recently) have alleles that are common and cannot be in complete or even moderate LD with rare variants. If a genetic variant is associated with fitness, selection can drive one allele to a lower frequency [32]. If the effect of a SNP has a large effect on fitness, the frequencies of the causal variant will be low. It is very unlikely that SNPs that have been identified in the many GWASs to date will explain all genetic variation since it misses the contribution to the variance of rare variants since these are not "tagged" by the genotyped SNPs. This explains, for instance, the discrepancy between the family heritability of height of around 0.7 to 0.8 and the lower SNP- or marker-based estimates of 0.4 [33, 34].

This discussion is also linked to the potential inflation of family estimates but also the term *still-missing heritability*, which refers to the genomic variants that are not thoroughly tagged by SNPs. This has led Visscher and others to argue that we can learn from animal studies, since this body of work explains that in livestock (and likely humans) certain causal variants are in fact rare and in poor LD with common SNPs [10]. For this reason, the field is now moving in the direction of incorporating rare variants into predictions. For instance, a 2018 study by Ganna and colleagues [35] quantified the impact of rare and ultra-rare coding variation on 13 quantitative traits and 10 diseases. They find an impact of rare deleterious coding variants on complex traits concluding that there is likely widespread pleiotropic risk.

5.3.5   Missing and hidden heritability in prediction of phenotypes from genetic markers (SNPs)

The variation that we are attempting to explain in the phenotypes that we study is a combination of genetic and environmental factors and their interaction. The use of a polygenic score is thus one way to quantify the genetic factors. Recall the missing heritability discussion from chapter 1 (section 1.6) where we discussed the unexpectedly low predictive power that came from GWASs and the rise of thinking in terms of polygenicity. Remember that missing heritability is the gap between the comparatively large heritability from twin study estimates and GWAS heritability estimates, whereas hidden heritability is the discrepancy between SNP-based heritability (Yang and colleagues [33] GREML models) and GWAS heritability. Since the first SNP discoveries only explained a fraction of the heritability [36], a series of studies emerged investigating non-additive genetic effects

[37], epistatic effects [38], heterogeneity and/or gene-environment interaction [11], and inflated estimates from twin studies due to shared environmental factors [39] and the role of rare, non-genotyped variants [34].

The only way to achieve the upper-level estimate of h^2 is if we were able to identify all genetic variants affecting the trait and estimate their effects without any error. As we note throughout this book, error may creep into the analyses through multiple factors such as lack of accurate or harmonized phenotype measurement, need for repeated measures or, as we describe in the next chapter, due to interaction with environmental factors [11, 12]. An example of strong increases in PGS R^2 is the evolution of subsequent GWASs such as in the study of type 2 diabetes (see chapter 7) and years of education. The first meta-GWAS on educational attainment [40], for instance, produced three significant hits, the second 74 significant hits [41], and the third in 2018, more than 1,100. Together with the inclusion of more genetic variants, the main difference between the three studies was the sample size, which increased from around 125,000 individuals in the first study to over 1.1 million. In parallel, the R^2 increased from around 2% in the original GWAS to around 7–10% in the 2018 study. In comparison, SNP-based estimates from whole-genome studies are around 20–25% [11, 40] and represent the upper bound of what we can expect to discover with additive models.

5.3.6   Trade-off between prediction and understanding biological mechanisms

It is important to keep in mind that with PGSs, we are dealing with a quantitative construct and that the underlying biology of a phenotype is highly complex. Due to polygenicity, common variants in single genes are weak predictors. Yet single genes are vital to understanding biological and causal function. It is the knowledge derived from studying the cell-regulating functioning of these single genes that allow us to move beyond correlations to understand the mechanisms linking genotype to phenotype. The trade-off with polygenic scores is that by virtue of having to combine information from many SNPs in PGSs, we often move further away from the specific biology underpinning the phenotype. This has implications for many of the quantitative analyses we conduct. As we demonstrate in chapter 10, you obtain more predictive results when you include all of the SNPs in the calculation of PGS for traits that are highly polygenic. Yet by doing this, you lose the biological specificity. We should be clear, however, that it is not simply a strict dichotomy between biological specificity and SNP thresholds. It is true that if we construct a score where we understand the mechanisms (i.e., only use SNPs that have a causal pathway that is understood), that score would have fewer SNPs and poorer performance. However, if we construct a score only from genome-wide significant SNPs, our understanding of the underlying mechanisms will still not be better than for a score that uses all of the SNPs. The lack of biological specificity is arguably a property of all polygenic scores if we adopt PGSs that include only scores with many independent SNPs.

The trade-off is thus between maximizing prediction and understanding biological mechanisms. In a 2018 review of the genetics of obesity in the *Lancet*, for instance, Goodarzi [13] summarizes how although over 300 SNPs have been isolated in relation to BMI, hip–waist ratio, and other adiposity traits, the absence of a serious biological functional understanding of obesity has prevented clinically relevant weight loss interventions. Many studies thus now strive to move beyond prediction to more precision, such as interventions in disease risk prediction, gene-environment interaction analysis, or Mendelian Randomization regressions. We return to specific issues in the mechanism-prediction trade-off in the discussion of the application of scores.

5.4   Shared genetic architecture of phenotypes

Until now we have discussed the PGSs for a specific trait in relative isolation, yet, particularly with complex phenotypes, PGSs for a single trait are habitually correlated with multiple phenotypes. As outlined in figure 5.1, when working with PGSs it is important to grasp that there is often a *shared genetic architecture* underlying many phenotypes. Although far from exhaustive, in this section we describe some of the main techniques used to disentangle this shared genetic architecture.

5.4.1   Predicting other phenotypes

PGSs often have a shared genetic architecture, and many diseases and traits have a shared etiology. Schizophrenia and bipolar disorder, for example, are intertwined diseases. Understanding their co-occurrence is vital when studying them or designing potential treatments. The PGSs for schizophrenia, for instance, have been used to predict bipolar disorder [42]. This study showed that there is to some extent a shared genetic etiology between the two phenotypes, suggesting that the same genes are associated with both outcomes. In contrast, the score for schizophrenia did not predict non-psychiatric health conditions, such as coronary artery disease, Crohn's disease, hypertension, or type 1 or 2 diabetes.

Reproductive traits have also been demonstrated to be highly interrelated [43]. A large-scale GWAS published in 2016 studied two reproductive behavioral traits of age at first birth (AFB) and number of children ever born (NEB) [44]. PGSs were used to examine their association with a variety of fertility and nonfertility traits. These included age at menarche, age at menopause, age at voice breaking (for boys), and age at first sexual intercourse. Although the PGS for number of children ever born had a relatively low predictive power, when entered into a regression model to predict childlessness, it had striking results. The PGS for NEB could predict the probability of remaining childless at the end of the reproductive period with an increase of one standard deviation of the PGS associated with a decrease of around 9% in the probability to remain childless for women [44, 45]. The PGS for (a later) age at first birth was likewise associated with both a later age at menarche and later age at natural menopause [44]. Biological functional work also suggested shared etiologies between reproductive traits (and infertility traits such as endometriosis).

5.4.2 Phenotypic and genetic correlation

Many phenotypes are also highly genetically correlated. Here it is important to distinguish between phenotypic and genotypic correlations. Although it may occur, a phenotypic correlation does not automatically imply a genetic correlation, even if the phenotypes are partly heritable. A genetic correlation also does not imply biological causation. In this section we focus on genetic correlation or overlap between phenotypes. *Genetic correlation* is an estimate of the proportion of additive genetic effect shared between a pair of traits. Consider, for example, two heritable traits such as schizophrenia and bipolar disorder, which usually have a high phenotypic correlation.[1] With genetic correlations, we are interested in examining whether there is also a genetic correlation or in other words whether the two traits share the same genes.

The most common method that is used to examine genetic overlap is LD score regression, developed by Bulik-Sullivan et al. in 2015 [46]. In chapter 12 we demonstrate how to estimate genetic correlations from GWAS summary statistics using the LDSC package (https://github.com/bulik/ldsc) (see appendix 1). LDSC exploits the LD structure of the data to estimate the degree of genetic correlation. The approach initially required GWAS summary statistics for all SNPs from GWASs and a reference sample from which the LD could be estimated in order to estimate the LD score regression. The method is written formally based on the following relationship:

$$E[z_{1j}z_{2j}] = \frac{\sqrt{N_1 N_2}}{M} \ell_j \rho_g + intercept,$$

Where z_{kj} is the Z-score of SNP j from the GWAS of trait k ($k = 1, \ldots, 20$), N_k is the sample size of the GWAS of trait k, l_j is the LD Score of SNP j, M the number of SNPs included in the GWAS, ρ_g the genetic covariance between traits 1 and 2, with the regression intercept represented by *intercept*. The slope from the regression of $\hat{z}_{1j}z_{2j}$ on $\sqrt{N_1 N_2}\ell_j$ can be used to estimate the genetic covariance between the two traits. It is also possible to estimate the heritabilities of the two traits, h^2_{g1} and h^2_{g2} from the univariate LD score regressions of traits 1 and 2. It therefore follows that an estimate of the genetic correlation is:

$$\hat{r}_g = \frac{\widehat{\rho_g}}{\sqrt{\hat{h}^2_{g1}\hat{h}^2_{g2}}}$$

In chapter 12 we demonstrate how to estimate these LD scores and the interpretation of the results. In that chapter we also show how to obtain genetic correlations via the website LDHub (http://ldsc.broadinstitute.org/ldhub/) [47]. This is an online database that can be used as a web interface for LD score regression. The site is constantly being updated but includes the SNP heritability of hundreds of traits and genetic correlation results. You are also able to download hundreds of genetic overlaps between traits.

Figure 5.2 provides an example of genetic correlations across multiple traits. Here we show our 2016 study where we used LD score regression to examine the correlation between reproductive behavior phenotypes (age at first birth [AFB], number of children ever born [NEB]) and 27 related phenotypic correlations. This included traits that were developmental or related to fertility (e.g., age at menarche, menopause, voice breaking, polycystic ovary syndrome [PCOS], age at first sexual intercourse, birth weight), behavioral (years of education, three smoking traits), personality and neuropsychiatric (e.g., neuroticism, schizophrenia, well-being, autism), cardiometabolic (e.g., LDL cholesterol, triglycerides, type 2 diabetes), and anthropometric (BMI, height, waist–hip ratio). As figure 5.2 shows, AFB was positively correlated mainly with human development and behavioral traits, while being negatively correlated with PCOS and the cardiometabolic and anthropometric traits. Once multiple testing was controlled for, NEB was only significantly and negatively correlated with years of education and age at first sexual intercourse. The two most striking and significant correlations were with AFB and age at first sexual intercourse and years of education. Years of education had in fact a 0.70 correlation with AFB, which we explore there and in related papers. Although LD score regression is a powerful tool to identify possible relationships between traits, it does not allow us to establish causal directions or relationships or to adjust for potential mediating factors. The relationship between many of the traits is highly complex with potential bidirectional mechanisms. We explore some of these relationships later in chapter 13, on Mendelian Randomization, and in chapter 15, which delves into future research directions.

5.4.3 Pleiotropy

Pleiotropy refers to the phenomenon of a single gene affecting multiple traits. It is derived from the Greek term *pleion*, which refers to *more*, and *tropos*, meaning *way*. Pleiotropic genes thus refer to those genes that exhibit multiple effects on phenotypes. If a mutation occurs in a pleiotropic gene, for instance, it could affect several phenotypes simultaneously. This is attributed to the fact that gene coding is used by many cells or different targets that have the same signaling function. The topic of pleiotropy was introduced over 100 years ago by German geneticist Ludwig Plate in 1910 [48]. It influenced many fields of evolutionary biology as well as physiological and medical genetics. Since 1910, the meaning of the term has evolved, particularly with the introduction of the molecular genetic data that we examine in this book. It has largely been studied in the area of senescence, which refers to physiological changes in individuals as they age.

Paaby and Rockman outline several different types of pleiotropy, noting that the discussion is often plagued by conceptual difficulties regarding the various meanings of pleiotropy and how to study these mechanisms [49]. In this introductory textbook we are only

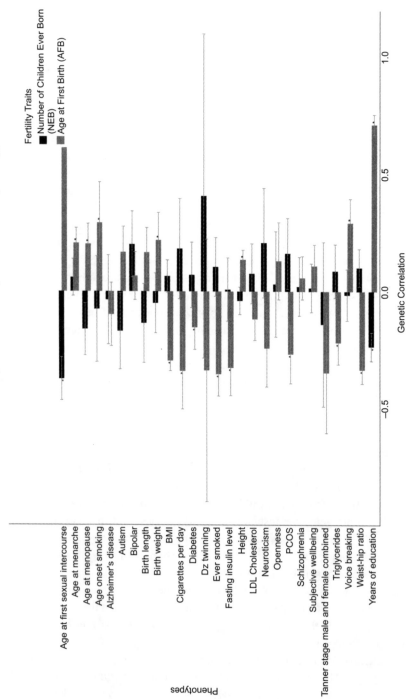

Figure 5.2

Genetic overlap between reproductive behavior (AFB, NEB) and 27 other phenotypes.

Source: Produced by the authors; see also reference 45.

Notes: Results for LD score regressions. Gray bars represent 95% confidence intervals. Asterisks indicate statistically significant results after controlling for multiple testing.

able to superficially introduce the different types of pleiotropy, many of which are studied at the molecular biology level. *Molecular-gene pleiotropy* studies the number of functions a molecular gene has, such as when a gene interacts with multiple proteins and catalyzes multiple reactions. This is, for instance, biochemical research into protein-protein interactors in a gene and the number of reactions that it catalyzes. *Developmental pleiotropy* is when mutations—not the molecular genes—are the unit of study. Here researchers often examine the genetic and evolutionary autonomy of different aspects of the phenotype, independent from fitness. Here key questions often include the examination of molecular pleiotropy and the relative importance of cis-regulatory[2] versus protein-coding variants, extending far beyond the scope of this textbook. *Selectional pleiotropy* is the study of when the phenotype has multiple effects on fitness. A key feature for this type of pleiotropy is that traits are considered as being defined by selection and not the intrinsic attributes of the individual. Examples harken back to some of the basic evolutionary texts that propose an *antagonistic pleiotropy* model examining the evolution of aging or mutations that form the basis of sexually antagonistic pleiotropy and pleiotropic trade-off that underlies adaptation [50]. Some have argued that antagonistic pleiotropy is common for genetic disorders [51]. A common example of antagonistic pleiotropy—when multiple effects of genes have opposing effects on fitness—is sickle cell disease.

Figure 5.3 provides a very simple genotype-phenotype map that illustrates additive pleiotropic effects. In this graph the genes G1, G2, and G3 represent the different genes that are shown to contribute to the three different phenotypes P1, P2, and P3. For instance, G1 influences both P1 and P2, G2 influences both P2 and P3, and G3 impacts P2 and P3. Note that pleiotropy is often used synonymously to refer to genetic correlation. However, for more precision, it is useful to distinguish between *direct pleiotropy* and *indirect pleiotropy*. All of the previous examples refer to direct pleiotropy, when a gene has a direct causal effect on multiple phenotypes. This is in parallel to the common cause model discussed previously. Indirect pleiotropy refers to a gene having a causal effect on P1, which in turn causally influences P2. This refers to the mediation model discussed in chapter 2, where P1 would be a mediator between genes and P2. In both scenarios, we would observe a genetic correlation between two phenotypes; however, the mechanisms leading to this observation are genuinely different and in the latter model, there may be no biological link between genes and P2 in spite of the fact that we observe an association.

Two recent studies used PGSs to study the pleiotropy between educational attainment and number of children ever born in Iceland and the United States [52, 53]. Both find that the PGS for educational attainment significantly predicts the number of children and that the genetic covariance based on regression models can be used to quantify the expected evolutionary change. Not surprisingly, direct evidence for evolutionary change is—whilst significant—very small. For example, in the United States it is a reduction of one week of education per generation as a result of natural selection. Even when the results are rescaled

GENOTYPE

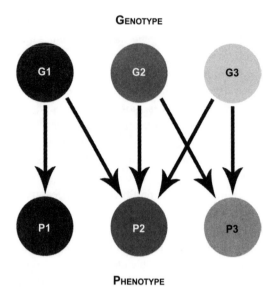

PHENOTYPE

Figure 5.3
Genotype-phenotype map showing additive pleiotropy effects.
Source: *CC BY-SA* 4.0 [54].

to account for missing heritability, the genetic selection predicts changes in education of no more than around 1.5 months. As we noted elsewhere [12], it is vital to consider gene-environment interaction such as gains in educational expansion and consider that changes are slow and would need to be stable and persist across several generations. The afore-mentioned studies also do not consider mortality selection.

It is now agreed that *pleiotropy is ubiquitous*. Pickrell and colleagues [55] examined 42 phenotypes to demonstrate pleiotropy and show that several loci were associated with a large number of traits. They then used these loci to identify the phenotypes that shared multiple genetic causes and developed a method to identify pairs of traits that have a causal relationship. Here they showed how BMI causally results in an increase in triglyc-eride levels. Others have examined the entire GWAS catalog to determine the preva-lence of pleiotropy, with 44% of the genes reported in the catalog being associated with more than one phenotype [56]. These authors showed that the degree of pleiotropy was positively scaled with a gene's average effect size and negatively with the variance of effects sizes in genes with a given number of associated phenotypes. As touched upon briefly in chapter 1, the knowledge that pleiotropy is ubiquitous has led others such as Boyle et al. [57] to argue that gene regulatory networks are so interconnected and that all genes affect the functions of core disease-related genes in the form of an omigenic model. Gratten and Visscher [58] have argued that this "pervasive pleiotropy" has real implications for personalized medicine and genome editing, which we return to in chapters 14 and 15.

5.4.4 Multitrait analysis

GWASs have often prioritized phenotypes that have been easy to measure consistently across various cohorts. In many cases it may be difficult to harmonize or measure the phenotype of interest. For this reason, the proxy-phenotype method was introduced by Rietveld and colleagues [59]. The *proxy-phenotype method* identifies common genetic variants in a two-stage approach by first conducting a GWAS on a phenotype and then second, by using an independent sample to measure the association of SNPs found in that original GWAS with related phenotypes. The aforementioned authors did this with educational attainment, linking it in the second stage with cognitive performance, memory, and absence of dementia.

This was extended by the *multitrait analysis of GWAS (MTAG)* by Turley and colleagues in 2018 [60], which is a method that allows for the joint analysis of summary statistics derived from GWASs for different traits. Here the authors demonstrate how applying MTAG to the GWAS results for depressive symptoms, neuroticism, and subjective well-being yielded more associated loci that were not isolated in the original individual GWAS and increased the variance explained by PGSs to levels that matched theoretical expectations. As compared to the 32, 9, and 13 genome-wide significant loci identified in the single-trait GWAS for the traits mentioned above, MTAG increased the number of associated loci to 64, 37, and 49, respectively. The increase is particularly relevant for the neuroticism GWAS that had a smaller sample size. In chapter 12 we provide an example application of MTAG.

5.5 Causal modeling with polygenic scores

Earlier in this book in chapter 2, we outlined the various types of multivariate causal models that are possible in our introductory statistical chapter. PGSs can be treated as standard, continuous variables in regression models and are for many phenotypes by now well-powered for prediction analysis already in relatively small samples ($N < 1000$) (see box 5.2). In this section, we provide an overview of some of the central applications that we explore later in this book. These include examining genetic confounding, gene-environment interaction, and Mendelian Randomization.

5.5.1 Genetic confounding

Genetic confounding is the case when an extraneous variable or variables explain at least partly the association (or lack thereof) between a PGS and the phenotype. In 2000, Eric Turkheimer, one of the founders of behavioral genetics, outlined the three laws of behavioral

Box 5.2
What target sample size do I need to apply a PGS?

Most will want to know whether their target data are sufficiently powered to examine a particular research question. Various articles have discussed the sample sizes that are required to achieve a given R^2 or AUC (area under the curve) in the target sample. The central paper published on this topic is by Dudbridge [2], who provided multiple estimates of the power and predictive accuracy of PGSs in relation to various scenarios of different types of phenotypes and data availability. There are also R packages to engage in these types of calculations.[3] It can be useful to perform a PGS analysis to examine the phenotypic variance explained R^2 by the association p-value. See box 4.1 for a detailed discussion of genetic power calculations.

---

genetics [61]. While the "everything is heritable" lesson is important to learn, it is equally important to acknowledge that "everything is environmental." If we study various diseases, behaviors, and traits, they are to a large extent and often more associated with nongenetic factors. Researchers have consistently shown that socioeconomic circumstances are the most robust and replicated predictors of health, cognitive ability, and others across multiple phenotypes [62–64].

5.5.2 Mendelian Randomization

As we elaborated upon in chapter 2, there are multiple strategies to estimate a causal effect. The most optimal situation would be to conduct a randomized controlled trial. However, with many of the outcomes that we study, this is simply infeasible and unethical. One alternative design that aims to approximate this is the instrumental variable approach, which in this area of research has been coined Mendelian Randomization. Since we have an entire chapter dedicated to this topic (chapter 13) we only briefly touch upon it here. *Mendelian Randomization (MR)* is a technique that has been developed to test whether there are causal relationships among variables using genetic information. For example, does high cholesterol cause high blood pressure? As we note in chapter 13, MR has proven to be very effective using PGSs. The technique relies on a few assumptions that are important to keep in mind. In MR it is essential to minimize the risk of including "noise" due to direct pleiotropy in PGSs by eliciting the genes that have a strong biological effect on the trait of interest. For this reason, if PGSs are used in MR, it is not recommended to use high p-value thresholds, which may violate the assumptions needed for the methods. For a detailed discussion of this issue, refer to the discussion by Hemani et al. [65] and chapter 13.

5.5.3 Controlling for confounders

The second approach to estimating a causal effect is to attempt to *control for potential confounders* in our models. Both family and molecular genetic studies have shown that genetic influences across many phenotypes are shared. Although the nature of these shared influences remains unclear, they likely confound phenotypic associations to some extent via direct pleiotropy.

To describe this, we use two examples from the literature: the association between schizophrenia and age at parent's first birth [66] and having children both very young (i.e., teen pregnancy) and very old has different socioeconomic as well as mental health outcomes for the offspring compared to average ages at childbirth. Individuals on both the lower and higher tails of the age at first birth distribution show, for example, a higher likelihood that the children will be diagnosed with schizophrenia. This has considerable implications since age at first birth has been delayed by around 4–6 years across many countries [67]. The question is whether the next generations will have a higher prevalence of schizophrenia or, given the U-shaped association, whether this relationship between age of parents at childbirth and schizophrenia is causal. If genes were to confound this relationship, the association may be specific to certain populations. Therefore, shifts in the distribution of parental age at birth across time or between populations may not affect the prevalence of the disease. There are of course many other explanations for the link of parent's age at childhood and their offspring's health and well-being. This includes differences in resources and socioeconomic status, stability of relationships, and education, which has been shown to be higher in older parents and thus impact children's later outcomes [68].

One hypothesis is that genes confound the relationship between age at first birth of parents and the development of schizophrenia in children. Polygenicity is key in this hypothesis. Genetically, parents might be located on a spectrum of genetic predisposition to develop schizophrenia. A disposition for schizophrenia may, for example, lead to abrupt and risky sexual behavior associated with teenage pregnancy or to problems in finding a mate, which delays or prevents childbirth. Since parents transmit their genes to their children, children of older or younger parents may carry a high genetic predisposition for schizophrenia more often than others and therefore also be diagnosed more often. Following this hypothesis, we expect that the parent–child association between age at first birth and schizophrenia is not causal but confounded by genes. But how might we test this hypothesis?

The application of PGSs makes this relatively straightforward. Several studies have analyzed the distribution of the PGS for schizophrenia across ages of first birth [66, 69]. These studies show that the distribution of the schizophrenia risk in parents has the same shape as in the development of schizophrenia in their offspring. Both teenage parents and parents older than 35 show an elevated PGS for schizophrenia. This suggests that genetic inheritance confounds the relationship between parental fertility timing and children's mental health, at least in part. Therefore, models that do not consider genetic inheritance when studying this association may have biased estimates. More precisely, they potentially overestimate the causal effect of parental age at childbirth on an offspring's mental health status. Statements about the consequences of birth postponement on population health might also be misleading.

5.5.4 Gene-environment interaction and heterogeneity

Gene-environment interaction is a core and emerging topic in this area of research. Because chapter 6 presents detailed theories, a discussion of gene-environment interplay, many examples, and methodological issues, we only briefly summarize the main points in relation to PGSs here. In chapter 11 we also provide several applied examples that will allow you to understand how to technically deal with key problems.

It is first useful to reiterate the meaning of an "interaction" with respect to gene-environment research in the context of PGS applications. Here we distinguish between direct effects of genes on a phenotype and interaction effects with the environment. The first instance implies that holding environmental conditions constant, genes cause—when varying between individuals—differences in phenotypes. The second scenario describes a different reaction to an environmental exposure, based on different genotypes. Interaction means that genetic effects on a phenotype will be different in different environments. These aspects are clarified in chapter 6.

Domingue and colleagues, for example, illustrated a heterogeneous reaction to spouse loss in the United States using the Health and Retirement Study [70]. As with other stressful events in life, the loss of a spouse can lead to depression. The degree to which depressive symptoms occur and the duration they remain, however, differs between individuals. As we elaborate upon in chapter 6, they test a theory called the diathesis-stress model, namely that a genetic predisposition for subjective well-being [71] might buffer the detrimental effect of spouse loss. They show that, while there is a general increase of depressive symptoms after spouse loss, U.S. adults with higher PGSs for well-being indeed experienced fewer depressive symptoms compared to those with lower genetic scores for well-being. In another study, Domingue and colleagues [72] show that the influence of the PGS for smoking on smoking increased across birth cohorts.

Recommendations for using PGSs in gene-environment interaction studies can be complicated and quite nuanced. In chapter 6, in table 6.2, we list the multiple challenges, why they are problematic, and potential solutions, which we do not repeat here. In this type of research, we are interested in modeling the genetic effects that have a differential effect in different environments. Yet, the ability to specify different environments remains challenging. In theory, it would be ideal to run a GWAS, taking into account possible environmental interactions. In practice, these studies would be underpowered. Even in large samples such as the UK Biobank, which has 500,000 individuals, it can be difficult to differentiate on certain key environmental factors since the sample is selective, consisting of healthier and higher socioeconomic status individuals [73].

In the absence of the ability to do this, we can also isolate which SNPs should be included in the analysis. The selection of SNPs is discussed in detail in chapter 10, where we demonstrate how to create and validate PGSs. Rosenquist and colleagues [74] use a single variant in the FTO gene to show that birth cohort interacts with genetic predisposition to obesity. It was possible to use one variant in that paper since FTO has an unusually high effect variant. Unlimited food availability in many industrialized countries together with an increasingly sedentary lifestyle, means that alleles for obesity are potentially "riskier" among more recent birth cohorts. Conversely, a study by Barcellos et al. [75] shows that increasing education has an effect on reducing health inequality. Using a PGS that contains all alleles from the 2018 GWAS on educational attainment, they tested whether the PGS moderated the effects of education on health. Using the natural experiment of the one-year increase in compulsory education in England and Wales in 1972, they found that education reduced the health gap in unhealthy body size for those at both the top and bottom terciles of the genetic risk for obesity.

5.6   Conclusion

The bulk of what most applied researchers will perform is likely to apply PGSs to multiple phenotypes and across various environments. Knowing where to start among this expansive and dynamic research area can be difficult. The goal of this chapter was to not only define PGSs and provide the context of how they emerged but to arm researchers with an understanding of some of the more practical steps they need to undertake. For this reason, in figure 5.1 and the accompanying discussion, we provided a flowchart starting from data to validation and processing, generating a PGS and using it for prediction in addition to ways in which to think about shared genetic architecture and modeling applications. Our aim was to also flag particular challenges and problematic areas in addition to providing potential solutions and further reading, which we summarized in table 5.1.

PGSs are a useful tool to summarize genetic information in one variable to apply to further statistical analyses. We attempted to present a balanced view on their usage but also potential restrictions. Perhaps the most problematic is that by virtue of being based on GWASs and the historical lack of diversity in the populations studied to date, they have been historically less widely applicable to a variety of populations and groups. The hope is that this will change over the next few years. We also noted that PGSs still remain proxies of the true heritability of traits, given the missing, still-missing, and hidden heritability issues. As sample sizes of GWASs increase, together with the move to look beyond common variants, this field will continue to expand. Beyond a lack of diversity, another important deficit in current GWAS designs is that signals of SNPs discovered in GWASs can be overstatements, since as Kong and colleagues [76] recently showed, they also tag effects of parental nurturing.

Although some researchers remain focused on the R^2, maximizing prediction is not always the ultimate and useful goal for certain research questions. Understanding the underlying biology and function of the main genetic markers may bring us further than statistical solutions and predictions. Since pleiotropy is ubiquitous, PGSs also often have a shared genetic architecture. Exploring correlated phenotypes, predicting related phenotypes, or multitrait analysis can be fruitful avenues to take in this respect. We also anticipate considerable gains in the upcoming period including better measurement of phenotypes, or what is known as "deep phenotyping" from multiple means (e.g., patient records) and machine learning algorithms to optimize prediction. Granted tremendous advancements in PGSs, it remains unclear whether it is possible to create a genome-wide PGS that will be sufficiently able to identify individuals at a clinically significantly increased risk. PGSs are increasingly used in combination with clinical measures for screening, intervention, and life planning [18], but there remains considerable controversy. We turn to the use of PGSs in clinical applications in our final discussion of ethics (chapter 14) and future directions (chapter 15).

Exercises

1. What is the difference between direct and indirect pleiotropy—can you draw the causal models? See also chapter 2.

2. What are the trade-offs between prediction and understanding biological mechanisms (table 5.1) for each application integrating PGS in causal modeling (as discussed in section 5.5)?

3. Which of the challenges outlined in table 5.1 do you expect to be resolved, and which will remain to be solved in future research?

4. Go to the LD Hub (http://ldsc.broadinstitute.org/ldhub/) site. Examine the site and look at SNP heritability and genetic correlation results. (In chapter 12 we will focus on graphing these results.)

Further reading

Belsky, D. W., and S. Israel. Integrating genetics and social science: Genetic risk scores. *Biodemography and social biology* **60**, 137–155 (2014).

Euesden, J., C. M. Lewis, and P. F. O'Reilly. PRSice: Polygenic risk score software. *Bioinformatics* **31**, 1466–1468 (2014).

Maier, R. M. et al. Embracing polygenicity: A review of methods and tolls for psychiatric genetics research. *Psychological Medicine* **48**(7), 1055–1067 (2018).

Maier, R. M. et al. Improving genetic prediction by leveraging genetic correlations among human diseases and traits. *Nature Communications* **9**, 989 (2018).

Martin, A.R. et al. Current clinical use of polygenic scores will risk exacerbating health disparities. *Nature Genetics* **51**, 584–591 (2019).

O'Connor, L. J. et al. Polygenicity of complex traits is explained by negative selection. *American Journal of Human Genetics* (2019), doi: https://doi.org/10.1101/420497.

Vilhjálmsson, Bjarni J. et al. Modeling linkage disequilibrium increases accuracy of polygenic risk scores. *American Jounral of Human Genetics* **97**(4), 576–592 (2015).

Ware, E. B. et al. Heterogeneity in polygenic scores for common health traits. *bioRxiv* (2017) (available at https://www.biorxiv.org/content/10.1101/106062v1).

References

1. H. D. Daetwyler, B. Villanueva, and J. A. Woolliams, Accuracy of predicting the genetic risk of disease using a genome-wide approach. *PLoS One* **3**, e3395 (2008).

2. F. Dudbridge, Power and predictive accuracy of polygenic risk scores. *PLOS Genetics* **9**(3), e1003348 (2013).

3. N. R. Wray et al., Pitfalls of predicting complex traits from SNPs. *Nat. Rev. Genet.* **14**, 507–515 (2013).

4. A. R. Martin et al., Human demographic history impacts genetic risk prediction across diverse populations. *Am. J. Hum. Genet.* **100**, 635–649 (2017).

5. J. J. Lee et al., Gene discovery and polygenic prediction from a genome-wide association study of educational attainment in 1.1 million individuals. *Nat. Genet.* **50**, 1112–1121 (2018).

6. F. M. De La Vega and C. D. Bustamante, Polygenic risk scores: A biased prediction? *Genome Med.* **10**, 100 (2018).

7. T. G. Belgard, I. Jankovic, J. K. Lowe, and D. H. Geschwind, Population structure confounds autism genetic classifier. *Mol. Psychiatry* **19**, 405–407 (2014).

8. R. Makowsky et al., Beyond missing heritability: Prediction of complex traits. *PLoS Genet.* **7**, e1002051 (2011).

9. S. H. Lee, N. R. Wray, M. E. Goddard, and P. M. Visscher, Estimating missing heritability for disease from genome-wide association studies. *Am. J. Hum. Genet.* **88**, 294–305 (2011).

10. K. E. Kemper, H. D. Daetwyler, P. M. Visscher, and M. E. Goddard, Comparing linkage and association analyses in sheep points to a better way of doing GWAS. *Genet. Res. (Camb).* **94**, 191–203 (2012).

11. F. C. Tropf et al., Hidden heritability due to heterogeneity across seven populations. *Nat. Hum. Behav.* **1**, 757–765 (2017).

12. A. Courtiol, F. C. Tropf, and M. C. Mills, When genes and environment disagree: Making sense of trends in recent human evolution. *Proc Natl Acad Sci USA* **113**, 7693–7695 (2016).

13. M. O. Goodarzi, Genetics of obesity: What genetic association studies have taught us about the biology of obesity and its complications. *Lancet Diabetes Endocrinol.* **6**, 223–236 (2018).

14. S. M. Purcell et al., Common polygenic variation contributes to risk of schizophrenia and bipolar disorder. *Nature* **460**, 748–752 (2009).

15. G. Hemani et al., Inference of the genetic architecture underlying BMI and height with the use of 20,240 sibling pairs. *Am. J. Hum. Genet.* **93**, 865–875 (2013).

16. L. E. Duncan and M. C. Keller, A critical review of the first 10 years of candidate gene-by-environment interaction research in psychiatry. *Am. J. Psychiatry* **168**, 1041–1049 (2011).

17. M. C. Keller, Gene × environment interaction studies have not properly controlled for potential confounders: The problem and the (simple) solution. *Biol. Psychiatry* **75**, 18–24 (2014).

18. A. Torkamani, N. E. Wineinger, and E. J. Topol, The personal and clinical utility of polygenic risk scores. *Nat. Rev. Genet.* (2018), doi:10.1038/s41576-018-0018-x.

19. R. Plomin, C. M. A. Haworth, and O. S. P. Davis, Common disorders are quantitative traits. *Nat. Rev. Genet.* **10**, 872–878 (2009).

20. A. G. Cardno and I. I. Gottesman, Twin studies of schizophrenia: From bow-and-arrow concordances to star wars Mx and functional genomics. *Am. J. Med. Genet.—Semin. Med. Genet.* (2000), doi:10.1002/(SICI)1096-8628(200021)97:1<12::AID-AJMG3>3.0.CO;2-U.

21. P. F. Sullivan, K. S. Kendler, and M. C. Neale, Schizophrenia as a complex trait. *Arch. Gen. Psychiatry* (2003), doi:10.1001/archpsyc.60.12.1187.

22. I. I. Gottesman and J. Shields, Genetic theorizing and schizophrenia. *Br. J. Psychiatry* **122**, 15–30 (1973).

23. E. Essen-Möller, Evidence for polygenic inheritance in schizophrenia? *Acta Psychiatr. Scand.* **55**, 202–207 (1977).

24. N. Risch and K. Merikangas, The future of genetic studies of complex human diseases. *Science* **273**, 1516–1517 (1996).

25. M. C. O'Donovan et al., Identification of loci associated with schizophrenia by genome-wide association and follow-up. *Nat. Genet.* **40**, 1053–1055 (2008).

26. H. Stefansson et al., Common variants conferring risk of schizophrenia. *Nature* (2009), doi:10.1038/nature08186.

27. R. A. Fisher, The correlation between relatives on the supposition of Mendelian inheritance. *Trans. R. Soc. Edinburgh* **52**, 399–433 (1919).

28. S. Ripke et al., Genome-wide association analysis identifies 13 new risk loci for schizophrenia. *Nat. Genet.* **45**, 1150–1159 (2013).

29. M. E. Goddard, N. R. Wray, K. Verbyla, and P. M. Visscher, Estimating effects and making predictions from genome-wide marker data. *Stat. Sci.* **24**, 517–529 (2009).

30. J. Euesden, C. M. Lewis, and P. F. O'Reilly, PRSice: Polygenic risk score software. *Bioinformatics* **31**, btu848-1468 (2014).

31. E. Skafidas et al., Predicting the diagnosis of autism spectrum disorder using gene pathway analysis. *Mol. Psychiatry* **19**, 504–510 (2014).

32. L. B. Barreiro et al., Natural selection has driven population differentiation in modern humans. *Nat. Genet.* **40**, 340–345 (2008).

33. J. Yang et al., Common SNPs explain a large proportion of the heritability for human height. *Nat. Genet.* **42**, 565–569 (2010).

34. J. Yang et al., Genetic variance estimation with imputed variants finds negligible missing heritability for human height and body mass index. *Nat. Genet.* **47**, 1114–1120 (2015).

35. A. Ganna et al., Quantifying the Impact of Rare and Ultra-rare Coding Variation across the Phenotypic Spectrum. *Am. J. Hum. Genet.* **102**, 1204–1211 (2018).

36. T. A. Manolio et al., Finding the missing heritability of complex diseases. *Nature* **461**, 747–753 (2009).

37. Z. Zhu et al., Dominance genetic variation contributes little to the missing heritability for human complex traits. *Am. J. Hum. Genet.* **96**, 377–385 (2015).

38. O. Zuk and E. Hechter, The mystery of missing heritability: Genetic interactions create phantom heritability. *Proc. Natl. Acad. Sci.* **109**, 1193–1198 (2012).

39. J. Felson, What can we learn from twin studies? A comprehensive evaluation of the equal environments assumption. *Soc. Sci. Res.* **43**, 184–199 (2014).

40. C. A. Rietveld et al., GWAS of 126,559 individuals identifies genetic variants associated with educational attainment. *Science* **340**, 1467–1471 (2013).

41. A. Okbay et al., Genome-wide association study identifies 74 loci associated with educational attainment. *Nature* (2016), doi:10.1038/nature17671.

42. S. M. Purcell et al., Common polygenic variation contributes to risk of schizophrenia and bipolar disorder. *Nature* **460**, 748–752 (2009).

43. G. W. Montgomery, K. T. Zondervan, and D. R. Nyholt, The future for genetic studies in reproduction. *Mol. Hum. Reprod.* **20**, 1–14 (2014).

44. N. Barban et al., Genome-wide analysis identifies 12 loci influencing human reproductive behavior. *Nat. Genet.* **48**, 1–7 (2016).

45. Melinda C. Mills, Nicola Barban, and Felix C. Tropf, The sociogenomics of polygenic scores of reproductive behavior and their relationship to other fertility traits. *Russell Sage Found. J. Soc. Sci.* **4**, 122 (2018).

46. B. K. Bulik-Sullivan et al., LD score regression distinguishes confounding from polygenicity in genome-wide association studies. *Nat. Genet.* **47**, 291–295 (2015).

47. J. Zheng et al., LD Hub: a centralized database and web interface to perform LD score regression that maximizes the potential of summary level GWAS data for SNP heritability and genetic correlation analysis. *Bioinformatics* **33**, 272–279 (2017).

48. F. W. Stearns, One hundred years of pleiotropy: A retrospective. *Genetics* **186**, 767–773 (2010).

49. A. B. Paaby and M. V. Rockman, The many faces of pleiotropy. *Trends Genet.* **29**, 66–73 (2013).

50. G. C. Williams, Pleiotropy, natural selection, and the evolution of senescence. *Evolution (N.Y.)* **11**, 398–411. (1957).

51. A. J. Carter and A. Q. Nguyen, Antagonistic pleiotropy as a widespread mechanism for the maintenance of polymorphic disease alleles. *BMC Med. Genet.* **12**, 160 (2011).

52. A. Kong et al., Selection against variants in the genome associated with educational attainment. *Proc. Natl. Acad. Sci. USA* **114**, E727–E732 (2017).

53. J. P. Beauchamp, Genetic evidence for natural selection in humans in the contemporary United States. *Proc. Natl. Acad. Sci. USA* **113**, 7774–7779 (2016).

54. Alphillips6, Genotype by phenotype. *CC BY-SA 4.0* (2019).

55. J. K. Pickrell et al., Detection and interpretation of shared genetic influences on 42 human traits. *Nat. Genet.* **48**, 709–717 (2017).

56. K. Chesmore, J. Bartlett, and S. M. Williams, The ubiquity of pleiotropy in human disease. *Hum. Genet.* **137**, 39–44 (2018).

57. E. A. Boyle, Y. I. Li, and J. K. Pritchard, An expanded view of complex traits: From polygenic to omnigenic. *Cell* **169**, 1177–1186 (2017).

58. J. Gratten and P. M. Visscher, Genetic pleiotropy in complex traits and diseases: Implications for genomic medicine. *Genome Med.* **8**, 78 (2016).

59. C. A. Rietveld et al., Common genetic variants associated with cognitive performance identified using the proxy-phenotype method. *Proc. Natl. Acad. Sci. USA* **111**, 13790–13794 (2014).

60. P. Turley et al., Multi-trait analysis of genome-wide association summary statistics using MTAG. *Nat. Genet.* **50**, 229–237 (2018).

61. E. Turkheimer, Three laws of behavior genetics and what they mean. *Curr. Dir. Psychol. Sci.* **9**, 160–164 (2000).

62. E. Turkheimer et al., Socioeconomic status modifies heritability of IQ in young children. *Psychol. Sci.* **14**, 623–628 (2003).

63. E. Krapohl et al., Phenome-wide analysis of genome-wide polygenic scores. *Mol. Psychiatry* (2015), doi: 10.1038/mp.2015.126.

64. M. Economou and G. Pappas, New global map of Crohn's disease: Genetic, environmental, and socioeconomic correlations. *Inflamm. Bowel Dis.* **14**, 709–720 (2008).

65. G. Hemani, J. Bowden, and G. Davey Smith, Evaluating the potential role of pleiotropy in Mendelian randomization studies. *Hum. Mol. Genet.* **27**, R195–R208 (2018).

66. D. Mehta et al., Evidence for genetic overlap between schizophrenia and age at first birth in women. *JAMA Psychiatry* **73**, 497–505 (2016).

67. M. C. Mills, R. R. Rindfuss, P. McDonald, and E. te Velde, Why do people postpone parenthood? Reasons and social policy incentives. *Hum. Reprod. Update* **17**, 848–860 (2011).

68. A. Barbuscia and M. C. Mills, Cognitive development in children up to age 11 years born after ART—a longitudinal cohort study. *Hum. Reprod.* **32**, 1482–1488 (2017).

69. G. Ni, J. Gratten, N. R. Wray, and S. H. Lee, Age at first birth in women is genetically associated with increased risk of schizophrenia. *Sci. Rep.* **8**, 10168 (2018).

70. B. W. Domingue, H. Liu, A. Okbay, and D. W. Belsky, Genetic heterogeneity in depressive symptoms following the death of a spouse: Polygenic score analysis of the U.S. Health and retirement study. *Am. J. Psychiatry* (2017), doi:10.1176/appi.ajp.2017.16111209.

71. A. Okbay et al., Genetic variants associated with subjective well-being, depressive symptoms, and neuroticism identified through genome-wide analyses. *Nat. Genet.* (2016), doi: 10.1038/ng.3552.

72. B. W. Domingue, D. Conley, J. Fletcher, and J. D. Boardman, Cohort effects in the genetic influence on smoking. *Behav. Genet.* (2016), doi: 10.1007/s10519-015-9731-9.

73. A. Fry et al., Comparison of sociodemographic and health-related characteristics of UK Biobank participants with those of the general population. *Am. J. Epidemiol.* **186**, 1026–1034 (2017).

74. J. N. Rosenquist et al., Cohort of birth modifies the association between FTO genotype and BMI. *Proc. Natl. Acad. Sci. USA* **112**, 354–359 (2015).

75. S. H. Barcellos, L. S. Carvalho, and P. Turley, Education can reduce health differences related to genetic risk of obesity. *Proc. Natl. Acad. Sci. USA* **115**, E9765–E9772 (2018).

76. A. Kong et al., The nature of nurture: Effects of parental genotypes. *Science* **359**, 424–428 (2018).

77. S. M. Sodini, K. E. Kemper, N. R. Wray, and M. Trzaskowski, Comparison of genotypic and phenotypic correlations: Cheverud's Conjecture in humans. *Genetics* **209**, 941–948 (2018).

6

Gene-Environment Interplay

Objectives

- Understand and differentiate between different types of *gene-environment interplay* of *gene-environment (G×E) interaction* and *gene-environment correlation (rGE)*
- Understand the multiple ways of defining the *environment*, including by multilevels, domains, and temporal aspects
- Recognize the *history of G×E studies* and common errors, from classic approaches to candidate gene and more recent genome-wide approaches
- Comprehend and differentiate between the *central theoretical G×E models* of diathesis-stress, differential susceptibility, bioecological (social compensation), and social control
- Differentiate between *different types of rGE*, including passive, evocative (reactive), and active and comprehend why rGE models are important and basic research designs
- Grasp potential *future directions* in this area of research

6.1   Introduction: What is gene-environment (G×E) interplay?

Gene-environment interplay is the phenomenon of gene-environment interaction (G×E) and **gene-environment correlation (rGE)**. The majority of research in this area examines *gene-environment interaction (G×E)*, which studies whether the effect of the genotype on the phenotype varies across different environments. This effect can be causal or noncausal. Gene-environment correlation (rGE) is the process by which an individual's genotype influences or is associated with exposure to the environment or in other words, how genes and the environment operate in tandem. The correlation thus measures whether there are different allele frequencies in different environments. This area of research has gained increased attention since researchers in the medical sciences, epidemiology, and social sciences are often interested in how a particular genotype could predispose certain subgroups in the population to various environmental exposures. Geneticists and biologists, conversely, more often focus on how environment may be related to the expression of a gene and lead to a particular disease or trait.

Many complex traits such as cardiovascular disease, cancer, diabetes, and psychiatric disorders have been shown to be strongly affected by both genetic and environmental factors [1, 2]. Recall that particularly for complex behavioral traits, polygenic scores (PGSs) derived from genome-wide association studies (GWASs) do not often capture a large percentage of the phenotypic variance. Most complex diseases and behaviors have a strong environmental component that has been well established in the nongenetic literature. Pioneering researchers in this area have shown that understanding complex traits requires not only information about genetic risks but also the importance in accounting for the social and natural environment of individuals [3–6]. Some studies, for instance, reveal that certain positive genetic predispositions are realized in high-resource or stress-free environments, whereas negative predilections are exacerbated in negative enviornments [7, 8]. By studying G×E, our aim is to identify genetic vulnerabilities or strengths that are either realized or suppressed in particular environments [9].

The aim of this chapter is to provide readers with an overview of the main concepts of this area of research, including the most prominent theoretical models. In chapter 11 we provide empirical examples and discuss additional methodological challenges in this area of research. A simple G×E model includes a phenotype or trait (T), a genetic factor (G), an environmental factor (E), and often potential confounders (C). The previous chapters have discussed genetic factors (G) in detail, which in this book focuses on using genetic loci identified by GWASs and PGSs. In the next section we move from our focus on genetics to define the multifaceted term of environment and the interdependence of environmental factors. We then provide a brief history of G×E research, starting with classic approaches, through to the controversial and often unreplicated candidate gene (cG×E) studies followed by genome-wide G×E approaches. The next section maps the four key theoretical G×E models followed by a summary of the main challenges in this area of research and potential solutions. We provide an introduction of the different types of rGE and research designs in section 6.5 and reasons why this area of research remains difficult to study yet is still vital to consider. We conclude with a discussion of future directions.

6.2   Defining the environment in G×E research

When one thinks of vernacular uses of the word *environment*, images such as exposure to pollutants or sunlight first come to mind. In the context of genetic research, environment (E), however, can adopt multiple forms. The environment is best characterized as a multilevel, multidomain, and multitemporal (life course, longitudinal) framework that is the upstream processes that may influence the trait under study [10]. In genetics, E is in practice everything that that is nongenetic. Recall from our previous statistical chapter that causal designs are often a main focus of this type of research. An **exogeneous variable** is one whose value is determined by factors outside of the causal system under study [11]. In this type of research, E is often represented as an exogenous environmental variable such as air pollution, high altitude or a change in policy (e.g., taxation of smoking, mandatory years of schooling) or a measure of some form of exposure (e.g., to junk food or high caloric environment).

6.2.1 Nature and scope of E: Multilevel, multidomain, and multitemporal

Of crucial importance to this area of research is attention to the scope, measurement, and definition of the environment [10]. A classic definition of environmental factors in the medical literature is: "The environmental risk factor can be an exposure, either physical (e.g., radiation, temperature), chemical (e.g., polycyclic aromatic hydrocarbons), or biological (e.g., a virus); a behaviour pattern (e.g., late age at first pregnancy); or a 'life event' (e.g., job loss, injury)" ([12], p. 764). As Boardman and colleagues argue [10], these types of definitions can be extended to account for behavior at a higher aggregated group level to account for the social, political, and cultural environment. A definition such as the one listed above, demarcates the environment as a set of proximate environmental moderators (see box 6.1) of associations. A *proximate cause* is an event that is either the closest to or, in fact, responsible for causing the result that we observe.

When examining the interplay of genotypes and the environment, the focus is on the role that is played by an individual's location within natural, social, and cultural structures as a fundamental determinant of both their vulnerability and their exposures. Empirical sociology has in particular provided strong theoretical and measurement models to transcend physical environment definitions to encompass the environment as multilevel, multidomain, and multitemporal. *Multilevel environment* refers to the supra-individual context in which individuals are "nested" or grouped within different levels of analysis.

Box 6.1
E as a moderator of the relationship of G and T (trait)

Moderation was discussed previously in chapter 2 (see section 2.4.2, figure 2.10). It is defined as what occurs when the relationship between two variables depends on a third variable. In our case, it is when the relationship between genes and a trait (T) depends on the environment. E is the moderator. We often measure a moderating variable by adding an interaction term to a regression model to see if that variable affects the direction or strength of the relationship between G and T. Put another way, the moderator E is a third variable that affects the zero-order correlation between G and T or the value of the slope of T (dependent variable or outcome) on G (independent variable or covariate). More detail on this specification is provided later in this chapter with applied examples using computer code and interpretation of coefficients in part III of this book.

These levels include countries, states, provinces or regions, neighborhoods, schools, and families. *Multidomain environment* denotes the often multiple and parallel environments that interact in several spheres of people's lives. These include the natural environment (e.g., altitude, temperature) but also social, economic, cultural, and institutional environments (e.g., health, social, or employment policy). If we recognize that we also have a *multitemporal environment*, we acknowledge that there are changes over time both within individuals (i.e., as they age *across* the life course) but also birth cohort (i.e., when they are born) and historical period effects (i.e., historical period they live in) within populations [13]. Here it is also useful to draw on the *life course perspective*, which recognizes that environments change across the life course. For instance, gestation occurs in the uterine environment, which is affected by the mother's behavior (e.g., smoking, diet during pregnancy), whereas in childhood and adolescence important environmental components are an individual's parents, school, peer group, and neighborhood. In adulthood, individuals are impacted by higher educational institutions, the workplace, their partner, and family unit. The life course perspective goes beyond examining the individual in exclusion to embrace linked lives (partners, children, or families) and a move from one trait or event to multiple sequences across the life course [14–16]. This links with previous findings that have shown that heritability often increases with age, such as the stronger heritability of the FTO gene (related to obesity) over the life course [17].

6.2.2   Interdependence of environmental risk factors

An aspect that makes it particularly challenging to study G×E is that environmental risk factors are rarely independent from one another. Social characteristics, such as lower socioeconomic status, smoking or poor air quality, often cluster in geographical areas such as neighborhoods, schools, or the workplace. For example, imagine that your aim is to examine the impact of stressful life events or maternal smoking during pregnancy on the outcome of birth weight and later childhood development. If you would only examine stressful life events and maternal smoking, you would miss the crucial fact that both health-related behaviors and the social risks that might lead to higher levels of maternal smoking or exposure to stress are derived from the same source of clustered lower socioeconomic status.

This type of deeper contextual understanding is crucial for the correct interpretation of G×E and genetic associations that differ between groups on key measures such as ethnicity or socioeconomic status. If complex traits such as health are primarily driven by physical and social features of the neighborhood or environment, genes may have little to do with the individual differences that we observe in some groups. Various studies have shown that certain genetic effects are less pronounced in resource-poor environments. In one of the most influential early examples in the field, Turkheimer et al. [7] found that the heritability of cognitive test scores was almost zero for those who came from poor environments but, importantly, heritability increased dramatically in tandem with gains in socioeconomic resources. In other words, those from higher-resource environments were able to realize their genetic potential whereas those from low-resource environments could not. Genetic factors related to cognitive performance was thus suppressed or not realized in the most disadvantaged groups.

Another example is the study of the relationship between apolipoprotein E-allele (APOE) and change in cognitive function, most often associated with an increased risk for Alzheimer's disease [18]. Boardman and colleagues tested whether the relationship between APOE and change in cognitive function varied in contexts with a higher or lower level of social disorder. They hypothesized that social contexts with high levels of disadvantage or disorder may dominate more subtle genetic effects on outcomes. The authors found that the genetic effect of the APOE-E allele was in fact the weakest in socially disorganized neighborhoods and strongest in socially organized ones [19]. These distinctions are crucial for the correct interpretation of genetic effects and G×E between different groups, particularly the most disadvantaged groups. Since ethnic minorities are often concentrated in more disadvantaged neighborhoods (e.g., in the United States), it is not only misleading but patently incorrect to conclude that ethnicity or race would be the cause of detrimental or lower genetic outcomes. Putting aside the problem that we discussed in chapter 4 that most GWASs are derived from European-ancestry populations, and PGSs cannot be applied outside of different ancestry groups, if the proper genetic scores were applied, environmental structures may still result in differential exposure and the weakening or watering down of genetic effects. Furthermore, as described in relation to population stratification in chapter 3, we know that genetic variation is often related to geographic location [20].

6.3 A brief history of G×E research

G×E refers to the case where we examine the moderation of genetic effects across various environments. These interactions are most often studied in one of three ways. First, heritability-environment (H×E) interaction or classic models estimate the relative contribution of genes to trait variance across different environments. Second, candidate G×E designs focus on environmental moderation of the association between a particular allele and a trait. This is what has been referred to as candidate gene (cG×E) or allele-by-environment interaction. Some have distinguished these two approaches as "latent" (H×E) versus "measured" (cG×E). The third and now most commonly used method employs PGSs identified by GWASs, and applies them across various environmental contexts.

6.3.1 Classic approaches

As we touched upon earlier, before the availability of genome-wide data and approaches, behavior genetic methods with family-based data (e.g., twins, adopted children, parents, siblings) were used to estimate genetic and environmental influences on a phenotype.

This partitioned the heritability or proportion of the total variance in a specific sample from a population that was attributed to genetic variation, which we described in detail in previous chapters. Particularly the extended multivariate models in this classic research advanced the field in several ways. First, they established whether two variables that have a high phenotypic correlation could share the same underlying genetic basis. Second, they pointed to sources of the nonshared environment and latent constructs. It was difficult, however, for these models to estimate the ways in which genetic influences and the environmental worked jointly. Family models are still very relevant for this type of research to answer particular questions. There is mounting evidence, for instance, that genetic effects, particularly on behavioral phenotypes in GWASs, might be biased or substantially attenuated after controlling for family fixed effects when using samples of siblings.

6.3.2   Candidate gene cG×E approaches

There are two types of candidate gene approaches. First, there were early candidate gene studies in the 2000s, that focussed on predefined loci of interest based on what was thought to be a priori knowledge of the loci's biological function or impact on the trait being examined. More thorough reviews of this research have been explored elsewhere, particularly in the area of psychology where these studies were more common [21, 22]. Second, candidate genes can also be selected from top hits from GWA studies. There was some early success in this area using the APOE-E allele in relation to Alzheimer's disease and the FTO gene related to obesity [23].

Candidate gene studies were introduced in the early 2000s since costs and technology restricted both the sample size of genetic studies but also often only the genotyping of a small number of loci. Early studies often examined plausible neurobiological processes such as neurotransmission since approved pharmaceutical therapies targeted these pathways. In fact, 89.2% of candidate gene studies in the first decade of this research examined genes involved in neurotransmission [22]. Many studies repeated virtually identical narratives regarding the functioning of these neurotransmitters, yet it soon became apparent that most of the findings were false-positives [21]. Some focused on examining genetic main effects but many well-known studies focused on G×E.

Duncan et al. [22] examined 103 cG×E studies that had six or more replication attempts. They found that virtually all lacked unequivocal support and one received no support at all. The false-positive results of candidate gene studies were primarily attributed to the following reasons. First, *candidate gene hypotheses were wrong* and in principle most should have warranted a null finding. Research stemmed from a clear biological basis that focused on drugs that were deemed to work. Here it is important to maintain a distinction between whether a gene is associated with a trait and whether the gene is involved in the biology of that trait, which had not been established. GWAS has revealed that the majority of the associated variants are not in protein-coding regions (exons), which is where most of the candidate genes were selected. Rather, they appear to be in intergenic and intronic regions that are less well understood [24]. We note, however, that there is a subtlety to this point, which is that the noncoding variants may well exert their effects through their influence on the expression of protein-coding genes. Second, *statistical significance norms* increased the risk of false positives, a topic we covered already in detail in chapter 2 [25]. Third, most of the studies were *seriously underpowered* to detect any association with such a small effect. Duncan and colleagues [22] elaborate that when comparing previous candidate gene studies, even the variants that had the strongest associations from GWASs were considerably smaller than those hypothesized by candidate gene studies. Fourth, there was a *strong publication bias* toward positive results. Remarkably, this led to the editor of the journal *Behaviour Genetics* to write in a 2012 editorial: "Behaviour genetics literature has become confusing and it now seems likely that many of the published findings of the last decade are wrong or misleading and have not contributed to real advances in knowledge" [26, p. 1].

All of these points refer to the culprit of *researcher degrees of freedom* [25]. This refers to the phenomenon of the research process where researchers make key decisions in the course of collecting and analysing the data. It includes how much data should be collected, whether some observations should be excluded, which conditions should be combined, inclusion of control variables, and transformation of specific measures. Since researchers rarely make these decisions beforehand, while exploring various analytical techniques, there is an inherent drive that leads to a research outcome that "works" and favors statistical significance. This thus increases the likelihood that at least one of these many analytical attempts produces a false positive finding.

A detailed review of these studies may be found elsewhere [21, 22]. Perhaps the most infamous cG×E study is the 2002 *Science* publication of Caspi [27], who examined whether the carriers of a short allele in the serotonin receptor gene (*5HTTLPR*) were sensitive to stressful life events. Carriers of the short allele were compared to those who had two long alleles at that locus, who appeared to be protected from the deleterious effects of strain and stress. The study gained considerable attention both initially due to the pioneering approach of cG×E but then later due to a lack of replication [21, 28]. The research was inherently appealing for many since it seemingly corroborated the importance of the environment in predicting genetic effects.

6.3.3 Genome-wide polygenic score G×E approaches

As we noted in previous chapters, with the arrival of the GWAS, reduction in costs, and the technical ability to genotype more than a small selection of variants, most G×E work now applies PGSs from GWASs. Many studies have been published using this approach since around 2014. Within psychology, some have adopted a case-control design. For example, multiple studies examine the relationship between the genetic predisposition toward major depressive disorder (MDD) and childhood trauma. Some found a significant interaction between the PGS for MDD and childhood trauma in predicting depression [29, 30].

Peyrot and colleagues [29], for instance, found that individuals who had a higher PGS for MDD and had experienced childhood trauma were more likely to develop depression (MDD) than those with a low PGS and no history of trauma. A growing number of studies use the compulsory schooling age reform in the United Kingdom, which demanded that students complete an additional year of school, as a natural experiment. Barcellos et al. (2018) [31], for example, examined whether genetic makeup moderates the effects of education on health outcomes. A series of studies by Boardman and colleagues has shown how the heritability of smoking was significantly reduced in U.S. states when restrictive policies on the sale of cigarettes and higher taxes were introduced at the state level [32]. We explore some of these examples in more detail when we describe the central theoretical models.

6.4   Conceptual G×E models

G×E interaction is often described with the aid of four key conceptual models, summarized in table 6.1. These theories are also often specified and distinguished by the functional form of the relationship between Genotype (G), Environment (E), and Trait (T). We also illustrate the theories using their functional forms in figure 6.1. It is perhaps important to note at the outset that these models were not rigorously specified or formalized to be mutually exclusive. For this reason, they can at times overlap and have blurred and fluid empirical applications. Others are not separate or exclusive theoretical models per se, but rather models that describe the positive (e.g., compensation model) or negative (e.g., diathesis-stress or triggering model) complement of the other model.

6.4.1   Diathesis-stress, vulnerability, or contextual triggering model
The majority of G×E research applies the diathesis-stress model, also interchangeably known as the vulnerability or contextual triggering model. The *diathesis-stress model*, developed by Monroe and Simons [33], proposes that genetic differences associated with *negative outcomes in risky environments* will have either an attenuated relationship or be entirely muted in low-risk environments. The model posits that the genetic propensity for a trait lies dormant until it is triggered by some sort of stressor or environmental exposure. The word *diathesis* originates from the Greek term for a predisposition but in this area of research it is used to refer to a tendency to suffer from a particular condition. Here diathesis is often represented by genetic or biological predictors.

Stressors or triggers in this theory are harmful or adverse conditions and can range from major stressful life events (e.g., death of spouse, divorce) to minor or chronic conditions or what Shanahan and Hofer [36] term *contextual triggering*. As we describe shortly in our discussion of gene-environment correlation, diathesis may even influence whether the environment is experienced by an individual in the first instance. Most research has focused on studying how genetic influences might be moderated by adversity, particularly

Table 6.1
Summary of theoretical and conceptual models underlying G×E studies.

Theory	Brief summary	Original article and further reading	Example of an empirical article that uses this theory
Diathesis-stress, also known as vulnerability, contextual triggering	A predisposition (i.e., a diathesis) for the phenotype lies dormant until triggered by environment (e.g., stressor).	Monroe and Simons (1991) [33]	South and Krueger (2008) [34]
Bioecological or social compensation model	Genetic influences are maximized in stable and adaptive environments that permit positive, stable interactions (proximal processes) between individuals and their environment, enabling them to reach their genetic potential.	Bronfenbrenner and Ceci (1994) [35]; Shanahan and Hofer (2005) [36]	Turkheimer et al. (2003) [7]
	Social compensation: environment is free of stress or has positive enriching properties.		
Differential susceptibility	Plasticity varies by individual with some (known as orchids) being more susceptible to (i.e., genetically influenced by) the effect of both positive and negative environments whereas others (known as dandelions) are more resilient.	Belsky and Pluess (2009) [6]	South and Krueger (2013) [8]
Social control or social push	Genetic influences are filtered and dampened in particular environments.	Shanahan and Hofer (2005) [36]	Boardman (2009) [32]; Dick et al. (2007) [37]; Liu and Guo (2015) [38]
	Social control: social norms and structural constraints.		

Source: Adapted from South et al. (2017) [39].

inspired by the early Caspi study [27] and related work. Extensions have been made, however, in diverse areas, including academic achievement [40].

6.4.2 Bioecological or social compensation model

Although they are often presented as separate theories, conceptually the bioecological or social compensation model is the actually the flipside or mirror of the previous theory. This theory often focuses on the environmental context of where individuals live, work, or interact (e.g., religious locations, schools) [35]. In contrast to the previous theory that focuses on negative and adverse conditions, this theory assumes that low-risk or highly stable environments allow positive and enduring interactions. These are called *proximal processes* of interaction between individuals and their environment. These interactions in

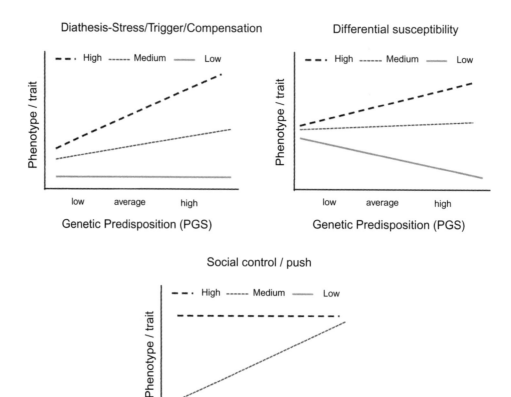

Figure 6.1
Gene-environment interaction models.
Source: Adapted from Liu and Guo [38].
Note: Lines represent genetic predisposition measured as PGSs by three groups of high, medium, and low PGS. The magnitude of the genetic association is indicated by the slope, with a steeper slope indicating a greater genetic association.

turn allow individuals to realize their genetic potential. This is similar to what Shanahan and Hofer [36] refer to as "social enhancement." Studies in this area often focus on the adaptation and buffering of context for individuals to reach their genetic potential. A buffering or social enhancement environment could be, for instance, parents who stimulate healthy eating, physical exercise, have many books in the household, or do not smoke. Or, those with the more "risky" alleles of two polymorphisms on the FTO (i.e., obesity-related) gene could avoid having this translate into a high BMI if they compensated, for instance, by being highly physically active.

A body of research that has used this theory examines the inheritance of cognitive ability and intelligence and how the genetic influence of these measures varies by socioeconomic status. This stems from the aforementioned study by Turkheimer and colleagues [7], who showed that the genetic influences on intelligence were only realized among those with the highest socioeconomic status. Those from low-risk environments (e.g., high socioeconomic status) and higher PGSs for IQ or cognitive ability would be more likely to realize their genetic potential. Even those in the lower PGS group (i.e., lower cognitive ability genetic score) that were in low-risk environments would do better than those from high-risk environments. Those from high-risk environments would have the lowest levels of educational achievement, regardless of their genetic score. This finding has been successfully replicated in several studies, such as Tucker-Drob and colleagues' work on the relationship of cognitive [41] and math ability [42] in young children. The original finding by Turkheimer et al., however, has evaded replication in several non-American contexts such as the United Kingdom [43] and the Netherlands [44]. As we have noted elsewhere, the differences in the moderation of genetic influences on IQ may be related to environmental variation between countries, historical periods, and levels of socioeconomic status and inequality [13, 45].

6.4.3   Differential susceptibility model

The differential susceptibility model [6] is also a variation on the diathesis-stress model. Both theories suggest that individuals are differentially susceptible to environmental influences. The diathesis-stress model describes this distinction almost exclusively in relation to negative influences. As the middle panel of figure 6.1 illustrates, the **differential susceptibility model**, sometimes also referred to as the plasticity model, hypothesizes that there are groups sensitive to both negative and positive environments. Some have also referred to this as the biological sensitivity to context model or the orchid–dandelion hypothesis [46]. Anyone who has attempted to grow an orchid knows that it requires a delicate mix of dedicated expert care to survive whereas dandelions thrive in any environment. Very risky environments (e.g., low socioeconomic status) allow for the expression of genetic vulnerabilities to poor outcomes and the low-risk or highly enriched or compensated environments (e.g., high socioeconomic status) allow genetic predispositions to be realized and flourish.

This theory also supposes that individuals have a certain level of plasticity or adaptability to their environment, with some being more susceptible to both positive and negative environments. This is characterized by crossover interactions; those who have the highest likelihood of the outcome when they have a high PGS (genetic risk) and high environmental stressor also have the lowest likelihood of the outcome when there is a low genetic risk and low environmental risk. Others have focused on the nuances of distinguishing between the diathesis-stress and differential susceptibility model in more detail [47]. An excellent discussion can also be found by Belsky et al. [48]. An example of this model is South and Krueger's [8] work on the "orchid" effect in how physical health differs in relation to an individual's marital relationship quality. The authors found that the heritability of subjective health was the highest among both highly distressed and highly satisfied couples.

6.4.4   Social control or social push model

The **social control model**, sometimes also referred to as the social push model, hypothesizes that genetic associations are attenuated or dampened in the presence of socially restrictive environmental contexts. In other words, the model does not emphasize direct environmental effects on the genetic contribution but rather that the genetic propensity can be weakened in the case of strong environmental influences. This could include social norms, such as age norms of the timing of life events, parental monitoring, or religious norms that restrict certain social behavior and environmental exposure such as smoking or drinking.

Although introduced separately by Shanahan and Hofer [36], as we noted earlier, these theories often conceptually overlap and have similar results in that both involve the presence of an environmental context that is protective of potential detrimental genetic predispositions. In the social control model, the environment is comprised of constraints imposed by structural processes and social norms. In the social compensation model, the environment is notable for the actual absence of stress or manifestation of enriching aspects [39]. The diathesis-stress model focuses on the combination of genetic predispositions and the presence of stress. The social control and compensation models focus on the environmental circumstances that might dampen or inhibit lower genetic predispositions resulting in a detrimental outcome or trait. Some have shown how adolescent smoking decreases with higher levels of parental monitoring [37]. Others examined the genetics of alcohol metabolism and the extent to which certain alleles affect alcohol consumption and that relationship to religion, family environment, and childhood adversity [49]. Social control can also take the form of policies such as a study that demonstrated that the heritability of smoking was significantly reduced in U.S. states that introduced restrictive policies on the sale of cigarettes and higher taxes [32].

6.4.5   Research designs to study G×E

Table 6.2 provides an overview of the main challenges of G×E research and potential solutions. In part III of this book, and particularly in chapter 11, we cover the details of model estimation and the interpretation of interaction parameters. Here we also address some of the vexing methodological issues noted at the bottom of table 6.2, such as sensitivity to scaling effects, basic statistical tests, properly controlling for confounders, and whether you require a main effect. In this chapter we only discuss the initial conceptual issues, with the methodological challenges and solutions explored in more detail in chapter 11.

Table 6.2
Summary of challenges and potential solutions for G×E analysis.

Challenge	Why is this problematic?	Potential solutions	Example of study, further explanation
Theoretical model underlying G×E unclear	Lack of understanding of causal mechanisms leads to black box explanation	Extensively review nongenetic literature on topic	
Articulating plausible biological pathway for G×E associations	Lack of understanding of causal mechanisms leads to black box explanation	Engage in or refer to downstream biological analyses related to G	Duncan and Keller (2011) [21] discuss various shortcomings of the literature
Candidate-gene (single-variant) approach	Lacks power, agnostic to biological processes, based on an a priori theory that has not been biologically proven, often small selection of SNPs	Conduct or include examinations on gene-based or set-based models Replicate in other samples with similar phenotypic measures	Duncan and Keller (2011) [21] discuss various shortcomings of the literature Frayling et al. (2007)—FTO gene [23]
Inconsistent measure of exposure or phenotype	Various studies used existing and only roughly comparable measures Harmonization of easiest and broadest measure that approximates E to obtain largest sample possible Environmental conditions are time-varying and manifold	Collaborate on common core exposures and definitions before data collection Prioritize precision of measures Engage in sensitivity analyses with different specifications Move beyond static or time-constant studies	
Environment viewed as proximate environmental moderator	Environments defined as obesity or maternal smoking are far downstream from social environmental factors that structure exposure	Contextualize individually based risk factors and examine what puts people at risk	Boardman et al. (2013) [10]
Too much emphasis on individual environment	Does not account for group-level behavioral, normative, and cultural processes that shape traits and behavior	Distinguish between the actions of people and the circumstances in which these actions occur	Boardman et al. (2013) [10]
Genetic participants from selective environments	Samples largely from healthy, higher-socioeconomic-status, and older populations in specific Western countries	Recognize selective sample in interpretation of results	Pampel et al. (2010) [50] show socioeconomic disparities in health behavior

(continued)

Table 6.2 (continued)

Challenge	Why is this problematic?	Potential solutions	Example of study, further explanation
	Specific disease profiles and few from low-resource environments to compare environments	Move to use representative population samples Draw from social science literature to recognize unique environments and life course (age) effects	Mills and Rahal (2019) [51] show that 72% of GWAS findings emanate from three countries (U.S., U.K., Iceland); datasets are often selective on geography, age (older), sex (more females)
Only European-descent populations considered	Cannot use genetic scores or generalize results beyond European ancestry groups	Locate consortiums or GWASs on other ancestry populations; consider admixture analysis; wait for the field to catch up	Mills and Rahal (2019) [51] show that all GWASs (2007–2018) range between 72% and 96% European ancestry, currently 88% in 2017
Lack of power and large-enough sample size	Need sufficiently large sample to estimate magnitude and strength of association	Engage in a power analysis to understand sample size required to detect G×E effect	See box 4.1 for genetic power calculators
SNP-based analyses lack power	Single-step analysis subject to multiple comparisons burden due to large number of SNPs considered at once	Conduct more efficient two-step tests	Gauderman et al. (2017) [52]
Interaction is dependent on scaling metric	Only multiplicative scale considered	Evaluate interaction on both multiplicative and additive scales	Gauderman et al. (2017) [52]
Detected interactions driven by confounders	Interactions are artefact driven by confounders (e.g., ethnicity, gender, age, socioeconomic status) rather than genetic or environmental variables Researchers incorrectly enter confounders as covariates in general linear models	Researchers need to enter the covariate-by-environment and the covariate-by-gene interaction in the same model that tests G×E interaction	Keller (2014) [53] explains the problem and how to solve it
Standard GWAS results may mask G×E	Genetic discovery is mostly based on GWAS meta-analyses of individuals from various environments. Only robust associations that replicate across multiple countries and birth cohorts are seen as valid.	Potential solutions include the search for a exogenous environmental interaction variables—also in reference to the GWAS discovery or conducting large-scale discovery studies in specific environments	Conley 2017 [54] explains the issue quite intuitively, and Tropf et al. 2017 [13] show evidence that G×E across countries and birth cohorts partly explain hidden heritability and produce a reduced while robust heritability estimate

Notes: GWAS = genome-wide association study; SNP = single-nucleotide polymorphism; G×E = gene-environment interaction.

The first challenge is specifying the theoretical model, which is often very difficult to achieve without an extensive literature review and strong understanding of the trait, main predictors, and variation across environmental conditions. A second and often insurmountable challenge for many newcomers (and even seasoned researchers) is the inability to articulate a plausible biological pathway for these models. Here it is essential to work directly with biologists or become versed in downstream biological and functional analyses not covered in this book. A third challenge as noted extensively in this chapter is that researchers should be reticent in adopting a candidate-gene approach for some topics and be careful to replicate research. A fourth challenge is an inconsistent measure of exposure or the phenotype under study. Studies can sometimes defy replication due to the lack of harmonization of the original phenotype measurement. As we noted in our discussion of E, a related problem is that environmental conditions can be observed on multiple levels and across various time periods.

A fifth issue is that the environment is often viewed as a proximate environmental moderator, and thus distant from the actual factors. Here it is important to understand the specific contexts of individuals and what places them at risk in that particular environment. A sixth challenge is that in some cases too much emphasis is placed on the individual environment, missing larger phenomenon such as couple or group-level behavior and larger cultural processes. Here we encourage researchers to distinguish between the individual actions of people and the circumstances in which the actions occur. Table 6.2 then addresses two points related to the lack of diversity in this research domain, discussed in detail in chapter 4. Here we note issues in the sample selectivity of GWAS research in terms of the healthy volunteer bias and extreme concentration of subjects coming from a handful of countries and overrepresented by those of European ancestry.

The final remaining five points are methodological issues. It is essential to determine whether you have a sample size that is large enough to even detect a G×E association, which we discussed in chapters 2 and 4 (see also box 4.1). Since a single-step analysis may be subject to a multiple comparisons burden (i.e., due to many SNPs compared at once), researchers should also consider a two-step test. Since only the multiplicative scale is often used in analyses, it is likewise important to evaluate the interaction on the additive scale to ensure that the interaction is not attributed to the scaling metric. A common error is that interaction effects are actually driven by confounders. To avoid this problem, it is necessary to enter the covariate-by-environment and the covariate-by-gene interaction in the same model that tests for the G×E interaction. Finally, the standard GWAS results used as a basis for the PGSs in these G×E models may mask environmental effects (see chapter 4).

6.5 Gene-environment correlation (rGE)

There is often considerable confusion over the distinction between gene-environment interaction (G×E) and gene-environment correlation (rGE). Gene-environment correlation (rGE) is a process by which genes and the environment operate together and change in tandem. In other words, rGE is the phenomenon by which an individual's genotype influences or is associated with exposure to the environment. It has been a longstanding area of study in psychology and psychiatry [55, 56]. It is also reminiscent of the well-established life course models in sociology of individual-environment correlations that describe how an individual's behavior, ability, or personality shape his or her environment [57]. These studies often try to uncover causal mechanisms underpinning how genetic predispositions control or influence environmental exposure. These genetic variants are often seen to indirectly influence environmental exposure via particular behaviors. Three main rGE processes were first described by Plomin and colleagues in 1977 [55] and categorized as passive, evocative, and active.

6.5.1   Passive gene-environment correlation (rGE)

Passive gene-environment correlation (rGE) is the association between the genotype a child inherits from her or his parents and the environment in which the child is raised. Parents not only pass on genetic material to their children but also create a home environment that is influenced by their own heritable characteristics. This idea was empirically examined, for instance, in a *Science* paper in the form of what the authors termed "nontransmitted alleles" in relation to educational attainment, reproductive behavior, and health [58]. In other words, although parents pass on their genetic material through sexual reproduction and genetic recombination (see chapter 1, section 1.2), the remainder of their nontransmitted genetic material still shapes the environment in which their children reside. Passive rGE accounts for the observed correlations between the child's behavior and their environment. For example, consider children who are aggressive and are also physically disciplined (e.g., spanked) by their parents. If these parents are also more aggressive than average they create an environment that includes physical discipline but also transmit a genetic risk of aggressive behavior to their children. This increases the likelihood that their children will be more aggressive and, as parents, will be more likely to use physical discipline [59]. Another example is the assortative mating of highly educated parents. If both parents are highly educated, they create an environment of learning, monitoring of homework, and higher educational expectations. They likewise transmit a higher joint genetic risk of obtaining higher education (often correlated with cognitive ability).

A challenge with studying passive rGE is the fact that the result may represent a spurious relationship between the environment and trait. A **spurious relationship** is when two more variables that are not causally related to one another are mistakenly believed to be related due to presence of a third omitted (or confounding) variable. If we take the first example, parents who impart aggressive behavior are at an increased risk of physically disciplining or even abusing their children. This physical maltreatment may be a genetic risk that parents transmit to their children and not a causal risk factor of children's own aggressive behavior.

6.5.2   Evocative (or reactive) rGE

Evocative (also known as reactive) gene-environment correlation (rGE) is when an individual's heritable traits evoke reactions from others in the environment. Children who are genetically prone to certain personality traits, such as introversion and related impositions such as being shy, for instance, may appear unapproachable or aloof to peers who as a result reinforce that trait through their reactions. Or, consider the relationship between having a higher genetic propensity for depression and partnership (marital) conflict. The genetic association and marital conflict may emerge as a result of problems in engaging with a depressed partner and not actually represent a causal effect of marital conflict and risk of depression.

6.5.3   Active rGE

Active gene-environment correlation (rGE), or niche creation, is when individuals actively select or create environments that are associated with their own genetic predispositions. Individuals who enjoy risk-taking or have extroverted personalities may actively seek out those environments. Those who feel more comfortable in controlled environments or ones that are highly regulated by social norms (e.g., religious norms, tax office) will be more likely to seek out peers and environments that are conducive to their preferences.

6.5.4   Why are models of rGE important?

Although rGE models are less studied than gene-environment interaction models and difficult to empirically examine, rGE models are important for several reasons. First, they examine the possibility that we may not fully understand the relationship between environmental risk exposure and the traits that we study. The fact that genes and environment may be correlated means that simple causal models of the association between environmental exposures (e.g., during childhood) and traits or outcomes are potentially confounded. Second, our causal understandings are often simplistic and do not measure reciprocal causal relationships over time, such as how genetically influenced behavior evokes responses which then reinforces that behavior. This reflects the example provided previously of a shy child with an introverted personality who may act invisible and evoke responses of being ignored by their peers, which in turn reinforces that behavior. Third, they inform us about how the trait we are studying develops over time, relating to the multitemporal dimension discussed previously. Life course and developmental pathways that lead to certain traits are increasingly recognized as important. Studies of rGE move toward understanding the reciprocal effects of individuals on their environments over time [60]. This may explain the results of studies that show how heritability changes with age by providing the explanation that behavior that is genetically influenced affects the way the environment interacts with the individual. Research in this area is providing increasing evidence that the pathway from genes to environment involves environment. In other words, there are no particular genes for environments but rather genes may evoke behavior and ability, which in turn shape an individual's environment.

6.5.5   Research designs to study rGE

It is often difficult to determine whether the relationships we observe between environmental exposure and the trait we study are causal since it is often the result of a complex and reciprocal process. Although this goes beyond the auspices of this introductory book, several researchers have developed unique research designs to study these processes, such as quasi-experimental designs [61, 62]. Others have used designs such as including parents whose children were conceived via assisted reproductive technologies where parents were biologically related but also used sperm, egg, or embryo donation [63]. They then compared the association between parent and child behavior of genetically versus unrelated pairs to estimate passive rGE. Others have used quasi-experimental designs and statistical matching of sibling pairs to rule out the possibility that evocative rGE confounds associations between men's marital status and antisocial behavior [64]. rGE models remain a challenging frontier since the magnitude of rGE varies across context, life course stage, involves reciprocal interactions with individuals and their environments, and varies within families and within populations [65].

6.6   Conclusion and future directions

6.6.1   Why haven't many G×Es been identified?

Detecting and replicating G×Es has been one of the main challenges in this area of research. As discussed previously, a fundamental challenge has been the lack of replicability related to low statistical power and the massive sample sizes required to detect interaction effects of even a low to moderate magnitude. Second, we often assume that there has been measurement error in both G and E, which in turn reduces the overall ability to detect real associations. In relation to G, we acknowledge that the causal genetic locus is rarely directly measured or understood. When using PGSs with multiple genetic loci, measurement error and a lack of understanding of the biological basis of the genetic predictors are even more pronounced. It is also important to note that we often rely on linkage disequilibrium (LD) and locate a region, but many times not the actual causal gene. As noted in chapter 3, the size of haplotype blocks varies and LD is sometimes very high with certain variants. The move to sequencing and study of more diverse populations will likely help in this respect. Third, in relation to E, as noted earlier, there are often difficulties in measuring multiple aspects of the environment and a high correlation of environmental factors. The response is often a relentless drive to increase sample size. However, a more cost-effective and precision measurement approach to not only E but also the traits we study is clearly a more efficient solution. Fourth, although quite implausible, it may be that G×E is theoretically pleasing but does not exist. Since we are generally examining only common genetic variants that have appeared in the immediate past, selection pressure could have weeded out all large G×E effects, leaving us with the true finding that there is no G×E. Even if G×E is important for rare variants or rare environmental circumstances, our current methods may render it virtually impossible to detect. One suggestion for future avenues of research is to stratify the analyses to identify subgroups that may pick up G×E. Aschard and colleagues showed age-specific per-allele odds of SNPs associated with breast cancer, for example, which would allow a focus on those who would most benefit from genetic screening for this disease [66].

Exercises

1. If a genetic predisposition would only lead to high blood pressure if someone drinks too much alcohol, which of the three gene-environment interaction models would describe this process the best? Try to draw the model as well.

2. Think of an example for the differential susceptibility gene-environment interaction model.

3. If someone with a genetic predisposition for smoking grows up in a family of smokers, does this describe gene-environment interaction or gene-environment correlation?

4. Provide an example for each of the following types of active, passive, and evocative gene-environment correlation.

Further reading

For a more in-depth reviews and reflection on the following topics, see:

Belsky, J., M. Pluess, and K. F. Widaman. Confirmatory and competitive evaluation of alternative gene-environment interaction hypotheses. *J. of Child Psy. and Psy.* **52**(1), 1135–1143 (2013).

Boardman, J. D., D. Daw, and J. Freese. Defining the environment in gene-environment interaction research: Lessons from social epidemiology. *Am. J of Pub. Health* **103**, S64–S74 (2013).

Del Giudice, M. Statistical tests of differential susceptibility: Performance, limitations, and improvements. *Dev. and Psycho.* **29**(4), 1267–1278 (2017).

Mitchell, C. et al. Genetic differential sensitivity to social environments: Implications for research. *Ann. Rev. of Psy.* **65**, 41–70 (2014).

Roisman, G. I. et al. Distinguishing differential susceptibility from diathesis-stress: Recommendations for evaluating interaction effects. *Dev. and Psycho.* **24**(2), 389–400 (2012).

Shanahan, M., and S. Hofer. Social context in gene–environment interactions: Retrospect and prospect. *J. Gerontol. Ser. B Psychol. Sci. Soc. Sci.* **60**, 65–76 (2005).

For a detailed review of the problems in candidate (cG×E) research, particularly in the area of psychological research, see references 21 and 22.

References

1. M. C. Nelson, P. Gordon-Larsen, K. E. North, and L. S. Adair, Body mass index gain, fast food, and physical activity: Effects of shared environments over time. *Obesity* **14**, 701–709 (2006).

2. J. Boardman et al., Is the gene-environment interaction paradigm relevant to genome-wide studies? The case of education and body mass index. *Demography* **51**, 119–139 (2014).

3. J. D. Boardman, B. W. Domingue, and J. M. Fletcher, How social and genetic factors predict friendship networks. *Proc. Natl. Acad. Sci. USA* **109**, 17377–17381 (2012).

4. J. D. Boardman et al., Do schools moderate the genetic determinants of smoking? *Behav. Genet.* **38**, 234–246 (2008).

5. B. W. Domingue et al., The social genome of friends and schoolmates in the National Longitudinal Study of Adolescent to Adult Health. *Proc. Natl. Acad. Sci. USA* **115**(4), 702–707 (2018).

6. J. Belsky and M. Pluess, Beyond diathesis stress: Differential susceptibility to environmental influences. *Psychol. Bull.* **135**, 885–908 (2009).

7. E. Turkheimer et al., Socioeconomic status modifies heritability of IQ in young children. *Psychol. Sci.* **14**, 623–628 (2003).

8. S. C. South and R. F. Krueger, Marital satisfaction and physical health: Evidence for an orchid effect. *Psychol. Sci.* **24**, 373–378 (2013).

9. E. B. Bookman et al., Gene-environment interplay in common complex diseases: Forging an integrative model—Recommendations from an NIH workshop. *Genet. Epidemiol.* **35**, 217–225 (2011).

10. J. D. Boardman, J. Daw, and J. Freese, Defining the environment in gene-environment research: Lessons from social epidemiology. *Am. J. Public Health* **103**, 64–72 (2013).

11. J. Pearl, *Causality: Models, reasoning, and inference* (Cambridge: Cambridge University Press, 2000).

12. R. Ottman, Gene-environment interaction: Definitions and study designs. *Prev. Med.* **25**, 764–770 (1996).

13. F. C. Tropf et al., Hidden heritability due to heterogeneity across seven populations. *Nat. Hum. Behav.* **1**, 757–765 (2017).

14. N. Barban and F. C. Billari, Classifying life course trajectories: A comparison of latent class and sequence analysis. *J. R. Stat. Soc. Ser. C (Applied Stat.)* **61**, 765–784 (2012).

15. M. C. Mills, *Introducing survival and event history analysis* (Los Angeles: Sage, 2011).

16. M. C. Mills and H.-P. Blossfeld, Globalization, uncertainty and the early life course: A theoretical framework. In H. P. Blossfield, E. Klijzing, M. Mills, and K. Kurz (eds.), *Globalization, uncertainty and youth in society*, 1–24 (London: Routledge, 2005).

17. C. M. A. Haworth et al., Increasing heritability of BMI and stronger associations with the FTO gene over childhood. *Obesity* **16**, 2663–2668 (2008).

18. J. Kim, J. M. Basak, and D. M. Holtzman, The role of apolipoprotein E in Alzheimer's disease. *Neuron* **63**, 287–303 (2009).

19. J. D. Boardman et al., Social disorder, APOE-E4 genotype, and change in cognitive function among older adults living in Chicago. *Soc. Sci. Med.* **74**, 1584–1590 (2012).

20. J. Novembre et al., Genes mirror geography within Europe. *Nature* **456**, 98–101 (2008).

21. L. E. Duncan and M. C. Keller, A critical review of the first 10 years of candidate gene-by-environment interaction research in psychiatry. *Am. J. Psych.* **168**, 1041–1049 (2011).

22. L. E. Duncan, A. R. Pollastri, and J. W. Smoller, Mind the gap: Why many geneticists and psychological scientists have discrepant views about gene-environment interaction (G×E) research. *Am. Psychol.* **69**, 249–268 (2014).

23. T. M. Frayling et al., A common variant in the FTO gene is associated with body mass index and predisposes to childhood and adult obesity. *Science* **316**, 889–894 (2007).

24. L. A. Hindorff et al., Potential etiologic and functional implications of genome-wide association loci for human diseases and traits. *Proc. Natl. Acad. Sci. USA* **106**, 9362–9367 (2009).

25. J. P. Simmons, L. D. Nelson, and U. Simonsohn, False-positive psychology. *Psychol. Sci.* **22**, 1359–1366 (2011).

26. J. Hewitt, Editorial policy on candidate gene association and candidate gene-by-environment interaction studies of complex traits. *Behav. Genet.* **41**, 1–2 (2012).

27. A. Caspi, Role of genotype in the cycle of violence in maltreated children. *Science* **297**, 851–854 (2002).

28. D. M. Dick, An interdisciplinary approach to studying gene-environment interactions: from twin studies to gene identification and back. *Res. Hum. Dev.* **8**, 211–226 (2011).

29. W. J. Peyrot et al., Effect of polygenic risk scores on depression in childhood trauma. *Br. J. Psych.* **205**, 113–119 (2014).

30. N. Mullins et al., Polygenic interactions with environmental adversity in the aetiology of major depressive disorder. *Psychol. Med.* **46**, 759–770 (2016).

31. S. H. Barcellos, L. S. Carvalho, and P. Turley, Education can reduce health differences related to genetic risk of obesity. *Proc. Natl. Acad. Sci. USA* **115**, E9765–E9772 (2018).

32. J. D. Boardman, State-level moderation of genetic tendencies to smoke. *Am. J. Pub. Health* **99**, 480–486 (2009).

33. S. M. Monroe and A. D. Simons, Diathesis-stress theories in the context of life stress research: Implications for the depressive disorders. *Psychol. Bull.* **110**, 406–425 (1991).

34. S. C. South and R. F. Krueger, Marital quality moderates genetic and environmental influences on the internalizing spectrum. *J. Abnorm. Psychol.* **117**, 826–837 (2008).

35. U. Bronfenbrenner and S. J. Ceci, Nature-nurture reconceptualized in developmental perspective: A bioecological model. *Psychol. Rev.* **101**(4), 568–586 (1994).

36. M. Shanahan and S. Hofer, Social context in gene-environment interactions: Retrospect and prospect. *J. Gerontol. Ser. B Psychol. Sci. Soc. Sci.* **60**, 65–76 (2005).

37. D. M. Dick et al., Parental monitoring moderates the importance of genetic and environmental influences on adolescent smoking. *J. Abnorm. Psychol.* **116**, 213–218 (2007).

38. H. Liu and G. Guo, Lifetime socioeconomic status, historical context, and genetic inheritance in shaping body mass in middle and late adulthood. *Am. Sociol. Rev.* (2015), doi:10.1177/0003122415590627.

39. S. C. South, N. R. Hamdi, and R. F. Krueger, Biometric modeling of gene-environment interplay: The intersection of theory and method and applications for social inequality. *J. Pers.* **85**, 22–37 (2017).

40. J. Jaekel, M. Pluess, J. Belsky, and D. Wolke, Effects of maternal sensitivity on low birth weight children's academic achievement: a test of differential susceptibility versus diathesis stress. *J. Child Psychol. Psych.* **56**, 693–701 (2015).

41. E. M. Tucker-Drob, M. Rhemtulla, K. P. Harden, E. Turkheimer, and D. Fask, Emergence of a gene × socioeconomic status interaction on infant mental ability between 10 months and 2 years. *Psychol. Sci.* **22**, 125–133 (2011).

42. M. Rhemtulla and E. M. Tucker-Drob, Gene-by-socioeconomic status interaction on school readiness. *Behav. Genet.* **42**, 549–558 (2012).

43. K. B. Hanscombe et al., Socioeconomic status (SES) and children's intelligence (IQ): In a UK-representative sample SES moderates the environmental, not genetic, effect on IQ. *PLoS One* **7**, e30320 (2012).

44. S. van der Sluis, G. Willemsen, E. J. C. de Geus, D. I. Boomsma, and D. Posthuma, Gene-environment interaction in adults' IQ scores: Measures of past and present environment. *Behav. Genet.* **38**, 348–360 (2008).

45. A. Courtiol, F. C. Tropf, and M. C. Mills, When genes and environment disagree: Making sense of trends in recent human evolution. *Proc. Natl. Acad. Sci. USA* **113**, 7693–7695 (2016).

46. B. J. Ellis and W. T. Boyce, Biological sensitivity to context. *Curr. Dir. Psychol. Sci.* **17**, 183–187 (2008).

47. G. I. Roisman et al., Distinguishing differential susceptibility from diathesis-stress: Recommendations for evaluating interaction effects. *Dev. Psychopathol.* **24**, 389–409 (2012).

48. J. Belsky, M. J. Bakermans-Kranenburg, and M. H. van IJzendoorn, For better and for worse: Differential susceptability to environmental influences. *Curr. Dir. Psychol. Sci.* **16**, 300–304 (2007).

49. T. L. Wall, S. E. Luczak, and S. Hiller-Sturmhöfel, Biology, genetics, and environment: Underlying factors influencing alcohol metabolism. *Alco. Res.* **38**, 59–68 (2016).

50. F. B. Pampel, Krueger, R.F., and J. T. Denney, Socioeconomic disparities in health behaviors. *Annu. Rev. Sociol.* **36**, 349–370 (2010).

51. M. C. Mills and C. Rahal, A scientometric review of genome-wide association studies. *Commun. Biol.* **2** (2019), doi:10.1038/s42003-018-0261-x.

52. W. J. Gauderman et al., Update on the State of the Science for Analytical Methods for Gene-Environment Interactions. *Am. J. Epidemiol.* **186**, 762–770 (2017).

53. M. C. Keller, Gene×environment interaction studies have not properly controlled for potential confounders: The problem and the (simple) solution. *Biol. Pysch.* **75**, 18–24 (2019).

54. D. Conley, The challenges of G×E: Commentary on "Genetic endowments, parental resources and adult health: Evidence from the Young Finns Study." *Soc. Sci. Med.* **188**, 201–203 (2017).

55. R. Plomin, J. C. DeFries, and J. C. Loehlin, Genotype-environment interaction and correlation in the analysis of human behavior. *Psychol. Bull.* **84**, 309–322 (1977).

56. L. J. Kendler and K. S. Eaves, Models for the joint effect of genotype and environment on liability to psychiatric illness. *Am. J. Psych.* **143**, 279–289 (1986).

57. G. H. J. Elder and M. J. Shanahan, The life course and human development. In *Handbook of child psychology*, vol. 1: *Theoretical models of human development* (New York: Wiley, 1998), pp. 665–715.

58. A. Kong et al., The nature of nurture: Effects of parental genotypes. *Science* **359**, 424–428 (2018).

59. L. F. DiLalla and I. I. Gottesman, Biological and genetic contributors to violence—Widom's untold tale. *Psychol. Bull.* **109**, 125–129; discussion 130–132 (1991).

60. C. R. Beam and E. Turkheimer, Phenotype-environment correlations in longitudinal twin models. *Dev. Psychopathol.* **25**, 7–16 (2013).

61. S. R. Jaffee, L. B. Strait, and C. L. Odgers, From correlates to causes: Can quasi-experimental studies and statistical innovations bring us closer to identifying the causes of antisocial behavior? *Psychol. Bull.* **138**, 272–295 (2012).

62. E. M. Foster, Causal inference and developmental psychology. *Dev. Psychol.* **46**, 1454–1480 (2010).

63. F. Rice, G. Lewis, G. T. Harold, and A. Thapar, Examining the role of passive gene-environment correlation in childhood depression using a novel genetically sensitive design. *Dev. Psychopathol.* **25**, 37–50 (2013).

64. S. R. Jaffee, C. M. Lombardi, and R. L. Coley, Using complementary methods to test whether marriage limits men's antisocial behavior. *Dev. Psychopathol.* **25**, 65–77 (2013).

65. A. Knafo and S. R. Jaffee, Gene-environment correlation in developmental psychopathology. *Dev. Psychopathol.* **25**, 1–6 (2013).

66. H. Aschard, N. Zaitlen, S. Lindström, and P. Kraft, Variation in predictive ability of common genetic variants by established strata. *Epidemiology* **26**, 51–58 (2015).

II

Working with Genetic Data

7

Genetic Data and Analytical Challenges

Objectives

- Understand the *genotyping and sequencing technologies* that produce genomic data
- Comprehend *linkage disequilibrium* and *imputation* in relation to genomic data
- Understand the large drop in costs in genomic data and the *limitations of genotyping arrays* and gain a basic understanding of *next-generation sequencing*
- Know about the *most prominently used data* in human genetics for genome-wide association study discovery and the *sources that archive and distribute* this data
- Grasp the *different formats of genomic data* in the computer program PLINK
- Gain an introduction into the sample data used in this book
- Have a basic understanding of *data storage, transfer, size,* and *computational power* required

7.1   Introduction

Since 2005, but particularly since around 2015, there have been considerable advances in the collection, availability, technology, and the sample size of genetic data. The aim of this chapter is to provide an overview of genomic data. We first discuss the staggering and rapid developments in genotyping and sequencing arrays that are used to measure the genetic variants of a single person. We then provide a brief overview of some of the most prominently used human genetic data sources in this area of research and a brief explanation of where and how to obtain them. In the third section we offer a more detailed description of the different types of genetic data formats that you will encounter. Within this chapter we refer to the genomic data that we work with as simply as "data." As we noted in an earlier discussion, researchers often in this field use the term "cohort" to refer to different datasets. We do not adopt the term because it may cause unnecessary confusion, since the term is often used in demography and other sciences to represent birth cohort or a particular cohort study design, and in medical sciences to refer more broadly to groups of people in a more general sense. In this chapter readers will also encounter code in R and PLINK. Refer to appendix 1 for information on how to download these programs and appendix 2 for a description of the data used in this book.

7.2   Genotyping and sequencing array

7.2.1   Genotyping and sequencing technologies

The typical way that genomic information is collected is via *biological samples*, either from saliva or blood. The DNA is then extracted from the sample using biochemical methods and analyzed using a genotyping or sequencing platform. *Genotyping* most often refers to the use of genotyping microarrays, a technology that rapidly developed in the last decades, which are used for the measurement of hundreds of thousands to millions of genetic variants in a single person. A *DNA microarray* (also called SNP [single-nucleotide polymorphism] microarray) is a laboratory tool with the dimensions of a microscope slide that is printed with thousands of tiny spots in defined positions, with each spot containing a known short segment of DNA (called probe) that is complementary to different alleles of a given SNP. Once the DNA is extracted from the blood sample or from saliva, it is fragmented into small segments using biochemical methods. The DNA fragments pair with the complementary probes in the microarray.

There are multiple different probes for each SNP that in turn correspond to specific alleles.[1] For example, if a specific SNP has two common variants, C and T, there will be some probes designed to detect whether the sample has a C in a specific position of the DNA sequence while other probes are designed to detect if the individual has a T in that position of DNA. Since the DNA fragments are labelled with a fluorescent dye, the array can be scanned and then used to detect the genotype of a person. If there is approximately equal florescence in the two probes, we can conclude that the person is heterozygous at that SNP; if we detect fluorescence only in one probe, the individual is most likely homozygous at that SNP.

The process of determining the genotype based on microarray data is called genotype calling. Microarrays can be designed to detect specific SNPs of interest in particular genes or used to genotype variants spread across the entire genome, producing what are generally called genome-wide data. They can measure SNPs or other genetic variants such as copy number variations (CNVs) or indels (see chapter 1). Depending on the number of probes, microarrays are designed to detect genetic variations more common in one ancestral group or that exist in multiple ancestries. The Health and Retirement Study [1] (see box 7.1), for instance, which is a survey on aging based on a sample of U.S. respondents more than 50 years old, started genotyping its respondents in 2006, using a microarray produced by the company Illumina and called the HumanOmni2.5 Quad BeadChip. This array genotypes 2.5 million genetic variants across the genome for individuals of different ancestries.

Box 7.1
Description of the Health and Retirement Study

The Health and Retirement Study (HRS) is a nationally representative, longitudinal panel study of over 37,000 American individuals from around 23,000 households aged 50 and above, and their spouses. The study was launched in 1992 with data collected every two years. The survey contains detailed sociodemographic and health information in addition to a genetic sample. In 2006, the HRS initiated an enhanced face-to-face (EFTF) interview. In addition to the core interview, the EFTF interview includes a set of physical performance tests, anthropometric measurements, blood and saliva samples, and a self-administered questionnaire on psychosocial topics. The blood sample collected since 2006 provides information on a range of biomarkers as part of the EFTF interview. The HRS collected blood-based biomarkers on half the sample in 2006; the other half of the sample provided biomarker data in 2008. The first group was asked for another blood sample in 2010 and the second group gave repeated samples in 2012. Between 2006 and 2012, the HRS genotyped almost 20,000 respondents who provided DNA samples and signed consent forms. HRS subjects were genotyped using the Illumina HumanOmni2.5 Quad BeadChip, with coverage of approximately 2.5 million genetic variants. The HRS genetic sample produces several data products. The HRS makes many polygenic scores publicly available and constantly updated on its website, making it an excellent resource for researchers. At the time of writing this book, there was the HRS Polygenic Score—Release 3, with 42 scores that covered educational attainment, height, body mass index (BMI), blood pressure, Alzheimer's disease, smoking, depression, age at first birth, and number of children ever born, to name only a few. Other HRS genetic data require additional authorization to access. The HRS is sponsored by the National Institute on Aging (NIA U01AG009740) and is conducted by the University of Michigan.

7.2.2 Linkage disequilibrium and imputation

The completion of the first human genome was attributed to the initial widely adopted Sanger sequencing method. This did not allow for population whole genome sequencing. To solve this problem, researchers used the fact that humans have many LD blocks, which we discussed in chapter 3 (section 3.6). Recall that due to LD, there is a reduction in the number of genetic variants that need to be measured. The result is that, once we know the alleles of a few SNPs in one region, we can infer the neighboring alleles with a high degree of confidence. This process is called *imputation*, and it is a fundamental part of genetic analysis (see box 7.2). Common genetic variation is thus assessed by genotyping hundreds of thousands of variants across the genome using information from what is referred to as reference panels, such as **HapMap** (https://www.genome.gov/10001688/international-hapmap-project) [2], the Haplotype Reference Consortium (http://www.haplotype-reference-consortium.org/), and the 1000 Genomes Project (http://www.internationalgenome.org/) (see figure 7.1) [3].

After genotyping, the haplotype information was then used to examine ungenotyped sites [5]. Reference panels such as the HapMap project or the 1000 Genomes Project have been used to effectively create an atlas of human genetic variations that are used to infer the "missing genotypes" that are not directly measured using microarrays. For instance, the genetic

Box 7.2
Imputing genetic data

Imputation of genotypes is undertaken to estimate unmeasured or missing genotypes. The data that is imputed is often expressed as genotype dosage. Metrics are used to measure the imputation quality for each SNP. Imputation is paramount when genetic information is combined across different data sources that use different genotyping chips. Imputation exploits patterns of linkage disequilibrium, which are the correlations between alleles at pairs of SNPs. Imputation is a process that uses data from a reference panel to guess the variants that are not genotyped (i.e., unmeasured or missing). Different reference panels are available for imputation, such as the 1000 Genomes Project or HapMap (see the "Further reading and resources" section). It works by matching haplotypes from the sample to the haplotypes in a reference sample (see discussion of haplotypes in chapter 3). Because haplotypes in the sample are matched to multiple haplotypes in the reference panel, the imputation is not always unique. In the example shown here, the first variant that needs to be imputed can be matched with the reference panel: one-third of the reference haplotypes is a T and two-thirds a C. Imputation is a probabilistic procedure that takes into account statistical uncertainty. Alternatively, it is possible to convert to a "best guess" genotype, which, in this example, would be a C. Imputation for rare variants may be challenging as it is limited by the sample size of the reference sample and also by limited LD of rare variants with common variants and the larger number of rare variants. The main reference panels used are HapMap, the 1000 Genomes Project, and the Haplotype Reference Consortium (see the "Further reading and resources" section).

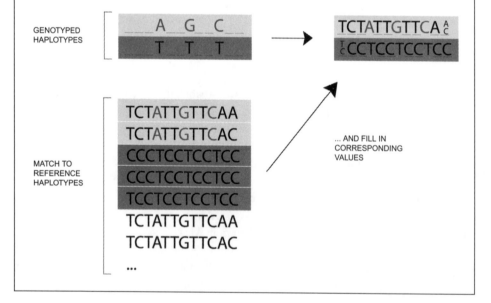

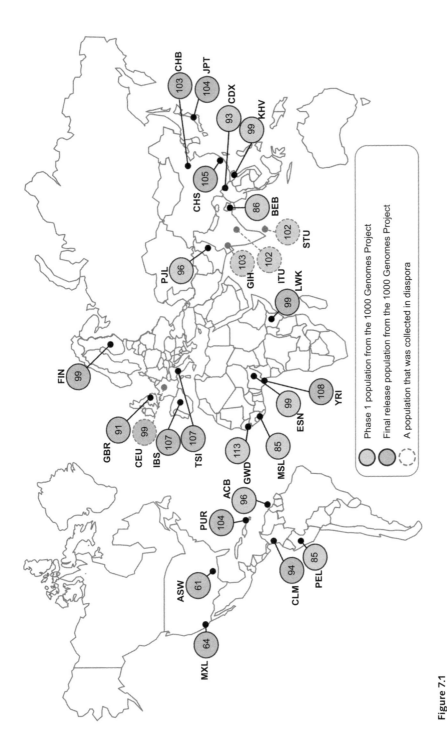

Figure 7.1

History and coverage of the 1000 Genomes data.

Source: Oleksyk et al. (2015) [4].

Notes: ACB = African Caribbeans in Barbados; ASW = Americans of African ancestry in the southwestern United States; BEB = Bengali from Bangladesh; CDX = Chinese Dai in Xishuangbanna, China; CEU = Utah residents with Northern and Western European ancestry; CHB = Han Chinese in Beijing; CHD = Chinese in metropolitan Denver; CHS = Southern Han Chinese; CLM = Colombians from Medellín, Colombia; ESN = Esan in Nigeria; FIN = Finnish in Finland; GBR = British in England and Scotland; GIH = Gujarati Indians in Houston; GWD = western divisions in the Gambia; IBS = Iberian population in Spain; ITU = Indian Telugu from the United Kingdom; JPT = Japanese in Tokyo; KHV = Kinh in Ho Chi Minh City, Vietnam; LWK = Luhya in Webuye, Kenya; MSL = Mende in Sierra Leone; MXL = people of Mexican ancestry from Los Angeles; PEL = Peruvians from Lima, Peru; PJL = Punjabi from Lahore, Pakistan; PUR = Puerto Ricans from Puerto Rico; STU = Sri Lankan Tamil from the United Kingdom; TSI = Tuscans in Italy; YRI = Yoruba in Ibadan, Nigeria.

data from the HRS (discussed previously) can be extended to 22 million genetic variants using imputation. Imputation is not always perfect, especially in regions where there are not many nearby SNPs. Or in the large UK Biobank study of approximately 500,000 individuals, phasing and imputation increased the number of testable variants over one hundredfold, to approximately 96 million variants (see Bycroft et al. 2018 in the "Further reading and resources" section). For this reason, imputed data come with an associated *imputation quality* that measures how confident we can be that the missing genotype is really, for example, CC or TC. We will discuss the importance of imputation quality later in part II of this book, where we show how to cope with quality control issues (see chapter 8, section 8.5).

7.2.3   Limitations of genotyping arrays and next-generation sequencing

There are several limitations to genotyping arrays. First, using "tag" SNPs that are high in LD with neighbor variants works only for common variants and moderately for rare ones. This is due to the constraints on LD imposed by allele frequency. Second, the probes on arrays only target SNPs and certain simple structural variants. This means that if the goal is to understand rare and complex structural variants, another approach is necessary. Recall from chapter 1 that *rare variants* are SNPs (or indels, CNVs) that are present in less than 1% of the population. These are rare mutations that therefore may not be present in the reference panels and thus cannot be easily detected using genotyping. Also, the fact that different ancestry groups have different variants may privilege the detection of genetic associations in loci that are more commonly measured with genotype, which is most likely variants that are common in individuals with European ancestry.

For this reason, sequencing data are increasingly more common. Sequencing data are rising in popularity due to reduction in prices and increased computational power. In particular, *next-generation genome sequencing* analyzes millions of small fragments of DNA in parallel in order to genotype the whole genome. Next-generation sequencing is still uncommon in large epidemiological studies due to its costs, but it is increasingly used clinically to detect the presence of rare genetic variants. An alternative to whole-genome sequencing is so-called *exome sequencing*, which is used to sequence the variations in coding regions rather than the entire genome. As of 2018, the UK Biobank, for instance, is exome sequencing its entire sample of 500,000 individuals using Illumina short-read technology. Using standard arrays, we capture between 2% and 2.5% of the total variation of the human genome [6]. These new methods of genome sequencing provide access to the remaining approximately 98% of the genome and the difficult-to-impute genetic variants. This includes the ability to examine how much rare variation contributes to certain phenotypes. Sequencing data allows researchers to examine potentially informative genetic variants outside of protein-coding regions. Some companies, such as Gencove (https://gencove.com/) or Oxford Nanopore Technologies (https://nanoporetech.com/applications/whole-genome-sequencing), are offering *ultra-low-coverage sequencing* as an alternative to microarray genotyping. This technology uses sequencing methods to genotype a limited number of variants (which is

Box 7.3
A GWAS case study of type 2 diabetes

It is not the case that conducting one GWAS will provide the definitive results that isolate all variants associated with a phenotype. As technology advances and ever larger sample sizes are available, authors often engage in new iterations of studying the same phenotype. Mark McCarthy and Anubha Mahajan wrote an interesting blog in 2018 outlining the GWAS journey in the study of type 2 diabetes [9]. The DIAGRAM consortium, for instance, has conducted multiple GWA studies of type 2 diabetes. The first study that they conducted was in 2007 as part of the Wellcome Trust Case Control Consortium and included 500,000 SNPs from around 5,000 individuals for a total of 2.5 billion genotypes. A publication in 2018 included almost 900,000 individuals from 32 different studies from European descent populations [10]. Only one decade later, they could now include 27 million SNPs and 900,000 individuals for a total of 25 trillion genotypes. The most recent study not only increased the sample size but also used more detailed imputation reference panels. In 2019, studies often use the Haplotype Reference Consortium (HRC) panel, which used around 30,000 genomes in around 65,000 human haplotypes combining data from multiple studies. Current studies used a sub-panel of the HRC, the 1000 Genomes Project, which contains considerably more detail on a few hundred European genomes. One way to imagine the change is to think of how digital photography has improved over time, from grainy pixels to now highly detailed pixelation. The use of the 1000 Genomes Project panel affords a more robust analysis and particularly allows the examination of the risk of low-frequency alleles. In the most recent study, this consortium detected 243 loci that reached genome-wide significance and an additional 160 secondary signals for a total of 403 significant signals across 243 loci. Another key difference was that they were able to detect more signals at lower ranges of the allele frequency spectrum (i.e., minor allele frequency [MAF] < 5%), which included rare variants. The authors argue, however, that most findings still reside in common shared variants. The 2018 study had both an increase in the sample size but also used what is known as fine-mapping of causal variants and an extension of fine-mapping through the integration of tissue-specific epigenomic information. In other words, it is now possible to examine in more detail which variants are "taking the lead" and actually driving the association. With a genome-wide heritability explaining 18% of the type 2 diabetes risk, the authors also highlighted 18 genes attributable to coding variants for validated therapeutic targets.

---

why it is called ultra-low-coverage) without being restricted to predefined SNPs. Recent studies have shown that even with low coverage, this technology can outperform standard genotype arrays especially in populations of non-European ancestry. There is also the term *fine-mapping*, which is a process to refine the lists of associated variants to a more credible set that is more likely to include the causal variant (see box 7.3).

7.2.4   Drop in costs per genome

Modern genetics has also been about the development of faster and cheaper genotyping and genome sequencing technology, which resulted in the remarkable drop in sequencing costs in the past few years. Many of the advances we discuss in this chapter and book are

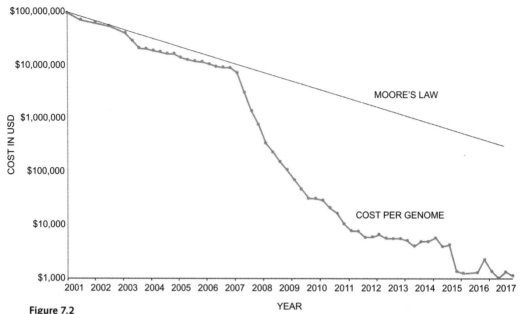

Figure 7.2
Drop in costs per genome versus drop in costs for computational power (Moore's law), 2001–2017.
Source: Adapted from Wetterstrand KA. DNA Sequencing Costs: Data from the NHGRI Genome Sequencing
Program (GSP), available at: www.genome.gov/sequencingcostsdata.

possible due to the staggering reductions in the costs of genome sequencing over the past
three decades (see also box 7.3 for a case study) [7, 8]. Figure 7.2 provides an overview of
the significant drop in price over just three decades, which is typically contrasted with the
dramatic drop in costs for computational power (Moore's law).

Moore's law is named after the cofounder of Intel, Gordon Moore, who observed that the
number of transistors per square inch on integrated circuits doubled every year since their
invention, while the costs were in turn halved. Although it is hard to definitively estimate the
total cost of sequencing of the first full human genome, by combining figures the NHGRI
(National Human Genome Research Institute) estimate the total cost was somewhere between
$500 million and $1 billion. This includes the broader umbrella costs as well as related technological development, model organism genome mapping, bioethics, and program management
[7]. This figure thus shows that the drop in costs per genome since 2006 has sharply dropped
in comparison to Moore's law predictions for drops in costs in computational power.

7.3   Overview of human genetic data for analysis

As with most types of data that we use in science, there is no central repository that
registers all possible data that are available to use in this type of research. Levels of open-
ness and availability also differ for each study and national funding requirements or

regulations. The move to open science and replication of research likewise means that data that were previously not released to external researchers may become increasingly available. This section provides a brief overview of the most prominently used sources of human genetic data in genome-wide association studies (GWASs) and key sources that archive and distribute this data.

7.3.1   Prominently used genetic data

To our knowledge, the most comprehensive overview of the most prominently used data in all GWASs for the past 13 years until late 2018 can be found in the accompanying Supplementary Material of the GWAS review article by Mills and Rahal (2019) [11]. Here we list the top 2,000 data sources that have been used in the largest GWASs from 2005 to October 2018. Readers can find the full list online at our GitHub site, https://github.com /crahal/GWASReview/blob/master/tables/Manually_Curated_Cohorts.csv. To generate this list, we manually extracted the most frequently used data across the majority of the largest 1,250 GWASs as of August 29, 2018, with the objective of providing the first systematic estimate of the frequency and identification of data sources used in GWAS. An overview of the top 10 datasets is shown in table 7.1 in addition to a description of some their key distinguishing features.

As the table shows, the most frequently used datasets have several commonalities. First, as we elaborated upon chapter 4, the most frequently used data emanates from high-income countries (the United States, the United Kingdom, Iceland, The Netherlands, Ireland, and Germany) that share similar rates of disease prevalence and population profiles. As noted earlier, large Asian ancestry data sources are mainly from Japan, China, and South Korea, with almost 8% of recorded GWASs involving Japanese participants representing over 14% of participants contributing to GWAS (see box 4.3). A second similarity is that most of the prominently used data sources in the table engaged in random probability or population sampling to gain as representative a sample as possible. We should note that some of the largest datasets that have emerged since 2018 and beyond in GWA studies are from the UK Biobank [12] or direct-to-consumer genetic companies such as 23andMe [13]. These studies are less representative and contain more healthy, older, and higher-socioeconomic-status individuals. Third, the most prominently used data sources are deeply and richly phenotyped across many traits, which likely made them more accessible for multiple needs. There are many data sources as well that are collected for specific diseases such as endometriosis, cardiovascular disease, or rare diseases.

Fourth, many of the most prominent datasets are older populations with disease diagnosis aimed at unravelling the pathways to disease and disability in old age. In this respect, they miss the longer-term development of disease and intervention possibilities that an asymptomatic younger population might afford (with the exception of some studies such as the 1958 British Birth Cohort or additional data collection in cohorts such as the Framingham Heart Study). Fifth, they are all prospective longitudinal datasets, following individuals or birth cohorts over a longer period, thus facilitating a life-course approach to understanding

Table 7.1
Most frequently utilized datasets across the largest GWASs, 2007–2018.

Cohorts	Count	N	Country of recruitment	Age range	Study design	Female (%)
Rotterdam Study (RS)	398	14,926	Netherlands	55–106	Prospective cohort	57
Cooperative Health Research in the Region of Augsburg (KORA)	255	18,079	Germany	24–75	Population-based	50
Framingham Heart Study (FHS)	207	15,447	U.S.	5–85*	Prospective cohort, three generation	54
Atherosclerosis Risk in Communities Study (ARIC)	204	15,792	U.S.	45–64	Prospective cohort, Community	55
Cardiovascular Health Study (CHS)	179	5,888	U.S.	65+	Prospective cohort	58
British 1958 Birth Cohort Study (1958 BC/NCDS)	156	17,634	U.K.	0+	Prospective birth cohort	48
U.K. Adult Twin Register (TwinsUK)	140	12,000	U.K., Ireland	18–97	Longitudinal entry at various times	84
European Prospective Investigation into Cancer CANCER (EPIC)	132	521,330***	10 EU countries	21–83**	Prospective cohort	71
Nurses Health Study (NHS)	129	121,700	U.S.	30–55	Prospective cohort	100
Study of Health in Pomerania (SHIP)	127	4,308	Germany	20–79	Prospective cohort	51

Source: Mills and Rahal (2019) [11], table 3.
Note: The top 10 most frequently utilized cohorts across the majority of the largest third of all GWASs as of August 29, 2018 (with studies ranked by *N*), manually extracted and harmonized. Additional fields (country of recruitment, age range, and study design) manually curated from web searches. * denotes originally 30–62 years; ** denotes variation by country; *** denotes full sample, including nongenotyped participants.

the pathways to certain diseases, disability, behavior, or mortality. Sixth, it is striking that all but one of these cohorts is comprised of predominantly female participants (ranging from 48% to 100%). This sex ratio imbalance is rarely addressed, yet sexual dimorphism or sex differences in disease are viewed as increasingly relevant, such as in autism [14] or reproductive traits [15]. Finally, although many started as focused hypothesis-driven clinical samples to study one type of disease, most have expanded to contain a breadth of phenotypes and there is a trend in data collection of adding new samples or generations over time.

Although the selectivity of the sample and lack of demographic diversity discussed above is less often discussed and addressed in this research, there is an increased move by funders and researchers to strengthen ancestral diversity. In 2018, the National Institute of Health (NIH) in the United States launched a national program called the All of Us Research Program (https://allofus.nih.gov/). The aim is to collect 1 million or more volunteers from those aged 18 and above with different types of health status and diverse backgrounds. To counter the diversity challenges in this field, they also aim to oversample underrepresented communities. Other initiatives include the H3Africa consortium (Human Heredity and Health in Africa, https://h3africa.org/) led by African scientists and includes 48 African projects of population-based genomic studies to build capacity and strengthen African-based research.

7.3.2 Sources that archive and distribute data

As we noted earlier in chapter 4, the majority of genomic data used in GWASs comes from samples in the United States and the United Kingdom and thus follow the data protocols in those countries. The most prominent large archive is the U.S.-based *dbGaP—Database of Genotypes and Phenotypes* (https://www.ncbi.nlm.nih.gov/gap), which curates and distributes genetic data. At the dbGaP website you can find information about how to access dbGaP data, resources that are available, and additional links. It contains additional data well beyond the auspices of this book as well. To start, it is useful to follow the numerous demo videos and overviews that are available that describe the process of applying for the data, how to access individual-level data, and set up separate accounts such as the eRA account. It can be somewhat confusing at first, but the information on the aforementioned website has detailed frequently asked questions (FAQs) and tutorials. Here you can also find more detailed information on downloading, decrypting, and extracting the data. Since this is a highly detailed process that is well documented and regularly updated at the aforementioned website, we do not repeat it here.

Currently, there is also a large amount of genomic data in the United Kingdom that researchers from around the world may use. Many U.K. longitudinal studies contain phenotypic, genotypic, and "omic" data and are funded by national research councils and thus required by the open science movement to allow access. Many, but not all, fall under the governance infrastructure called *METADAC: Managing Ethico-social, Technical and Administrative issues in Data Access*, which is governed by an interdisciplinary committee. The process of data sharing and governance is documented elsewhere [16], and readers can refer to the METADAC website (https://www.metadac.ac.uk/). As of early 2020, the largest openly available genetic data is the UK Biobank [17], which has its own data access process (https://www.ukbiobank.ac.uk/), generally accompanied with a payment to cover the processing of the data.

There are many additional datasets not mentioned here, and a full list of around 3,000 different datasets can be found on the GitHub site link [11]. Some publicly available U.S.-based genomic data that you can access includes prominent data such as the Health and Retirement Study (HRS, http://hrsonline.isr.umich.edu/; see box 7.1) [1], Add Health: National Longitudinal Study of Adolescent to Adult Health [18], the Wisconsin Longitudinal Study (WLS, https://www.ssc.wisc.edu/wlsresearch/) [19], or others such as the LifeLines Biobank in The Netherlands (https://www.lifelines.nl/). Data are also often distributed and released by local committees or consortiums that have collected the data directly via their own web-based application. As noted in the GWAS review article [11], the principal investigators responsible for obtaining the funding and collecting this data are often listed as coauthors on GWASs and central to these consortiums. Since there are many and varied ways that data is released for external use, researchers should investigate this for each individual data source.

In order to obtain access to genetic data, researchers are generally required to submit a research plan through their local IRB (institutional review board or independent ethics committee) but also often a committee, group, or website that manages data access. In most cases you need to specify only a small number of variables directly related to your research question with each application for a separate article or project. Many researchers in this area also increasingly encounter the question from journal editors to share their data and code. Virtually all genetic data adheres to national policies on data sharing, but due to the sensitive nature of genetic data access is very limited and strictly managed. It is possible to provide the code used to construct your variables and data analysis and then refer the journal editor to the data access protocol for the respective dataset. We now examine the actual genetic data that you will encounter and work with in the applied chapters in this book.

7.3.3 Obtaining GWAS summary statistics

GWAS consortia regularly publish their entire list of results, and it is virtually always a requirement of the journal where it is published. This allows other researchers to investigate the role of particular variants or to use them to construct polygenic scores in an independent genotyped sample. Summary results should include, at least, a list of SNPs with their rs number, chromosome number, genomic position, alleles, betas from the association analysis, or alternatively the Z-scores, and p-values of the association results. Often other summary statistics include additional information such as average allele frequency and heterogeneity statistics if the results are derived by a meta-analysis.

The NHGRI-EBI GWAS Catalog, which we described in chapter 4 contains some but not all of the many of the cataloged GWASs (see https://www.ebi.ac.uk/gwas/summary -statistics). The NHGRI-EBI GWAS Catalog provides a consistent, searchable and freely available database of published SNP-trait associations. It can be used to search for a particular trait or to examine the association between a genetic variant and possible traits. There you can also find links to some of the large consortiums which store their summary statistics on their own webpages: https://www.ebi.ac.uk/gwas/downloads/summary-statistics. As we have noted elsewhere, although it is now virtually always a publication requirement to release summary statistics, a smaller number of groups do not release them or only do so in exchange for authorship [11].

Web-based repositories have also been created that contain detailed information on thousands of publicly available GWASs. A recent research initiative led by Danielle Post-huma in the Netherlands created the *Atlas of GWAS Summary Statistics* (http://atlas.ctglab .nl) [20], containing, at the time of writing this book, summary statistics from over 4,000 different studies. Using this website, it is possible to select your phenotype of interest and download the entire list of association results. Data are harmonized, so it is possible to directly compare results from different studies. Moreover, they included their own GWAS results for many traits calculated from the UK Biobank.

The research group led by Benjamin Neale in the United States also created a database of GWAS statistics for 4,203 phenotypes available in the UK Biobank. Instead of using GWAS results from the GWAS produced by other groups, they ran a new GWAS on all (or virtually all) of the phenotypes available in the UK Biobank. This makes it arguably the largest resource for genomic data available to researchers. The advantage is that the GWAS analysis is conducted consistently for all the phenotypes after careful quality control (QC) procedures. The analyses also include 20 principal components and covariates (e.g., age, age2, sex, age*sex) (http://www.nealelab.is/uk-biobank/). They also generated sex-specific results and included all of the code that they used to run their analyses on GitHub (https://github.com/Nealelab/UK_Biobank_GWAS).

7.4   Different formats in genomics data

7.4.1   Genomics data is big data

Genomics data may initially appear to be rather unusual to those accustomed to working with epidemiological or social science data. Most readers will be familiar with the rectangular data structure, in which data are stored in a single file. In these types of files, each row typically contains the information of a single participant with each column providing information about a statistical variable (e.g., sex, age, disease status). This is widely used in statistical programs such as SPSS, Stata, or SAS. The dimension of this rectangular structure is $N \times K$, where N is the number of observations and K is the number of variables. For example, if we simulate a rectangular file in R as below and then examine it, you will see that the first column is an "id" (identification) variable of person 1 to 4. The second column is a binary covariate "sex" with the values of 1 and 2, with two additional variables, t1 and s1.

```
# simulate rectangular file
> rectangular <-data.frame (id=factor(1:4), sex =
c(1,2,1,2),
    t1 = c(19, 21, 82, 68), s1 = c(1,1,1,1))

> rectangular # examine data
      id    sex    t1    s1
1     1     1      19    1
2     2     2      21    1
3     3     1      82    1
4     4     2      68    1
```

Genomic data differ from some of the data many researchers might be familiar with due to the main distinction that typically we have more variables than observations. The widely used HRS genomic data, for instance, provides information on 22 million variables from around 20,000 individuals (see box 7.1). The number of variables is thus substantially higher than the number of observations, making the rectangular structure challenging to visually inspect. Excel, for instance, has a limit of 1,048,576 rows by 16,384 columns. Another limitation is the sheer file size of genomic data. Depending on whether the data contain only genotyped or also imputed data, the size of the genome-wide data files can be extremely large, usually on the order of several gigabytes or in some cases even terabytes. Genomic data is thus truly big data. In 2015, researchers warned that within the decade—when between 100 million to 2 billion human genomes are likely to be sequenced—data storage demands will far outstrip YouTube and Twitter's projected annual storage [21].

One way to think of genetic data is in terms of observations and variables, with the variable representing the genotype of a particular SNP. For example, if SNP rs9930506[2] has two variants, T and C, we can observe the genotypes (that can be TT; TC or CC) of all the respondents and store this information in a file. However, we also need to store the information of the SNP itself and map it on the human genome. rs9930506 is, in fact, a SNP that can be found in chromosome 16 at position 53796553 according to a predefined set of coordinates [Genome Reference Consortium Human Build 38 patch release 7]. This information needs to be stored in the data at the same time with the sample genotypes. In addition, you generally also often want to have additional information on the respondents, including sex, phenotype, and family relatedness with other respondents. Genetics existed long before DNA, and molecular biology and was based on the study of inheritance. Family relatedness therefore has historically been very relevant in genetic studies where the pedigree of families with hereditary disorders were used to study genetic transmission of diseases.

7.4.2   PLINK software and genotype formats

In 2007, software designed by Shaun Purcell and colleagues at the Broad Institute of Harvard and Massachusetts Institute of Technology in Boston was released. It soon became one of the most popular software applications to handle the growing mass of genetic data and to perform associations between (genome-wide) genotypes and phenotypes. The software is called PLINK and is updated often. In this book we use PLINK 1.9 and 2.0 (see appendix 1). PLINK can be used to handle genomic files, calculate statistics, and transform the data into different formats. We use both versions since at the time of writing this book, the 2.0 version of the software was still under development and some analyses are only available on PLINK 1.9. Version 1.9 performs analysis on genotyped data only, while version 2.0 can also be used with imputed data (see box 7.2). We will first start by describing the data structure for genotyped data and then extend the discussion to the PLINK 2.0 format.

Figure 7.3 provides an overview of the various types of files that you will most commonly use in PLINK, described in more detail in the following sections. Since text files are very time consuming to read, it is better to use binary files. The files that you will likely use most often are divided into three basic types, grouped in figure 7.3. The two

Text PLINK files

*.ped

```
FID     ID       F M S P -GENETIC INFO-
CH18526 NA18526  0 0 2 1 G G C C T T A A
CH18524 NA18524  0 0 1 1 G G C C T T A A
CH18529 NA18529  0 0 2 1 C G C C T T C A
CH18558 NA18558  0 0 1 1 G G C C G T A A
CH18532 NA18532  0 0 2 1 G G C C T T A A
```

*.map

```
Chr  SNP         SNP       Base-Pair
                 Position  Coordinate
8    rs17121574  12.7991   12799052
8    rs754238    12.8481   12840056
8    rs11203962  12.8484   12848438
8    rs6999231   12.8623   12862253
8    rs17178729  12.867    12867001
```

Binary PLINK files

*.bim

```
Chr  SNP         SNP       Base-Pair  Allele1 Allele2
                 Position  Coordinate
8    rs17121574  12.7991   12799052   G       G
8    rs754238    12.8481   12840056   G       G
8    rs11203962  12.8484   12848438   C       G
8    rs6999231   12.8623   12862253   G       G
8    rs17178729  12.867    12867001   G       G
```

*.fam

```
FID     ID       F M S P
CH18526 NA18526  0 0 2 1
CH18524 NA18524  0 0 1 1
CH18529 NA18529  0 0 2 1
CH18558 NA18558  0 0 1 1
CH18532 NA18532  0 0 2 1
```

*.bed

Binary version
of the SNP info
of the *.ped
file which is
only readable by
your computer

Covariates

```
FID     ID       Sex  Cohort  PC1      PC2       etc...
CH18526 NA18526  2    1       0.00542  -0.00876
CH18524 NA18524  1    1       0.04517  -0.00761
CH18529 NA18529  2    4       0.07776  -0.00231
CH18558 NA18558  1    3       0.00125  -0.00356
CH18532 NA18532  2    2       0.00456  -0.00651
```

Figure 7.3

Commonly used files in PLINK.

Note: FID = family identifier; ID = individual identifier; F = father ID; M = mother ID; S = sex of individual; P = phenotype outcome; Chr = chromosome; PC = principal component.

There are no headers in the file, and these are added here only for ease of interpretation.

text-format PLINK files contain information on individuals and their genotypes (.ped) and genetic markers (.map). The three binary PLINK files used most often are those that hold information on individual identifiers and their genotypes (.bed) and two readable binary text files that have material on individuals (.fam) and genetic markers (.bim). As we show in later chapters, you generally also include covariates, which necessitate a fourth set of files. For example, if you wanted to study type 2 diabetes, the .bed file contains the genotyped results of all individuals (e.g., if using a case-control study all patients and healthy controls). The .fam file contains the individual-related data (e.g., family interrelatedness with other individuals in the data, sex, type 2 diabetes diagnoses). The .bim file allows you to add the information about the actual physical position of the SNPs, and additional covariates could be included in the last file.

As shown in figure 7.3, the original PLINK 1.0 text format of genomic data is composed of a set of two files. The first file is the so-called *pedigree file*. A pedigree file, which in PLINK uses the suffix .ped, contains information on the sample (i.e., the list of individuals genotyped). Each line corresponds to an individual, and the first six columns provide information on this individual. In reality, the file does not contain header or variable names, but we have shown them here for ease of interpretation. The first two columns consist of a family identifier (FID) and an individual unique identifier (ID). Next to these we have information on the father (F) and the mother (M) identifier, which can be used to reconstruct the family pedigree. This information is not always present and very often only the information that is the unique individual identifier. Columns five and six contain information on the sex (S) and the phenotype (P) of interest. The remainder of the columns contain the genetic information. Each SNP consists of two columns indicating the individual genotype. For instance, in the example below, the genotype of the first individual (id NA18526) has GG as the first SNP, while the genotype of the third individual (id NA18529) is CG. A *.ped file* therefore has a large number of columns, exactly $6 + (K \times 2)$, where K is the number of SNPs genotyped. A .ped file can be opened in any text editor, although its dimension and the large number of columns may make reading it difficult.

Example.ped

```
COLUMN NUMBERS*
1    2      3 4 5 6 ————————GENETIC INFORMATION————————————

LABELS*
FID  ID     F M S P ————————GENETIC INFORMATION————————————

CH18526 NA18526 0 0 2 1 G G  C C  T T  A A  G G  G T  A G  G
T G  C C  T T  T T  C A  C C  A C  G G  C C
CH18524 NA18524 0 0 1 1 G G  C C  T T  A A  G G  A G  A A  G A
G G  C C  T T  C T  A A  C C  C C  G G  C C
```

```
CH18529 NA18529 0 0 2 1 C G C C T T C A G G G G T A G G
T G C C T T C T A A C C A C A G C C
CH18558 NA18558 0 0 1 1 G G C C G T A A G G G G A A G A
T G C C T T T T A A C C C C G G C C
CH18532 NA18532 0 0 2 1 G G C C T T A A G G G G A A G A
G G C C T T T T C C C C A C G G C C
CH18561 NA18561 0 0 1 1 G G G C G T C A G G A G T A G G
T G C C T T T T C A C C A C G G C C
CH18562 NA18562 0 0 1 1 G G C C T T A A G G G G A A G A
G G C C T T T T A A C C C C G G C C
CH18537 NA18537 0 0 2 2 G G G C G T C A G G G G A A A A
G G C C T T T T C C C C A A A G A C
CH18603 NA18603 0 0 1 2 G G C C T T A A G G G G T A G A
T G C C T T T T C C C C A A G G C C
CH18540 NA18540 0 0 2 1 G G C C T T A A G G G G T A G G
T G C C T T T T C A C C A C G G C C
```

Note: There is no header in the file, and these are added here for ease of interpretation.

A .ped file must be accompanied by a *.map file* in order to provide complete information on the genotype of a sample of individuals. A .map file provides information on which SNPs have been genotyped and how to locate them in the genome. The first column indicates the chromosome (Chr) number, the second is the SNP identifier (typically the rs number), while the third and fourth columns indicate the position of the SNP. The third, which is measured in centimorgans, is a measure of genetic distance based on recombination probability and therefore is not constant across the genome. One centimorgan equals a 1% chance that a marker at one genetic locus on a chromosome will be separated from a marker at a second locus due to crossing over in a single generation. The fourth column measures the base-pair coordinates or the genetic distance in base pairs, i.e., the number of molecules (letters) between variants. One centimorgan corresponds to around 1 million base pairs in humans on average. The centimorgan totals per chromosome are based on the Human Reference Genome. It is important to note that the location of SNPs can change based on the reference panel that is used. With the advancements in mapping the human genome, different releases of the Human Reference Genome have been published (see the "Further reading and resources" section). A reference genome is our navigation system in the exploration of the human genome. It can be used to map the location of SNPs across human DNA, but it needs to be updated to the most recent version. The current version of the Human Reference Genome is called GRCh38 and has been released by the National Center for Biotechnology Information.[3] The .map file has a dimension of K rows (number of SNPs) and 4 columns.

example.map

LABELS*			
Chr	SNP	SNP Position	Base-Pair Coordinate
8	rs17121574	12.7991	12799052
8	rs754238	12.8481	12848056
8	rs11203962	12.8484	12848438
8	rs6999231	12.8623	12862253
8	rs17178729	12.867	12867001
8	rs10105623	12.8683	12868315
8	rs2460915	12.8704	12870407
8	rs7835221	12.8781	12878098
8	rs2460911	12.8953	12895289
8	rs12156420	12.9146	12914557
8	rs17786052	12.9224	12922389
8	rs529983	12.9426	12942555
8	rs630969	12.9458	12945844
8	rs2460914	12.9581	12958068
8	rs607499	12.9619	12961886
8	rs634228	12.9633	12963283
8	rs556531	12.9893	12989321

*Note: There is no header in the file, and these are added here for ease of interpretation.

The combination of .ped and .map files can be used to describe the genotype of a sample of individuals.

7.4.3   PLINK binary files

PLINK .map and .ped files can be opened in a normal text editor but are inefficient in terms of data storage. As we noted earlier, a common way to store genetic data is in what is called binary files. In particular, a PLINK binary file compresses the genotype information included in the .ped file.

As illustrated in figure 7.3, a set of PLINK binary files consists of three files:

1.   A .bed file, which is not readable in a text editor and contains the information on the genotype in a compressed way.

2.   A .fam file indicating the information about individuals (equivalent to the first six columns of a .ped file, shown in figure 7.3).

3.   A .bim file indicating the information about the SNPs (virtually equivalent to a .map file, but with Allele1 and Allele2 columns; see figure 7.3).

As we will demonstrate in the chapters that follow, passing from a binary file to a .map/.ped PLINK file is straightforward.

7.5   Genetic formats for imputed data

The PLINK format described above is simple and straightforward but can only be used for genotyped data and is not suitable for *imputed genomic data* (see box 7.2). Imputation combines the information on genotyped SNPs to the information on the LD from a reference panel (for example, the 1000 Genomes Project or the Haplotype Reference Consortium). Although imputation can be very precise, it does come with some uncertainty. Imputed genotypes are associated with measures of imputation probabilities (sometimes called genotype calls) that give you an indication of how likely a certain genotype is based on the information from a reference panel. For instance, based on imputation, we can derive the genotype probabilities of the three possible genotypes of SNP rs2777888 as 28% CC, 52% TC, and 21% TT. A possible solution is to ignore the imputation uncertainty and opt for the most likely genotype (i.e., the genotype with the highest probability). This is a legitimate choice and often used in the calculation of polygenic scores (see chapter 10). Another alternative would be to use software that can handle imputation probabilities, which we turn to now.

7.5.1   PLINK 2.0

In 2018, the authors of PLINK released a new version of their software that is better suited to work with imputed data. The software, called PLINK 2.0, has a new data format for genetic data that updates the binary format of PLINK 1.0.

A set of PLINK 2.0 binary files is composed of three files:

1. A .pvar file indicating the information about the genetic markers (similar to a .bim file)

2. A .psam file indicating the information about the individuals in the sample (similar to a .fam sample file)

3. A .pgen file (nonreadable in a text editor) containing the information on the genotype probabilities in a compressed way.

PLINK 2.0 fields contain more information on the genetic variants included in the .psam file. In addition to the alleles (reference and alternative), additional information on the imputation quality (columns QUAL and INFO) is also provided. The .pgen file is a compressed binary file that cannot be read in a text editor but contains information on how likely it is that a genetic variant has a particular genotype. This information on imputation qualities and genotype probabilities are used extensively in quality control procedures and in GWASs to assess whether the results are driven by true association effects or by data anomalies.

Columns in a .pvar PLINK 2.0 file

1. IID (individual ID; required)
2. SID (source ID, when there are multiple samples for the same individual)
3. PAT (individual ID of father, "0" if unknown)
4. MAT (individual ID of mother, "0" if unknown)
5. SEX ("1" = male, "2" = female, "NA"/"0" = unknown)

Columns in a .psam file

1. POS (base-pair coordinate)
2. ID (variant ID; required)
3. REF (reference allele)
4. ALT (alternate alleles, comma-separated)
5. QUAL (phred-scaled quality score for whether the locus is variable at all)
6. FILTER ("PASS," "." or semicolon-separated list of failing filter codes)
7. INFO (semicolon-separated list of flags and key-value pairs, with types declared in header)
8. FORMAT (terminates header line parsing)
9. CM (centimorgan position)

7.5.2 Oxford file formats

Another popular format for genomic data is the format that is used by the collection of software designed by the statistical genomic group of the Department of Statistics from Oxford University and the Wellcome Trust Centre of Human Genetics, also based in Oxford. The format is sometimes referred to as Oxford file formats, and it is used in the software GTOOL and SNPTEST. Genomic data are stored in two files (similarly to the .ped and .map PLINK files): a genotype file and a sample file. The *genotype file* contains information on the genotype data on a one-line-per-SNP format, while columns represent individuals. This is exactly the opposite of a PLINK .ped file where rows indicate individuals and columns genotypes. An example of the Oxford genotype file format is shown below. The first five columns of each file contain information on the SNP identifier, the base-pair position of the SNP, the allele coded A, and the allele coded B. The next three numbers on the line give the probabilities of the three genotypes AA, AB, and BB at the SNP for the *first* individual in the cohort. The next three numbers should be the genotype probabilities for the *second* individual in the cohort. The next three numbers are for the *third* individual, and so on. All the probabilities *need to sum to 1*. If a SNP has been directly measured, then its genotype probabilities are (1, 0, 0).

In the example above, we report the genotypes of five hypothetical SNPs for two individuals. The first individual has the following genotypes (AA; GG; CC; CT; AG), while the second has the following (AA; GT; CT; CT; GG). The example reports a precise measure of the genotypes (without decimal points) indicating that these SNPs are directly measured, but this format can accommodate genotype likelihood, giving an indication on how likely it is to have a particular genotype. The dimension of a genotype file depends on the number of subjects N and the number of SNPs K because the file has K rows and $(N \times 3) + 5$ columns.

Example of Oxford genotype file:

SNPID	rs#	Base position	AlleleA	AlleleB	ProbInd1	ProbInd1
		pair				
SNP1	rs1	1000	A	C	1 0 0	1 0 0
SNP2	rs2	2000	G	T	1 0 0	0 1 0
SNP3	rs3	3000	C	T	1 0 0	0 1 0
SNP4	rs4	4000	C	T	0 1 0	0 1 0
SNP5	rs5	5000	A	G	0 1 0	0 0 1

Note: There is no header in the file and these are added here for ease of interpretation

The *sample file* consists of three parts: a header row detailing the names of the columns in the file, a row detailing the types of variables stored in each column, and a row for each individual detailing the information for that individual. The second line of the file details the type of variables included in each column. The first three entries are zeros, while other columns take a D if the variable is discrete or C if the variable is continuous. Phenotypes that are continuous variables are indicated with a P or B if they are binary (case-controls studies).

Example of an Oxford sample file:

ID _ 1	ID _ 2	missing	cov _ 1	cov _ 2	cov _ 3	cov _ 4	pheno1	bin1
0	0	0	D	D	C	C	P	B
1	1	0.007	1	2	0.0019	-0.008	1.233	1
2	2	0.009	1	2	0.0022	-0.001	6.234	0
3	3	0.005	1	2	0.0025	0.0028	6.121	1
4	4	0.007	2	1	0.0017	-0.011	3.234	1
5	5	0.004	3	2	-0.012	0.0236	2.786	0

7.5.3   The variant call format (VCF)

An additional format for genomic data that is very common in bioinformatics is the variant call format (VCF). This format can store genomic information for genotyped, imputed data, and even sequencing data. It is very flexible, because various types of information can be stored. It can be read by a text editor (although we do not recommended it for large files) or from the command terminal. VCF files contain all of the genetic information in a single file. The file has a large preamble of meta-information lines (prefixed with a double ## symbol), a header line (prefixed with a single # symbol), and data lines each containing information about the position in the genome and genotype information on samples for each position. VCF is the format for large genomic projects and is the format that has been used for releasing data from the 1000 Genomes Project (see figure 7.1). VCF files can be used to store any type of genetic variants, including CNVs, indels, and multiallelic SNPs. It can be read using different software and transformed in your preferred format using software such as PLINK.

Example of a preamble of a VCF file:

```
##fileformat=VCFv4.3
##fileDate=20090805
##source=myImputationProgramV3.1
##reference=file:///seq/references/1000GenomesPilot-NCBI36.
fasta ##contig=<ID=20,length=62435964,assembly=B36,md5=f126
cdf8a6e0c7f379d618ff66beb2da,species="Homo sapiens",
taxonomy=x> ##phasing=partial
##INFO=<ID=NS,Number=1,Type=Integer,Description="Number of
Samples With Data"> ##INFO=<ID=DP,Number=1,Type=Integer,
Description="Total Depth"> ##INFO=<ID=AF,Number=A,Type=Float,
Description="Allele Frequency"> ##INFO=<ID=AA,Number=1,
Type=String,Description="Ancestral Allele"> ##INFO=<ID=DB,
Number=0,Type=Flag,Description="dbSNP membership, build
129"> ##INFO=<ID=H2,Number=0,Type=Flag,Description="Hap
Map2 membership"> ##FILTER=<ID=q10,Description="Quality
below 10"> ##FILTER=<ID=s50,Description="Less than 50% of
samples have data"> ##FORMAT=<ID=GT,Number=1,Type=String,
Description="Genotype"> ##FORMAT=<ID=GQ,Number=1,Type=
Integer,Description="Genotype Quality"> ##FORMAT=<ID=DP,
Number=1,Type=Integer,Description="Read Depth"> ##FORMAT=<I
D=HQ,Number=2,Type=Integer,Description="Haplotype Quality">
```

Typical information stored in a VCF file:

	Name	Brief description (see the specification for details)
1	CHROM	The name of the sequence (typically a chromosome) on which the variation is being called. This sequence is usually known as the *reference sequence* (i.e., the sequence against which the given sample varies).
2	POS	The 1-based position of the variation on the given sequence.
3	ID	The identifier of the variation (e.g., a dbSNP rs identifier or, if unknown, a "."). Multiple identifiers should be separated by semicolons without white space.
4	REF	The reference base (or bases in the case of an indel) at the given position on the given reference sequence.
5	ALT	The list of alternative alleles at this position.
6	QUAL	A quality score associated with the inference of the given alleles.
7	FILTER	A flag indicating which of a given set of filters the variation has passed.
8	INFO	An extensible list of key-value pairs (fields) describing the variation. See below for some common fields. Multiple fields are separated by semicolons with optional values in the format: "<key>=[,data]".
9	FORMAT	An (optional) extensible list of fields for describing the samples. See below for some common fields.
+	SAMPLEs	For each (optional) sample described in the file, values are given for the fields listed in FORMAT.

7.6 Data used in this book

Analyses in this book are mainly based on two datasets, which are described in more detail in appendix 2 at the end of this book and available on the companion website to this book: http://www.intro-statistical-genetics.com. At the start of each chapter we note the data that will be used so that readers can ensure they are able to actively follow all exercises. For the practical exercises in part II of this book, chapters 8, 9, and 10 use a combination of publicly available data that you can download and additional data that we have simulated for an individual phenotype for BMI (see appendix 2). The simulations were conducted based on publicly available GWAS results for BMI. Based on the genotypes of the individuals, the effect sizes from the GWAS results, and an arbitrary heritability level, these simulations produce standardized phenotypes that behave in analyses according to the parameters we used. We use GCTA software for the simulations and refer to appendix 2 for the QC of the genotype data, the simulation syntax, and further details. In part II of this book, analyses include molecular genetic data using software packages such as PLINK, GCTA, and PRSice. You will learn how to clean data, generate polygenic scores, and run some basic analyses.

In part III of book, and specifically chapter 11, we turn to more advanced applications such as causal modeling and regression analysis using polygenic scores. Here we provide several practical examples based on real findings from the research literature using the publicly available data from the Health and Retirement Study (HRS) (see box 7.1). Examples include genetic confounding in education across generations or G×E interaction between birth cohorts and BMI. We describe how to obtain the HRS data in appendix 2 of this book. We provide all details as well as the R code to download the data directly on your computer in chapter 11, which is also available on the website for this book. We use HRS data since the data already have readily available polygenic scores and can be downloaded freely. Polygenic scores are an aggregated measure and not as sensitive as individual molecular genetic information. With the knowledge you will gain from part II of this book, including how to work with data and particularly chapter 10 on the construction of polygenic scores, you could apply for primary access to the genotype data and construct your own scores. This can be very useful, since the score construction might depend on the research interests (see also chapter 10) or the available scores. Some, for example, might be outdated because a new GWAS on the phenotype of interest has been published very recently.

7.7   Data transfer, storage, size, and computing power

7.7.1   Data storage

If it has not already been developed within your research team or university, it is essential that you follow a strict protocol for data storage and sharing that also fits your local, national, or regional regulations. A useful guide, for instance, is the U.S.-based protocols from the NIH (National Institute of Health) on genomic data sharing (see the "Further reading and resources" section). These guidelines ensure the participant's anonymity and represent higher standards in genetic research.

Although each data provider often has specific requirements that are contractually binding, a basic protocol that researchers can develop should encompass the following points:

1.   All data is anonymized and respondents are exclusively identified by an ID code. No researchers should have access at any time to participant names or identifying information. This means that the following material is removed: names, all geographical subdivisions that are smaller than a state/province or municipality, all elements of dates (except years), telephone or fax numbers, e-mail addresses, any identifying Social Security or medical record numbers, IP address numbers, and biometric identifiers.

2.   Only digitalized data is data (i.e., nothing on hard copy/paper).

3.   Data is stored anonymously on a secure, password-protected institutional server at your institution or, in some cases if agreed on, a secured cloud platform.

4.   Data is regularly backed up on a password-protected hard drive and stored securely in a locked filing cabinet.

5. No other member of your organization, student, or external user will have permission to access the data. In some cases, some data providers will allow limited permission for a small sample of the data to be used as part of the teaching module.

A detailed checklist for database curators is the document by Ekong et al. (2018) in our "Further reading and resources" section. Additional protocols need to be followed when working with particular types or methods of analyzing genetic data. Data from GWAS consortia, for example, are often stored in secure repositories and only meta-analytical association results are shared among the authors of that paper while research is being conducted. Often then only information about SNP association results will be available (marker name, position, reference allele and other allele, GWAS betas, SE, HapMap, or 1000 Genome allele frequency). To ensure greater anonymity, in some cases the minor allele frequency is substituted by HapMap or 1000 Genome (or any other eventual reference panel) allele frequency.

7.7.2 Data sharing, transfer across borders, and cloud storage

There are striking differences in national regulations for data sharing, which vary across many different countries and include a patchwork of Institutional Review Board (IRB) positions and regulations. As we discuss in more detail in chapter 14, on ethics, all researchers (including students) need to have their projects approved by their local IRB before engaging in research.

Often inherent to this type of research is data sharing and the use of multiple data sources in order to replicate results. Although there are some more developed models of genomics data sharing in particular areas of research, such as the International Cancer Genome Consortium, this is not always the case for other areas. In fact, a recent evaluation of genomics data sharing across multiple countries reveals complexity, contradiction, and confusion [22]. Data transfer to third countries outside of China, for instance, is often either prohibited or very difficult due to overlapping or complex data regulations. The United States also has a fragmented data protection regime with oversight across IRBs and data access committees [23]. Europe's recently introduced General Data Protection Regulation (GDPR) in May 2018 brought new restrictions related to the transfer of data across borders, complicated by additional unique country—and institutional—specific interpretations [24]. There are various bodies working toward shaping the interpretation of GDPR's rules to their national and local contexts. An area that has received relatively less attention is regulatory protection and data sharing across borders in relation to cloud-based storage providers. Many researchers store or share their data on these cloud-based storage providers. However, cloud providers often shift data across geographical locations with limited notification or oversight, so it is important to seek legal and technical advice to avoid problems [25].

7.7.3   Size of data and computational power

As we noted earlier, genomic data is truly big data, and in the coming years it will only get bigger. Computational demands can be very, very high in statistical genetic analyses compared to what you may be accustomed to from standard analyses using phenotypic data from a few thousand individuals. Demands depend on the genetic data that are used (genotype, imputed data, sequence data), the number of individuals in the data, and the statistical model that is applied. Within this book we use a smaller stylized data file for many of our examples. A simple rule of thumb is that if you are working directly with genetic data that contains information from more than 1,000 people, your laptop and desktop computer likely will not be able to handle it. If this is the case, you will very likely need to work on a cluster computer. Most universities and research institutions have these types of clusters available. As we show in the next chapter, you will then need to acquire a few extra skills, such as how to work with the command line but also how to work with job submission and management systems. Since every system differs, we do not cover this extensively in this book. There are many online videos and tutorials to which readers can refer or courses at your own institution [26].

To give a general indication of how large data might get, consider one of the largest publicly available datasets at the moment: the UK Biobank. If you store the full UK Biobank, the not-imputed data as of 2018 was 92 gigabytes (GB) and the imputed data was 2.1 terabytes (TB). This then scales in a linear fashion with the number of SNPs and individuals [27]. It is hard to estimate processing time exactly, since it depends heavily on the analysis that you are conducting. To provide a general indication, running QC analysis (discussed in the next chapter) on the scale of big data such as the UK Biobank, it will likely take days or weeks to run. For association analyses, running a standard BOLT-LMM association (https://data.broadinstitute.org/alkesgroup/BOLT-LMM /) on the full UK Biobank will take around 100 GB of RAM and several days to run, if it is given eight processors to use. This BOLT-LMM algorithm computes statistics for testing associations between phenotypes and genotypes using a linear mixed model [28, 29].

Another example is the estimation of SNP-heritability using Genome-wide Complex Trait Analysis (GCTA) software (see chapter 1 for a definition and chapter 9 for an application). For this analysis, first a matrix of pairwise genetic relatedness needs to be calculated. It its original version, this required around 5 GB of memory, for a dataset including only around 300,000 SNPs and around 4,000 individuals (AMD Opteron 2.8 GHz) [30]. Methods are frequently adapted incorporating algorithms that are more computationally efficient. A heritability analysis using Haseman-Elston regression, for example, requires less than 2 GB memory for 120,000 individuals in GCTA. A genetic relatedness matrix we estimated for around 35,000 people had a size of over 30 GB [31].

7.8   Conclusion

As with many chapters in this book, we recognize that we were only able to expose the tip of the iceberg in relation to the multiple aspects of genomic data. Readers should have a basic grasp of where this data comes from and the related genotyping sequencing technologies, LD and imputation. In light of the rapid developments in this area, particularly in exome sequencing and updating of computer programs such as PLINK, we anticipate that many new developments will take place even while this textbook is being printed. Our hope is that you have an overview of where and how you can obtain this data, which we aim to update as much as possible on the webpage that accompanies this book. Since working with genomic data may be fundamentally different from some of the data that you may have worked with in the past, we also attempted to provide an overview of how the different binary files are linked. Using the sample data in this book, you will be able to gain a basic understanding of how to conduct these types of analyses. You can then expand and apply this knowledge using larger and more diverse genetic sources, keeping in mind the often demanding data storage and computational power requirements. The next chapter now builds on what we have learned to now actively use and work with these genomic files.

Exercises

1. How many columns and rows would a .ped file have for a sample of 1,000 individuals, with 500,000 genotype variants?

2. Download the hapmap-ceu set of PLINK binary files from the exercises of chapter 8. See http://zzz.bwh.harvard.edu/plink/res.shtml and explore the .fam and .bim files in your computer, and describe the content of each column.

3. If not already available at your institution, use the "Further reading and resources" section below to draw up the data storage, processing and sharing guidelines that you will use when working with genetic data in the chapters that follow. Ensure that your work is also approved by your local IRB.

Further reading and resources

Further reading on data, imputation and software

Bycroft, C. et al. The UK Biobank resources with deep phenotyping and genomic data. *Nature* **562**, 203–209 (2018).

Chang, C. C. et al. Second-generation PLINK: Rising to the challenge of larger and richer datasets. *Giga-Science* **4** (2015) (available at https://doi.org/10.1186/s13742-015-0047-8).

Das, S. et al. Next-generation genotype imputation service and methods. *Nat. Gen.* **48**(10), 1284–1287 (2016).

Hoffmann, Thomas J., and John S. Witte. Strategies for imputing and analyzing rare variants in association studies. *Trends Genet.* **31**(10), 556–563 (2015).

Nielsen, R., J. S. Paul, A. Albrechtsen, and Y. S. Song. Genotype and SNP calling from next-generation sequencing data. *Nat. Rev. Gen.* **12**(6), 445–451 (2011).

Reference panels

Haplotype Reference Consortium: http://www.haplotype-reference-consortium.org/.

HapMap: https://www.genome.gov/10001688/international-hapmap-project.

1000 Genomes Project: http://www.internationalgenome.org/.

Additional imputation software

BEAGLE 5.0: https://faculty.washington.edu/browning/beagle/beagle.html.

IMPUTE4: https://jmarchini.org/impute-4/.

MACH: http://csg.sph.umich.edu/abecasis/mach/tour/imputation.html.

Minimac4: https://genome.sph.umich.edu/wiki/Minimac4.

Imputation and phasing servers

These are free genotype and imputation and phasing servers where you can upload GWAS data (e.g., in VCF or 23andME format) and receive imputed and phased genomes:

Michigan Imputation Server: https://imputationserver.sph.umich.edu/index.html#!.

Sanger Imputation Server: https://imputation.sanger.ac.uk/.

Data storage and sharing guidelines

Data storage and sharing rules and regulations evolve with different legislation and across countries and for many also need to heed new European GDPR regulations. Some examples are:

Ekong, R. et al. *Ethical data management: Checklist for gene/disease specific database curators.* The Human Variome Project Working Group WG08, CC 4.0 (2018) (available at https://osp.od.nih.gov/wp-content/uploads/NIH_Best_Practices_for_Controlled-Access_Data_Subject_to_the_NIH_GDS_Policy.pdf).

NIH guidelines: https://osp.od.nih.gov/wp-content/uploads/NIH_Best_Practices_for_Controlled-Access_Data_Subject_to_the_NIH_GDS_Policy.pdf.

Shabani, M., and P. Borry. Rules for processing genetic data for research purposes in view of the new EU General Data Protection Regulation. *Eur. J. of Hum. Gen.* **26**, 149–156 (2018).

References

1. A. Sonnega et al., Cohort profile: The Health and Retirement Study (HRS). *Int. J. Epidemiol.* **43**, 576–585 (2014).
2. K. A. Frazer et al., A sequence-based variation map of 8.27 million SNPs in inbred mouse strains. *Nature* **448**, 1050–1053 (2007).
3. R. A. Gibbs et al., A global reference for human genetic variation. *Nature* **526**, 68–74 (2015).
4. T. K. Oleksyk, V. Brukhin, and S. J. O'Brien, The Genome Russia project: Closing the largest remaining omission on the world genome map. *Gigascience* **4**, 53 (2015).
5. Z. Li et al., A partition-ligation-combination-subdivision EM algorithm for haplotype inference with multiallelic markers: Update of the SHEsis (http://analysis.bio-x.cn). *Cell Res.* **19**, 519–523 (2009).
6. O. Harismendy et al., Evaluation of next generation sequencing platforms for population targeted sequencing studies. *Gen. Biol.* **10**, R32 (2009).

7.   K. A. Wetterstrand, DNA sequencing costs: Data from the NHGRI Genome Sequencing Program (GSP) (2019) (available at https://www.genome.gov/sequencingcostsdata/).

8.   E. R. Mardis, A decade's perspective on DNA sequencing technology. *Nature* **470**, 198–203 (2011).

9.   M. McCarthy and A. Mahajan, Genome-wide association analyses in type 2 diabetes: The gift that keeps on giving. (2018) (available at http://mccarthy.well.ox.ac.uk/2018/10/gwas-gift-keeps-giving/).

10. A. Mahajan et al., Fine-mapping type 2 diabetes loci to single-variant resolution using high-density imputation and islet-specific epigenome maps. *Nat. Genet.* **50**, 1505–1513 (2018).

11. M. C. Mills and C. Rahal, A scientometric review of genome-wide association studies. *Commun. Biol.* **2** (2019), doi:10.1038/s42003-018-0261-x.

12. A. Fry et al., Comparison of sociodemographic and health-related characteristics of UK Biobank participants with those of the general population. *Am. J. Epidemiol.* **186**, 1026–1034 (2017).

13. K. Servick, Can 23andMe have it all? *Science* **349**, 1472–1477 (2015).

14. D. M. Werling, N. N. Parikshak, and D. H. Geschwind, Gene expression in human brain implicates sexually dimorphic pathways in autism spectrum disorders. *Nat. Commun.* **7**, 10717 (2016).

15. R. M. Verweij et al., Sexual dimorphism in the genetic influence on human childlessness. *Eur. J. Hum. Genet.* **25**, 1067–1074 (2017).

16. M. J. Murtagh et al., Better governance, better access: Practising responsible data sharing in the META-DAC governance infrastructure. *Hum. Gen.* **12**, 24 (2018).

17. C. Bycroft et al., The UK Biobank resource with deep phenotyping and genomic data. *Nature* **562**, 203–209 (2018).

18. K. M. Harris et al., The National Longitudinal Study of Adolescent to Adult Health: Research design (2009) (available at http://www.cpc.unc.edu/projects/addhealth/design).

19. P. Herd, D. Carr, and C. Roan, Cohort profile: Wisconsin longitudinal study (WLS). *Int. J. Epidemiol.* **43**, 34–41 (2014).

20. K. Watanabe et al., A global view of pleiotropy and genetic architecture in complex traits. *bioRxiv* (2019), doi:10.1101/500090.

21. E. Check Hayden, Genome researchers raise alarm over big data. *Nature* (2015), doi:10.1038/nature.2015.17912.

22. B. M. Knoppers and Y. Joly, Introduction: The why and whither of genomic data sharing. *Hum. Genet.* **137**, 569–574 (2018).

23. M. A. Majumder, United States: Law and policy concerning transfer of genomic data to third countries. *Hum. Genet.* **137**, 647–655 (2018).

24. M. Phillips, International data-sharing norms: From the OECD to the General Data Protection Regulation (GDPR). *Hum. Genet.* **137**, 575–582 (2018).

25. E. S. Dove, Y. Joly, A.-M. Tassé, and B. M. Knoppers, Genomic cloud computing: Legal and ethical points to consider. *Eur. J. Hum. Genet.* **23**, 1271–1278 (2015).

26. J. Yang et al., Concepts, estimation and interpretation of SNP-based heritability. *Nat. Genet.* **49**, 1304–1310 (2017).

27. UK Biobank, UK Biobank: Genotyping and Imputation Data Release, March 2018 (2018) (available at http://www.ukbiobank.ac.uk/wp-content/uploads/2018/03/UKB-Genotyping-and-Imputation-Data-Release-FAQ-v3-2.pdf).

28. P.-R. Loh et al., Mixed-model association for biobank-scale datasets. *Nat. Genet.* **50**, 906–908 (2018).

29. P.-R. Loh et al., Efficient Bayesian mixed-model analysis increases association power in large cohorts. *Nat. Genet.* **47**, 284–290 (2015).

30. J. Yang, S. H. Lee, M. E. Goddard, and P. M. Visscher, GCTA: A tool for genome-wide complex trait analysis. *Am. J. Hum. Genet.* **88**, 76–82 (2011).

31. F. C. Tropf et al., Hidden heritability due to heterogeneity across seven populations. *Nat. Hum. Behav.* **1**, 757–765 (2017).

8

Working with Genetic Data, Part I

Data Management, Descriptive Statistics, and Quality Control

Objectives

- Understand how to *use the command line*
- Open and work with *PLINK binary files*
- *Recode PLINK files* into other formats
- Understand the basics of *data management* to select information on particular markers or a subsample of individuals
- Derive information on *allele frequencies*, *phenotypes*, and *missing values*
- *Merge* different genetic files
- *Associate a phenotype* to a PLINK file
- Understand and perform *quality control* procedures at the individual, marker, and genome-wide association studies level

8.1   Introduction: Working with genetic data

The previous chapter introduced readers to the different types of genomic data. The aim of this chapter is to provide a gentle introduction of how to work with genetic data for those who are unaccustomed to working in a command line environment and have never used the computer program PLINK. We describe the command line and PLINK in more detail in the next section. After we outline the basics of using PLINK, such as calling PLINK, opening files, and importing data, we describe basic data management. This includes selecting individuals and markers and merging different genetic files. We then illustrate how to produce descriptive statistics including allele frequency, phenotypes, and missing values. Finally, we outline quality control (QC) of genetic data by individual and markers and QC for genome-wide association studies (GWASs), followed by a brief summary. Readers can choose to actively follow the exercises in this chapter on their own computer, and we provide an indication of how to do this in the next section.

8.2 Getting started with PLINK

8.2.1 The command line

As we described chapter 7, one of the most popular open-source free software programs for QC and GWAS analyses is PLINK [1, 2]. It was developed by Shaun Purcell and colleagues and facilitates multiple types of data handling and usage. Detailed instructions on how to install PLINK are available in appendix 1 of this book. There are also many online tutorials (http://zzz.bwh.harvard.edu/plink/tutorial.shtml), instructions, and forms of documentation for PLINK. This chapter provides a brief introduction to the basics and main commands that you will need; it is by no means exhaustive. We highly recommend that you refer directly to the ample online documentation and material that is available. Depending on your operating system (i.e., Windows, macOS, Linux/Unix), some commands and processes might differ. In this book we use examples for the macOS and Linux systems. Windows users can find more detail in appendix 1 and box 8.1. For readers who would like to actively follow the tutorial in this chapter, we advise that you to first engage in these four steps:

1. *Create a new directory* where you will store the data and PLINK. It could be, for example, /User/YourName/Chapter8 or E:\User\YourName\Chapter8 (see box 8.1).

2. *Install PLINK* on your computer (see http://www.cog-genomics.org/plink2/ and appendix 1).

3. *Download the example data* from the companion website of this book (http://www.intro-statistical-genetics.com).

   These are:

   - ALL.chr21.vcf.gz
   - BMI_pheno.txt
   - 1kg_EU_qc.bim, 1kg_EU_qc.bed, 1kg_EU_qc.fam
   - hapmap-ceu.bim, hapmap-ceu.bed, hapmap-ceu.fam
   - list.txt
   - 1kg_hm3.bim, 1kg_hm3.bed, 1kg_hm3.fam
   - individuals_failQC.txt
   - hello_world.sh

   The hapmap-ceua.zip files can be found at http://zzz.bwh.harvard.edu/plink/dist/hapmap-ceu.zip.

4. *Unzip the downloaded file(s)* into the new folder that you created. As with all zipped files, when you double-click, it will unzip in a folder called hapmap1. Within this folder you should see three files: hapmap-ceu.bed, hapmap-ceu.bim, and hapmap-ceu.fam.

Once PLINK is installed on your computer, you can execute all commands that we show in this book by typing them in the command line. The *command line* is an interface for typing commands directly to a computer's operating system. We need to use the command line because the usage of PLINK requires an active shell that waits for commands. One way to think of this is that the command line is a way of interacting with the computer program. You type a command in each of your successive lines of text (i.e., the command lines) and the *shell* serves as the command language interpreter or the program that handles your lines. The shell is like a language interpreter that accepts the text command and then converts it into the appropriate operating system functions. Those using Unix or familiar with the MS-DOS or Apple DOS interfaces of the 1970s and 1980s will already be accustomed to working in this type of environment. Due to the introduction of many graphical user interfaces (GUIs) with point-and-click, menu-driven actions, the fine art of the command line interfaces may have been lost for some. Box 8.1 provides a comparison of some of the command line differences between the operating systems.

Box 8.1

Differences between the command line in Windows, Mac OS Terminal, and Linux Shell

Most of the software presented in this book use the command line and thus no graphical user interface. There are some differences, however, in how the commands are executed between the different operating systems. Below is a list of some the differences between Windows, Mac OS, and Linux. Since both Mac OS and Linux are based on Unix, there are no differences in most of the commands between these two operating systems.

	Windows command line (CMD)	Mac OS terminal	Linux shell
Directory path	..\dir1\dir2\	../dir1/dir2/	../dir1/dir2/
List files and folder	dir	ls	ls
Call current location	dir	pwd	pwd
Move to directory	cd "path to the folder"	cd "path to the folder"	cd "path to the folder"
Get back to parent directory	cd ..	cd ..	cd ..
Get to the root directory	cd	cd /	cd /
Create a new directory	mkdir NewFolder	mkdir NewFolder	mkdir NewFolder
Remove directory	rmdir MyFolder	rm -r MyFolder	rm -r MyFolder
Rename directory	rmdir	mv oldName newName	mv oldName newName
Delete a file	del filename	rm fileName	rm fileName
Line break for commands	^	\	\
Execute plink	plink.exe	./plink	./plink
Execute GCTA	gcta.exe	./gcta	./gcta
Execute PRSice	Rscript PRSice.R or PRSice_win64.exe or PRSice_win32.exe	Rscript PRSice.R or ./PRSice_mac	Rscript PRSice.R or ./PRSice_linux

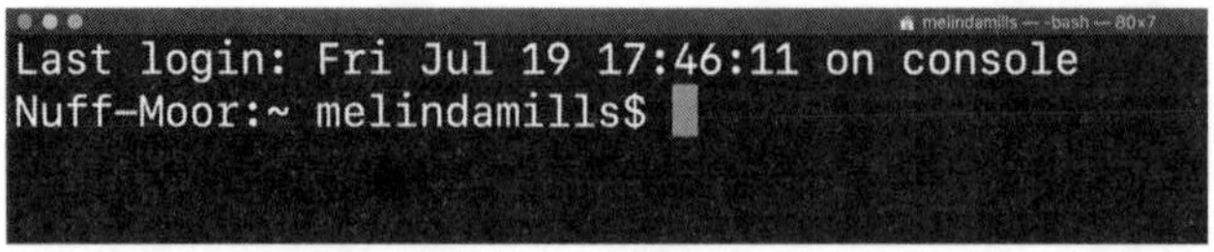

```
Last login: Fri Jul 19 17:46:11 on console
Nuff-Moor:~ melindamills$
```

Figure 8.1
The command line in macOS terminal window.

We discuss the command line for Windows users in more detail in appendix 1. Briefly, Windows 10 users can download and install Ubuntu, which will effectively allow them to carry out these commands as if in a Linux operating system. For *Mac users, to access the terminal or Unix command prompt*, you first need to open the *Utilities* folder. To do this click on the *Finder Icon* that is located at the bottom dock of your screen. Find *Applications* on the left panel of the Finder window then scroll down within the Applications window until you locate *Utilities* and then click it in order to open. On the right-hand panel you will find *Terminal*. When you double-click it, it should look like the image in figure 8.1.

For those following the practice session while reading this chapter, we suggest that you have the command line prompt ready now. The command line starts with a prompt just before the cursor ($ or >). In most cases the prompt of the current directory is displayed before the prompt. As with R, the current directory is important when using PLINK. By default, PLINK will load and save data and results files into that directory. To change the directory, use: cd. We outline the cardinal rules of PLINK in box 8.2.

8.2.2   Calling PLINK and the PLINK command line

As shown in figure 8.2, PLINK commands are composed of several arguments. Although the order of the arguments is arbitrary, a typical command begins by *calling PLINK*. Before doing this, however, you need to *ensure that you are in the correct directory* of where you stored PLINK. PLINK will be installed in the directory that you specified when you installed it (see appendix 1 and box 8.1). To find the current directory that you are using, type: pwd (i.e., print working directory). If PLINK is not in this directory, you will need to type the directory line of where you stored PLINK in the command line. This could be, for example: /usr/local/bin/plink. If you need to change directories, use the cd command, for example: cd /Users/yourname/plink.

In these examples we show running the commands after the cursor. When you engage in more advanced analyses, you will run PLINK from a script as you would in R or other programs.

As figure 8.2 shows, PLINK is invoked by typing ./plink on a Mac or Linux computer and plink.exe (on older versions of Windows computers)—given that PLINK is in your current working directory (see box 8.2). This is followed by the names of the genetic input file(s), then the actions we want to perform on these files, and ending by specifying the name of the output file. Note that these commands always begin with two dashes

Box 8.2
The 10 PLINK commandments for new users

---

1. PLINK is a command line program, so it needs to operate in that environment (i.e., DOS window or Unix terminal defined in text).

2. Put your files in the same directory as the ./plink (or plink.exe) file in order to do all analyses from one directory.

3. All results are written to files that have various different extensions.

4. Remember to always examine the LOG file, which is your console output for notes, warnings, and errors (see box 8.3).

5. PLINK has no cumulative memory. Each run that you make loads the data as if it was new and all previous filters and so on are lost.

6. Spelling and exact command syntax is extremely important. This includes the double dash or minus that you will use often (--).

7. You cannot combine all options with each other. PLINK does not always warn you, but for instance, basic haplotype tests cannot have covariates.

8. Directory paths are separated by the "slash" symbol (/ in Unix and OSx), while windows uses the "backslash" symbol (\). In Unix and macOS, the backslash is used to break different lines of code. (See box 8.1 for more differences between operating systems.)

9. PLINK is designed for human genetic data (which we cover in this book) and assumes the genome has 23 chromosomes with chr23 as the X chromosome. For those readers looking at other species, ensure you specify this (e.g., dog for 39 chromosomes).

10. It is well worth your time and effort to consult the excellent documentation that is available online at http://www.cog-genomics.org/plink2/.

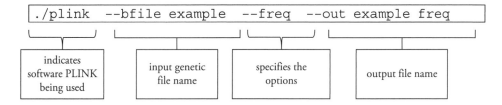

Figure 8.2
Anatomy of a PLINK command line.
Note: Some shells such as UNIX first show the path of the directory that contains the files.

(--). The command shown in figure 8.2 is only used as an instructive example for those following this tutorial, and you do not need to enter it.

The command in figure 8.2 shows the example of how to derive the allele frequencies of a genetic file. The *input file* name argument --bfile indicates the prefix[1] of the binary PLINK file we will use as input. As we touched upon in the last chapter, a binary file is computer-readable but not human-readable. All of our executable programs are stored in binary files, mostly in the form of numeric data files.

We call up the example --bfile, which consists of the three interlinked files of: example.bim, example.bed, and example.fam. The command --bfile (instead of --file) specifies that the input data are in a binary format. This is followed by the *options* that you would like to undertake. In our example, --freq instructs PLINK to calculate allele frequencies. There are many other options, such as --recode or --assoc for association analyses. Note that multiple options can be used within a single command line and the order is not relevant since PLINK implements a default order. Finally, you need to specify an *output file name*, which in the example below is --out example _ freq. Suffixes for the data file will be immediately added on in PLINK.

In this example, the command --freq would produce output with the suffix .afreq. If the name of the output file is not specified, the default name is plink (in lower case). New users should note that PLINK will overwrite files that have the same names without asking, so it is important to remain alert when using this command. A recommended practice is to always rename your output files to avoid confusion. You therefore do not use the default but engage in consistent *file naming conventions*. Keep file names short and meaningful, do not use spaces, and if using different version numbers sort in a way that the most recent version can be easily retrieved (2019v5example.txt or example _ v5 _ 2019.txt). It is also important to use meaningful directory names that you will be able to find (and remember) later. It is advisable that you use similar names for linked files, such as hapmap.ped and hapmap.map. Also, try to avoid using characters such as * : \ / <> | " ? [ ] ; = + & £ $ as they may be difficult to find or open, avoid using common words that are hard to distinguish later (e.g., version, draft), and try to avoid unnecessary repetition and redundancy in your file names and paths.

8.2.3   Running scripts in terminal

Since we are only running simple commands to familiarize readers on how to work with PLINK, we often type all commands simply at the command line. Once you engage in more advanced analyses, however, you will want to write your script and then run it in terminal. To run a script in terminal, you need to type: sh /path/to/file and then press enter.

For instance, type:

```
sh hello _ world.sh
```

And you will see the following output:

```
Hello World!
```

8.2.4   Opening PLINK files

The hapmap-ceu data   PLINK can be used to import data in different formats. One of the most commonly used formats, discussed already in detail in chapter 7, are the more compact *binary files* that consist of three files with the same name followed by the suffixes .bed, .bim, and .fam. The example data that we use is the hapmap-ceu data, which you should have already downloaded in your specified folder. CEU refers to Utah residents with Northern and Western ancestry and is one of the 11 populations in Hap-Map. The files for *hapmap-ceu data* are:

hapmap-ceu.bed   contains compressed information on genotype (not readable in
                                 text editor)
hapmap-ceu.fam   contains information about individuals
hapmap-ceu.bim   contains information about the SNPs

8.2.5   Recode binary files to create new readable dataset with .ped and .map files

As we noted in the preceding chapter, PLINK has its own format for genetic data that differs from the classical rectangular structure of statistical software such as Excel or Stata. By storing the data in different files, we have information on both the sample and the genetic variants. The first step in any data analysis is to get to know your data by producing some basic summary statistics. The three linked files above in the hapmap-ceu data are, however, in unreadable binary format. It is possible to transform a binary file into a human-readable set of files by using the option --recode by using the command below. Note that this command will only work if the working directory is the /User/Your-Name/Chapter8 directory (or one that you created earlier and if the PLINK binary file is compatible with your operating system). Windows users on older systems may get a warning output and can refer to appendix 1. By including the option --bfile in a PLINK command, we call up a binary file. We specifically say "call" and not open since it is not a human-readable file that you can open and observe.

```
./plink --bfile hapmap-ceu --recode --out hapmap-ceu
```

As the command above shows, the only difference is the plink.exe. A series of *outputs* that are printed directly on the terminal tell us what PLINK is doing and provides some basic descriptive information about the file.[2] Below we show the output printed to the terminal screen (divided into two parts). The first lines of the output shows the version number v1.90b6.7 and date of the software release (December 2, 2018), the PLINK version (in this case it is PLINK 1.90b6.7), and copyright. We then see that the log file is saved in default "plink.log" as we noted earlier. But also note that since we changed the name of the output file using the --out hapmap-ceu command, this will now be the output file.

```
./plink --bfile hapmap-ceu --recode --out hapmap-ceu
PLINK v1.90b6.7 64-bit (2 Dec 2018)              www.cog
-genomics.org/plink/1.9/
(C) 2005-2018 Shaun Purcell, Christopher Chang GNU
General Public License v3
Logging to hapmap-ceu.log.
Options in effect:
--bfile hapmap-ceu
--out hapmap-ceu
--recode
```

The second part of the output reports important information about the number of markers and individuals in the file. In the example below, we see that PLINK loaded 2,239,392 variants from the .bim file for 60 individuals (30 males and 30 females) from the .fam file. We also see that there are 60 founders and 0 nonfounders present. Founders are the individuals who do not have parents in the dataset. PLINK only uses founders to calculate allele frequencies and nonfounders (e.g., a sibling pair dataset) would be excluded. PLINK may also report a series of Notes, Warnings, and Errors when it detects potential problems that something might be wrong or is not standard in some way, but it does not stop the execution of the PLINK command. Refer to box 8.3 for a more detailed discussion and interpretation of the warnings and note in the output below. Another important statistic is the *genotyping rate* (shown as 0.992022 below), which is the average proportion of markers available for an individual (with nonmissing data) [3]. Later in this chapter we will see how this information can be used to assess the quality of the data.

```
16384 MB RAM detected; reserving 8192 MB for main workspace.
2239392 variants loaded from .bim file.
60 people (30 males, 30 females) loaded from .fam.
Using 1 thread (no multithreaded calculations invoked).
Before main variant filters, 60 founders and 0 nonfound-
ers present.
Calculating allele frequencies ... done.
Warning: 59 het. haploid genotypes present (see hapmap-
ceu.hh); many commands treat these as missing.
Total genotyping rate is 0.992022.
2239392 variants and 60 people pass filters and QC.
Note: No phenotypes present.
--recode ped to hapmap-ceu.ped+hapmap-ceu.map ... done.
```

Box 8.3
PLINK note, warnings, and error messages

PLINK has three types of errors and warnings that range in a gradient of severity from the light caution of a "Note," to "Warning," and to the more fatal "Error." A *Note* is simply PLINK informing you about something that might be useful to know, but it does not indicate an error or mistake. The note in the output above indicates that no phenotype is present, which for this example is not a problem since we are not performing an association analysis or loading phenotype data. For that we would need to use the --pheno command. A *Warning* alerts you to the fact that something is likely wrong but not fatal for your analysis. The Warning is: Warning: 59 het. haploid genotypes present (see hapmap-ceu.hh); many commands treat these as missing, which is generally related to male heterozygous calls in the X chromosome pseudoautosomal region. The pseudoautosomal regions (PAR1, PAR2) are homologous sequences on the X and Y chromosomes. It is strongly recommended that you address a warning particularly since in PLINK 2.0 it can be treated as an actual Error. The advice is to check the variants in the .hh file and if they are all near the beginning or end of the X chromosome use the command --split-x to solve this problem. An *Error* is a serious problem that causes PLINK to terminate. This can include messages such as "No input dataset," "Out of memory. Try the --memory and/or--parallel flags," or "All people removed." Refer to the PLINK website and ample resources that explain these issues and how to deal with them in more detail.

The advantage of using .ped and .map files instead of binary files is that we can visually inspect these files by opening them in a text editor. This can help us to understand the structure of the data but requires more memory and is a less computationally efficient storage method. We recommend that you use binary data when working with large files (for example, when you working with genome-wide data for thousands of individuals).

8.2.6 Import data from other formats

The option --make-bed can be used to *transform data from other formats* into a PLINK binary file. This creates a more compact version of the data that not only saves space but can speed up your analyses. The following command, for example, can be used to transform a .vcf file into PLINK binary files. A .vcf file refers to the 1000 Genomes Project text variant call format, which has variant information, including the sample ID and genotype call text file. We define and describe genotype calling in more detail shortly in the section on QC. Note that the current version of PLINK only works with genotyped data (see chapter 7). Imputed variants are automatically assigned to the most likely allele. As discussed in the previous chapter, PLINK 2.0 (currently in development at the time of writing this book) will efficiently store imputation quality information.

```
./plink --vcf ALL.chr21.vcf.gz --make-bed --out test_vcf
```

You should then see the following output below.

```
PLINK v1.90b6.7 64-bit (2 Dec 2018)           www.cog
-genomics.org/plink/1.9/
(C) 2005-2018 Shaun Purcell, Christopher Chang GNU Gen-
eral Public License v3
Logging to test_vcf.log.
Options in effect:
--make-bed
--out test_vcf
--vcf ALL.chr21.vcf.gz

16384 MB RAM detected; reserving 8192 MB for main
workspace.
--vcf: test_vcf-temporary.bed + test_vcf-temporary.bim +
test_vcf-temporary.fam
written.
1105538 variants loaded from .bim file.
2504 people (0 males, 0 females, 2504 ambiguous) loaded
from .fam.
Ambiguous sex IDs written to test_vcf.nosex .
Using 1 thread (no multithreaded calculations invoked).
Before main variant filters, 2504 founders and 0
nonfounders present.
Calculating allele frequencies ... done.
Total genotyping rate is 0.999787.
1105538 variants and 2504 people pass filters and QC.
Note: No phenotypes present.
--make-bed to test_vcf.bed + test_vcf.bim + test_vcf.
fam ... done.
```

PLINK can read data in various formats. We provide here a list of commands that can be used to import data from different software, including personal data from 23andMe. We show these only as an example.

Oxford format *(do not run this example)*

```
./plink    --gen oxford _ example.gen \
           --sample oxford _ example.sample \
           --make-bed --out oxford _ example
```

Bgen *(do not run this example)*

```
./plink    --bgen bgen _ example \
           --make-bed --out bgen _ example
```

23andme *(do not run this example)*

```
./plink    --23file 23andme _ example \
      --make-bed --out 23andme _ example
```

8.3   Data management

PLINK can also be used to manage and transform data. We might want, for instance, to restrict an analysis to a subset of individuals or to certain markers. If we need to merge datasets for an analysis, PLINK can be used to ensure that the reported alleles of a genetic variant are coded in the same way. We discuss ways to harmonize SNPs coded in different manners in another chapter. This is of particular importance when using data from different genotyping platforms. In this section we first describe how to select individuals and markers and then how to merge different genetic files.

8.3.1   Select individuals and markers

Select individuals   PLINK can be used to select or exclude certain individuals. This can be done by providing PLINK with a file with the IDs of individuals that will be included (using --keep) or excluded (using --remove) in the analysis. The file must be a space/tab-delimited text file with family IDs in the first column and within-family IDs in the second column.

The --keep option can be used to select individuals from the sample.

The --remove option does the inverse and excludes from the analysis the individuals listed in the file.

```
./plink    --bfile hapmap-ceu \
           --keep list.txt \
           --make-bed --out selectedIndividuals
```

```
PLINK v1.90b6.7 64-bit (2 Dec 2018)              www.cog
-genomics.org/plink/1.9/
(C) 2005-2018 Shaun Purcell, Christopher Chang GNU
General Public License v3
Logging to selectedIndividuals.log.
Options in effect:
--bfile hapmap-ceu
--keep list.txt
--make-bed
--out selectedIndividuals

16384 MB RAM detected; reserving 8192 MB for main
workspace.
2239392 variants loaded from .bim file.
60 people (30 males, 30 females) loaded from .fam.
Error: Failed to open list.txt.
```

Similarly, we can select or exclude entire families by using the `--keep-fam` and `--remove-fam` options. The threshold filters for missing rates and allele frequency are automatically set to exclude no one when you use the `--make-bed` option. You can specify these filters manually by using `--mind`, `--geno`, and `--maf` to exclude people. This is also when the commands `--extract/--exclude` and `--keep/--remove` can applied. For instance, if you would like to create a new file containing only individuals with high genotyping of at least 95% complete you can add the `--mind 0.05` command.

```
./plink --bfile hapmap-ceu --make-bed --mind 0.05 --out
highgeno
```

This in turn creates three new files: `highgeno.bed`, `highgeno.bim`, and `highgeno.fam`. Remember to never try to open the `.bed` file as it is unreadable but it is possible to view the `.fam` and `.bim` files. Recall that the `.fam` file is just the first six columns of the `.ped` file (see chapter 7).

Select markers   It may be that you are interested in a SNP, a set of SNPs, or a particular region. PLINK can also be used to extract the genotype information of specific variants into a separate and smaller file. It could be, for instance, that you are interested in a list of SNPs such as a list of HapMap 3 SNPs if you have 1000 Genomes data. We can do this by specifying the name of the SNPs we want to select using the --snps options or by providing a file (using the option --include) containing the marker names of the variants that we want to include in the new file. The option --exclude can be used to remove certain variants from the file. The following example illustrates how to select the genotype of a single variant. The variant rs9930506 is a SNP in the FTO gene that several studies have associated with increased BMI and body weight. It is possible to select this variant using the following commands. In the previous section we demonstrated how to read in data from the direct-to-consumer company 23andMe. If you have this data and are curious about how many risk alleles you carry on rs9930596, you could import in your own personal data and check using the commands below. rs9930596 is a genetic variant in the FTO gene, and the G allele has been associated with increased risk of obesity.

```
./plink    --bfile hapmap-ceu \
           --snps rs9930506 \
           --make-bed \
           --out rs9930506sample
```

The output below shows that only 1 variant passed the filters, which was the intention and that three separate files were created (.bed, .bim, and .fam).

```
PLINK v1.90b6.7 64-bit (2 Dec 2018) www.cog-genomics.org
/plink/1.9/
(C) 2005-2018 Shaun Purcell, Christopher Chang GNU General Public License v3
Logging to rs9930506sample.log.
Options in effect:
--bfile hapmap-ceu
--make-bed
--out rs9930506sample
--snps rs9930506

16384 MB RAM detected; reserving 8192 MB for main
workspace.
2239392 variants loaded from .bim file.
60 people (30 males, 30 females) loaded from .fam.
```

```
--snps: 1 variant remaining.
Using 1 thread (no multithreaded calculations invoked).
Before main variant filters, 60 founders and 0 nonfound-
ers present.
Calculating allele frequencies … done.
1 variant and 60 people pass filters and QC.
Note: No phenotypes present.
--make-bed to rs9930506sample.bed + rs9930506sample.bim +
rs9930506sample.fam
```

If we examine the new `rs9930506sample.bim` file, we see that it has one line for the variant:

```
16      rs9930506      0       52387966      G      A
```

8.3.2   Merge different genetic files and attaching a phenotype

Merging genetic files   During this type of analysis we often work with multiple files. It is frequently the case, for instance, that genomic data are stored by chromosome to avoid gigantic files that are hard to transfer over the internet and bear large computational require-ments. It can, therefore, often be the case that you have 22 separate autosomal files, the files for the sex chromosomes, and perhaps, a file for the mitochondrial DNA and want to merge everything into a single file. Or, you may have files from different subgroups, such as from different genotype centers. On other occasions, it may be necessary to merge files from dif-ferent studies to create a single file. Merging genetic files requires considerable care. Vari-ants measured in one file may not be measured in another fileand may have different alleles or base-pair positions. The command --bmerge in PLINK ensures that these issues are addressed properly. By default, the command --bmerge sets these types of mismatches to missing. It is possible, however, to make different specifications by using the option --merge-mode. If you need to merge several files at the same time, as in the case of merg-ing chromosome-specific files, use a file containing the names of the different genotype files and the option --merge-list. The following example merges two different genetic files, but note that it is only provided as an illustrative example (and thus not to be carried out).

```
./plink --bfile HapMap _ founders \
--bmerge HapMap _ nonfounders \
--make-bed --out merged _ file
```

Attaching a phenotype   A primary goal of genetic analysis is to study the association between a genotype and a phenotype. Phenotypes can be specified in a .fam file (see chapter 7). To "attach" a phenotype to a genetic file, we can use the --pheno command in PLINK as in the example above. Here we use two new files: 1kg _ EU _ qc and BMI _ pheno.txt. Looking first at the .fam file:

```
head 1kg _ EU _ qc.fam
```

With a small excerpt of the header:

```
0 HG00096 0 0 1 -9
0 HG00097 0 0 2 -9
0 HG00099 0 0 2 -9
0 HG00100 0 0 2 -9
```

Then the .bim file and a small excerpt:

```
head 1kg _ EU _ qc.bim
```

```
1      rs1048488    0      760912    C      T
1      rs3115850    0      761147    T      C
1      rs2519031    0      793947    G      A
1      rs4970383    0      838555    A      C
```

Looking at the phenotype file and a small excerpt:

```
head BMI _ pheno.txt
```

```
FID        IID          BMI
0          HG00096      25.022827
0          HG00097      24.853638
0          HG00099      23.689295
```

To merge the phenotype we can use:

```
./plink     --bfile 1kg _ EU _ qc\
            --pheno BMI _ pheno.txt \
            --make-bed --out 1kg _ EU _ BMI
```

With the output:

```
PLINK v1.90b6.7 64-bit (2 Dec 2018) www.cog-genomics.org
/plink/1.9/
(C) 2005-2018 Shaun Purcell, Christopher Chang GNU Gen-
eral Public License v3
Logging to 1kg _ EU _ BMI.log.
Options in effect:
--bfile 1kg _ EU _ qc
--make-bed
--out 1kg _ EU _ BMI
--pheno BMI _ pheno.txt

16384 MB RAM detected; reserving 8192 MB for main
workspace.
851065 variants loaded from .bim file.
379 people (178 males, 201 females) loaded from .fam.
379 phenotype values present after --pheno.
Using 1 thread (no multithreaded calculations invoked).
Before main variant filters, 379 founders and 0 nonfound-
ers present.
Calculating allele frequencies … done.
851065 variants and 379 people pass filters and QC.
Phenotype data is quantitative.
--make-bed to 1kg _ EU _ BMI.bed+1kg _ EU _ BMI.bim+1kg _ EU _
BMI.fam … done.
```

If we examine the .fam file, we see that instead of −9 (i.e., missing) the last column now has the quantitative (i.e., continuous) phenotypic values but the .bim file (not shown) remains the same.

```
head 1kg _ EU _ BMI.fam
```

```
0  HG00096  0  0  1  25.0228
0  HG00097  0  0  2  24.8536
0  HG00099  0  0  2  23.6893
0  HG00100  0  0  2  27.0162
```

8.4 Descriptive statistics

Just as with any type of analysis, it is important to understand your data and engage in initial descriptive analyses. PLINK can also be used to derive information about the data that you are using for the analysis or the variants genotyped.

8.4.1 Allele frequency
Allele frequencies can calculated by the --freq command in PLINK. The output file (with the suffix .frq) contains information on the alleles of the genotypes and the **minor allele frequency (MAF)** and the allele codes for each SNP.

```
./plink --bfile hapmap-ceu --freq --out Allele _ Frequency
head Allele _ Frequency.frq
```

Here we see the header of the output file Allele _ Frequency.frq:

CHR	SNP	A1	A2	MAF	NCHROBS
1	rs9629043	C	T	0.09322	118
1	rs11497407	A	G	0.008333	120
1	rs12565286	C	G	0.0678	118
1	rs11804171	A	T	0.03704	108
1	rs2977670	G	C	0.07143	112
1	rs2977656	T	C	0.008333	120
1	rs12138618	A	G	0.05833	120
1	rs3094315	G	A	0.1552	116
1	rs2073813	A	G	0.125	120

Here you will see that the columns are: CHR (Chromosome number or code for sex chromosomes); SNP (variant name, rsID for most of the SNPs); A1 (Allele 1, which is the usually the minor allele [i.e., with lower frequency]); A2 (Allele 2, usually the major allele), MAF (Allele 1 frequency); and NCHROBS (number of allele observations). It is also possible to do a variety of other analyses such as stratifying by a categorical variable using the --within option.

8.4.2 Missing values

Individual and variant missing values Examining missing values is important to assess the quality of the data. There are two types of missing values to consider. The first is by individuals. For each individual, we calculate the proportion of all variants that are missing. Second, genetic variants also may have missing values. For each genetic variant we calculate the proportion of individuals that have been genotyped and the proportion that have missing values. The command used by PLINK to derive this information is --missing, which produces two output files: a .imiss file for missing information by individual and a .lmiss file for missingness by SNP.

```
./plink --bfile hapmap-ceu --missing --out missing _ data
```

To reduce repetition we do not show the entire output of the file but only the last lines that report that two new files have been created.

```
--missing: Sample missing data report written to miss-
ing _ data.imiss, and
variant-based missing data report written to missing _
data.lmiss.
```

To examine the header (i.e., the top few lines) of the sample missing data, you can use the command below. If you do not use this and, for instance, type: more missing _ data.imiss you will be taken hostage by a long file which you can exit by typing "q" for quit.

```
head missing _ data.imiss
```

You will then see the following output. In the *individual-missing file* (.imiss), it is structured in a way that each row represents an individual. The columns refer to: FID (family ID), IID (within-family ID), MISS_PHENO (a Yes/No indicator of missing phenotype), N_MISS (number of missing genotype calls), N_GENO (number of potentially valid calls), and F_MISS (missing call rate).

FID	IID	MISS _ PHENO	N _ MISS	N _ GENO	F _ MISS
1334	NA12144	Y	15077	2239392	0.006733
1334	NA12145	Y	19791	2239392	0.008838
1334	NA12146	Y	13981	2239392	0.006243
1334	NA12239	Y	14072	2239392	0.006284
1340	NA06994	Y	16080	2239392	0.007181
1340	NA07000	Y	26113	2239392	0.01166
1340	NA07022	Y	17467	2239392	0.0078
1340	NA07056	Y	12133	2239392	0.005418
1341	NA07034	Y	20425	2239392	0.009121

To examine the *variant-missing file* (.lmiss), you can use the command:

```
head missing _ data.lmiss
```

Where you will see that the columns are represented by CHR (chromosome code), SNP (variant identifier), N_MISS (number of missing genotype calls, not including obligatory missings), N_GENO (number of potential value calls), and F_MISS (missing call rate). Columns that are only present in within-family data are CLST (cluster identifier) and N-CLST (cluster size).

CHR	SNP	N _ MISS	N _ GENO	F _ MISS
1	rs12565286	1	60	0.01667
1	rs12138618	0	60	0
1	rs3094315	2	60	0.03333
1	rs3131968	0	60	0
1	rs12562034	0	60	0
1	rs2905035	0	60	0
1	rs12124819	0	60	0
1	rs2980319	0	60	0
1	rs4040617	0	60	0

Filters   A number of additional filters can be applied to the data to extract particular sets of individuals. This is a non-exhaustive list of possible filters that can be applied to the data. For more information, refer to http://www.cog-genomics.org/plink/1.9/filter.

```
--filter-controls
```
Filters controls with a binary phenotype

```
--filter-males
```
Keeps only males (based on genotype data)

```
--filter-females       Keeps only females (based on genotype data)
--filter-founders      Keeps only founders (i.e., it excludes all samples with at
                       least one known parent)
--filter-nonfounders   The opposite of founders
```

```
./plink    --bfile hapmap-ceu \
           --filter-females \
           --make-bed
           --out hapmap _ filter _ females
```

To avoid repetition, we do not show the entire output but only the relevant lines that note that all females have been removed.

```
30 people removed due to gender filter (--filter-females).
```

8.5 Quality control of genetic data

As already touched upon briefly in chapter 4 in relation to GWASs, quality control (QC) is a standard procedure when working with genetic data. Poor study design and errors in genotype calling can introduce systematic bias in genetic research, in particular to association studies, undermining the validity of the results. Genetic studies should have a rigorous *quality control protocol* that is planned already at the initial design of the study (see chapter 4, where we discuss GWAS research design). To ensure that the QC protocol is independent from the results, the protocol is often pre-registered at an external repository such as the Open Science Framework (https://osf.io). Additionally, researchers from large GWAS consortia often aim to ensure that all the QC procedures are performed by independent centers, although this is quite variable. Quality control protocols are set up to minimize the risk of a general *inflation of associations* but also *false positive errors*, which occur when a variant that is not truly associated with the disease is significantly associated with the disease in a given study. Poor data quality can relate to multiple problems such as poor or inconsistent measurement of the phenotype, which reduces *statistical power* (i.e., the ability to detect a true association using a statistical procedure). A good QC protocol assures that data are comparable across studies and can be used in subsequent analyses. The three main types of QC that we now describe below are: (1) per-individual QC, (2) per-marker QC, and (3) genome-wide association meta-analysis QC. We will briefly describe the main procedures and how to implement them in PLINK.

8.5.1 Per-individual QC

The first step is to ensure that the individuals included in the sample have high-quality data. Per-individual QC of genome-wide data consists of setting up filters that remove individuals from the sample who may introduce bias in the analysis because of their low-quality data. Per-individual QC generally consists of five steps, namely identification of individuals:

1. with poor DNA quality (low call rate and missing genotype, distinguished below);

2. with high heterozygosity across autosomal chromosomes that indicates possible sample contamination or low levels of heterozygosity, which may be due to inbreeding;

3. with discordant sex information;

4. that are duplicated or related; and

5. from different ancestry groups.

Identification of individuals with poor genotype quality   A challenge in this type of analysis is missing data both from individuals who have high rates of their genotype missing (i.e., low genotype call rate) but also SNPs that are missing in a large proportion of individuals. Here we address the large variations in DNA quality that have an effect on *genotype call rate* and genotype accuracy. There are sometimes low-quality DNA samples that need to be removed from the sample. It is possible to detect this when we observe individuals that have a higher proportion of missing genotypes. Typically, individuals with more than 3–7% missing genotypes are removed from the analysis with the option --mind and a specification of the missing cut-off of, for example, 0.05 for a 5% missing rate. This can be specified in PLINK as follows:

```
./plink --bfile 1kg _ hm3 --mind 0.05 --make-bed --out
1kg _ hm3 _ mind005
```

Identification of heterozygosity across autosomal chromosomes   *Heterozygosity* refers to the carrying of two different alleles of a specific SNP (see chapter 1, section 1.5). The heterozygosity rate of an individual is the proportion of heterozygous genotypes. High levels of heterozygosity within an individual might be an indication of low sample quality whereas low levels of heterozygosity may be due to inbreeding. For this reason we exclude individuals with both high and low levels of heterozygosity in our analyses. Reasons for this may be due to inbreeding but also sample contamination. A basic command in PLINK to calculate heterozygosity is described below (option: --het).

```
./plink --bfile 1kg _ hm3 --het --out 1kg _ hm3 _ het
```

Two files are created: 1kg_hm3_het.het and .log. The heterozygosity for each sample is computed as the ratio of the number of heterozygote genotype calls to the total number of non-missing calls. Heterozygosity statistics can be inspected using standard software such as Excel or R. Outliers with very high or low values of mean heterozygosity can be filtered out using the --exclude option in a separate PLINK command. The rule is often to remove individuals who deviate ±3 standard deviations from the particular sample's heterozygosity mean [4]. Figure 8.3 shows the distribution of the heterozygosity statistic using the output from the previous PLINK command. The red lines indicate QC thresholds of ±3 standard deviations from the mean. Observations with higher or lower heterozygosity can be removed from successive analyses. To generate this histogram, use the following commands in RStudio (see appendix 1 for information on how to install R and RStudio if you have not done so already):

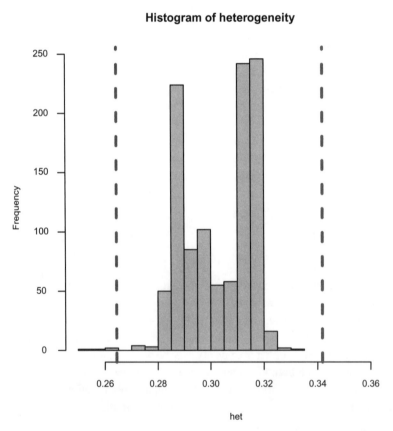

Figure 8.3
Histogram of heterozygosity statistic.

```
heterogeneity _ stats<-read.table("1kg _ hm3 _ het.het",
header=T)

attach(heterogeneity _ stats)
het<- (N.NM.-O.HOM.)/N.NM.
min<-mean(het)-3*sd(het)
max<-mean(het)+3*sd(het)

hist(het, col="lightblue", xlim=c(0.25,0.37),
main="Histogram of Heterogeneity")

abline(v = c(min, max), col="red", lwd=3, lty=2)
```

Identification of individuals with discordant sex information   Recall that males
only have one copy of the X chromosome and thus cannot be heterozygous for any marker
in sex chromosomes. Using genotype data from the X chromosome, it is thus possible to
check for discordance with ascertained sex. Typically, when a genotype-calling algorithm
detects a male heterozygote for an X chromosome maker it calls that genotype as missing.
Deriving correct sex information for the sample is important especially when analyses are
carried out stratified by sex. To calculate this, run the commands below including the
--check-sex option.

```
./plink --bfile hapmap-ceu --check-sex --out
hapmap _ sexcheck
```

Males should have an X chromosome inbreeding coefficient greater than 0.8, while it
should be smaller than 0.2 for females. PLINK compares the estimated inbreeding coefficient with ascertained sex and signal individuals with values outside these thresholds as
possible mismatches. Discordant sex information may be the result of errors in the lab and
if many individuals in your sample have this, you should scrutinize the data in more
detail. The X chromosome inbreeding coefficient F is reported in the file .sexcheck
obtained in the previous PLINK command.

```
head hapmap _ sexcheck.sexcheck
```

FID	IID	PEDSEX	SNPSEX	STATUS	F
1334	NA12144	1	1	OK	0.9999
1334	NA12145	2	2	OK	-0.06528
1334	NA12146	1	1	OK	0.9999
1334	NA12239	2	2	OK	0.05498
1340	NA06994	1	1	OK	0.9999
1340	NA07000	2	2	OK	-0.1001
1340	NA07022	1	1	OK	0.9998
1340	NA07056	2	2	OK	-0.03786
1341	NA07034	1	1	OK	0.9999

Identification of duplicated or related individuals   It is important to examine *inadvertent duplications* of individuals and *cryptic relatedness*, or in other words the possibility that individuals in your sample may be close relatives. To identify duplicate and related individuals, a measure called *identity by state (IBS)* can be calculated for each pair of individuals based on the average proportion of alleles shared in common at genotyped SNPs (only autosomal chromosomes). The degree of shared ancestry for a pair of individuals (identity by descent, IBD) can be estimated by using genome-wide SNPs and its derived IBS. Pairs of individuals who have an IBS equal to 1 are either monozygotic twins or are duplicates in the data. The expected IBD of siblings is 0.5, and 0.25 for second-degree relatives. Due to genotyping errors and population structure there is often variation around these theoretical values. Typically, pairs of individuals with IBS > 0.1875 are removed from the analysis.[3] Recent methods, based on linear mixed models (such BOLT-LMM), do not require us to remove related individuals as they take into account relatedness in their computation. Chapter 9 illustrates how to calculate relatedness and how to filter individuals based on IBS statistics.

Identification of individuals of divergent ancestry: Population stratification   As discussed in chapter 3, confounding due to population stratification is a major source of bias in association studies. For this reason, it is common to run analyses stratified by continental-level ancestry. A common practice to identify high-level ancestry groups is to run a Principal Component Analysis on a large number of variants. This is addressed in the next chapter in section 9.4 in more detail.

8.5.2   Per-marker QC
A second set of quality control analyses focuses on the data quality of variants. Several steps are taken sequentially to remove low-quality variants that might introduce bias in the study. Per-marker QC consists of five steps:

1. Exclusion of low call rate SNPs

2. Removal of SNPs with very low allele frequency (rare variants) in certain types of analyses

3. Identification and exclusion of variants with extreme deviation from the Hardy–Weinberg equilibrium (HWE)

4. In case-control studies, exclusion of SNPs with extremely different call rates between groups

5. In the case of imputed SNPs, exclusion from the study of variants with low imputation quality

It is important to remove suboptimal markers that may affect the analysis and thus increase the number of false positives. It is vital to pay attention to which markers are excluded and the potential LD aspects associated with this. Below, we give a minimal explanation of the steps required to perform per-marker QC analysis.

Low-call-rate SNPs   *Low call rates* refer to the situation where there are high rates of genotype missingness or in other words, SNPs that are missing in a large proportion of individuals. "Calling" in this context is used to represent the estimation of one unique SNP. Typically, markers with a *call rate* of less than 95% are excluded from the analysis. This can be specified with the option --geno in a PLINK command and a specification of the missing cut-off of, for example, 0.05 for a 5% missing rate.

```
./plink --bfile 1kg _ hm3 --geno 0.05 --make-bed --out
1kg _ hm3 _ geno
```

To avoid repetition, we do not show the detailed output for each of these commands. The following four files were created: 1kg _ hm3 _ mind005.bim, .fam, .bed, and .log.

SNPs with low minor allele frequency   Recall from chapter 1 (section 1.3.1) that minor allele frequency (MAF) is the frequency that the second most common allele at a site occurs in a given population. For example, in the HapMap project, SNPs with a MAF of 5% (0.05) or greater were targeted. SNPs that have low MAFs are rare. For this reason, analysts only include SNPs that are above a certain MAF threshold. There are two reasons to exclude MAFs with a low threshold. First, in the case of low MAFs, power is lacking to detect any real SNP-trait associations. Second, these SNPs are often more prone to genotyping errors. Low-frequency SNPs are more prone to bias due to genotyping error and low power to detect association and thus are usually removed from the study. We can exclude SNPs based on a MAF threshold of 1% with the option --maf 0.01.

```
./plink --bfile 1kg _ hm3 --maf 0.01 --make-bed --out
1kg _ hm3 _ maf
```

Two files are created, `1kg _ hm3 _ het.het` and `.log`. The choice of MAF threshold is highly dependent on sample size. If you have a large sample you can use a lower MAF threshold. Although there is no fixed rule, a large sample is considered a something like N = 100,000, MAF threshold 0.01, with a moderate sample N = 10,000, MAF threshold 0.05; below that would be considered a small sample. Typically SNPs with a MAF of less than 1–2% are excluded from the study, but studies with a small sample size may apply higher thresholds. Note also that this 1–2% has also been decreasing with advances in imputation.

Deviation from the Hardy–Weinberg equilibrium (HWE)   As described in detail in chapter 3 (section 3.5), the HWE assumes an infinitely large population, with no selection, mutation, or migration and that the genotype and allele frequencies are constant over generations if none of the conditions are violated. Deviation from the HWE can thus indicate genotyping error, evolution, and concerns regarding the relationship between the allele and genotype frequencies. Violation of HWE indicates that genotype frequencies are significantly different from expectations and the observed frequency should not be significantly different. For example, if the frequency of allele A = 0.20 and the frequency of allele T = 0.80, then the expected frequency of genotype AT is $2 * 0.2 * 0.8 = 0.32$. In GWASs, it is generally assumed that deviations from the HWE are the result of genotyping errors. The PLINK command for the HWE test is `--hwe` followed by a specification of the significance threshold for violation.

```
./plink --bfile 1kg _ hm3 --hwe 0.00001 --make-bed --out
1kg _ hm3 _ hwe
```

This produces four new files, `.log`, `.bim`, `.fam`, and `.bed`. Exclusion differs according to whether you have a binary or quantitative (i.e., continuous) trait. For binary traits the rule is often a HWE p-value $< 10^{-10}$ in cases and $< 10^{-6}$ in controls. For quantitative traits the recommendation is a HWE p-value $< 10^{-6}$ [4]. Different p-value thresholds have been applied in the literature and different references will have different levels.

Combining different QC filters   To remove all of the SNPs failing multiple QC filters, we can simultaneously apply the commands previously covered at both the individual and marker level. The file `individuals _ failQC.txt` includes all of the individuals that fail the individual level QC (e.g., extreme heterozygosity or related individuals).

```
./plink     --bfile 1kg _ hm3 \
            --mind 0.03 \
            --geno 0.05 \
            --maf 0.01 \
            --hwe 0.00001 \
            --exclude individuals _ failQC.txt \
            --make-bed --out 1kg _ hm3 _ QC
```

8.5.3 Genome-wide association meta-analysis QC

As our chapter on GWASs described in considerable detail, most studies are based on a meta-analysis of association results that combines several datasets using millions of genetic variants. To avoid type II errors (false positives), GWAS researchers expend substantial effort to identify outliers attributed to low-quality data during the final meta-analysis QC stage. Variants that are not robust to a series of checks are filtered out and not included in the analysis. Typical filters applied at this stage remove variants that have low allele frequency, low imputation quality, allele frequency diverging substantially from a reference sample, or that have association results driven by a specific study that cannot be replicated elsewhere. In some cases, it may appear to be a repetition of some of the previous steps (for instance removing SNPs with low minor allele frequency) that are generally filtered out before the meta-analysis.

There are several reasons, however, as to why these QC steps need to be repeated at the meta-analysis stage. First, variants that do not show any problem in a single study may have additional problems when pooled together in a multi-cohort study. As we noted in our GWAS chapter, these studies may be very complex and use the data from dozens of individual genetic studies from all over the world. A 2016 GWAS on human reproductive behavior, for instance, was based on the analysis of three phenotypes on sex-specific analyses of 63 different cohorts [5]. This implied an extensive check of hundreds of files containing the association results of millions of genetic variants. The QC in that study was performed by two independent centres (Based in Oxford, U.K., and Rotterdam, The Netherlands) that analysed the data and compared filters and diagnostic tools to ensure that the reported results were based on the best quality variants. QC for such a large project can often take several months. In this section, we summarize the main steps and diagnostic tests of the QC procedure for GWASs.

Filters   Although researchers do not want to discard "true" results, it is often the case that GWAS analysts prefer a conservative approach to ensure that low-quality markers are excluded from the analysis. These are possible filters that can be applied in the QC of genome-wide association results.

1. A first step in a GWAS is to *harmonize base pair positions* of the markers across files using a common reference. For instance, the US National Center for Biotechnology Information d 37.

2. Evaluating information on association results, especially if coming from a multitude of files, also needs to be carried out. Markers with missing information on effect allele or alternative allele, or with implausible values for effect estimates, standard errors, or *p*-values, are removed from the sample.

3. Markers that are not biallelic (pertaining to both alleles), or that are monomorphic (showing no variation or in other words opposite to polymorphic), are excluded from the list of final results.

4. Results from rare variants are usually problematic and may affect the results. Markers with MAF below 1% are typically removed. However, this threshold may depend on sample size of the study. Some small studies can more often have problematic results from variants with MAF smaller than 5%, while large biobanks may not have problems for variants that are much less common.

5. Imputation quality can affect the quality of association results. It is common to drop analysis markers with an imputation quality below 0.7.

Diagnostic checks   Following this, it is then common to run additional diagnostic checks for the SNPs remaining after applying the filters described above. A very common graph is the *allele frequency plot* (see figure 8.4). This checks whether the variants (1) are all coded in the same way and do not have errors in allele frequencies and strand orientations,[4] and (2) have similar allele frequencies across studies. It is common to produce a graph in which the allele frequencies in each study are plotted against the allele frequencies in a reference

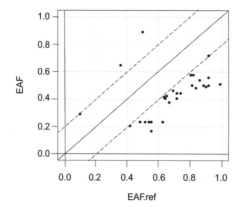

Figure 8.4
Allele frequency plots of age at first birth (AFB) by 1000 Genomes reference panel.
Source: Slightly adapted by authors from original EasyQC output for higher resolution and to fit black and white format.

sample, typically 1000 Genomes. Figure 8.4 shows an allele frequency plot that was used in the QC of our large GWAS consortium on human reproductive behavior [5]. The figure is a scatterplot with the expected allele frequencies of age at first birth (AFB) on the y-axis and the allele frequencies from a reference consortium (1000 Genomes) on the x-axis. Since we would need to plot around 10 million points, we plot only the points that diverge from the reference sample. In particular, here we only plot points that have a difference in allele frequency of 0.2 (hence the diagonal lines). In this case, it is possible to see that some variants have lower allele frequency than the reference sample.

A second diagnostic check that can be very useful to examine heterogeneity of results is the so-called *forest plot* (see figure 8.5). A forest plot is a common diagnostic tool used in meta-analysis. In this graph, taken from our same study discussed above, the association results of different studies for a given variant (rs2777888) are plotted with their confidence intervals across all data sources used in the study. This is a visual representation of the heterogeneity of the results. This plot is verification that all of the results have the same direction of effect. Larger studies should have smaller confidence intervals, giving more precise estimates. In this case it is, for example, the large samples from deCODE genetics and 23andMe that were included in our meta-analysis. The size of the box also represents the sample size of the data. Forest plots are typically combined with heterogeneity tests.

Additional QC diagnostic tools include plots of the reported p-values versus the p-values of the Z-scores (also called PZ plots), and Quantile-Quantile (Q-Q) plots where observed p-values are plotted against the theoretical quantiles we would observe in case of no association. Q-Q plots may reveal unaccounted for population stratification. We discuss Q-Q plots in more detail in chapter 12 (see section 12.1.3). It is also worth noting that we would expect inflation to show up in a Q-Q plot from a well-powered GWAS of a heritable trait. Another aspect to note when considering errors and bias in genotyped data is batch effects. *Batch effects* refer to measurement errors that arise due to laboratory conditions, reagent lots, or other differences. For an excellent review on this topic, refer to a 2010 study by Leek and colleagues [6].

8.6   Conclusion

This chapter provided a basic introduction of how to start working with genetic data using PLINK. Our aim is that you now have grasped the basics of how to work with this type of data using the command line, calling and opening PLINK and importing data. Data management is very important in these types of analyses and for this reason we attempted to outline some of the very basics about how to select individuals and markers and merge different genetic files or add a phenotype. You should now also be able to examine descriptive statistics including allele frequency and filtering for missing values. We concluded with one of the most important steps in this type of data analysis, which is QC of the individual-, marker-, and GWAS-level QC processes. In chapter 9 we will show you how to run more advanced analyses including association analysis, GWAS, identification of independent SNPS,

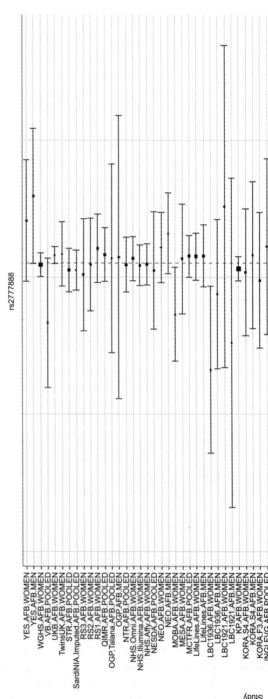

rs2777888

YES.AFB.WOMEN
YES.AFB.MEN
WGHS.AFB.WOMEN
VB.AFB.POOLED
UKB.AFB.WOMEN
TwinsUK.AFB.WOMEN
STR.AFB.POOLED
SardiNIA.Imputed.AFB.POOLED
RS3.AFB.WOMEN
RS2.AFB.WOMEN
RS1.AFB.WOMEN
QIMR.AFB.POOLED
OGP.Talana.AFB.POOLED
OGP.AFB.MEN
NTR.AFB.POOLED
NHS.Omni.AFB.WOMEN
NHS.Illumina.AFB.WOMEN
NHS.Affy.AFB.WOMEN
NESDA.AFB.POOLED
NEO.AFB.WOMEN
NEO.AFB.MEN
MOBA.AFB.WOMEN
MESA.AFB.WOMEN
MCTFR.AFB.POOLED
LifeLines.AFB.WOMEN
LifeLines.AFB.MEN
LBC1936.AFB.WOMEN
LBC1936.AFB.MEN
LBC1921.AFB.WOMEN
LBC1921.AFB.MEN
KP.AFB.WOMEN
KORA.S4.AFB.WOMEN
KORA.S4.AFB.MEN
KORA.F3.AFB.WOMEN
INGI.FVG.AFB.POOLED

Study

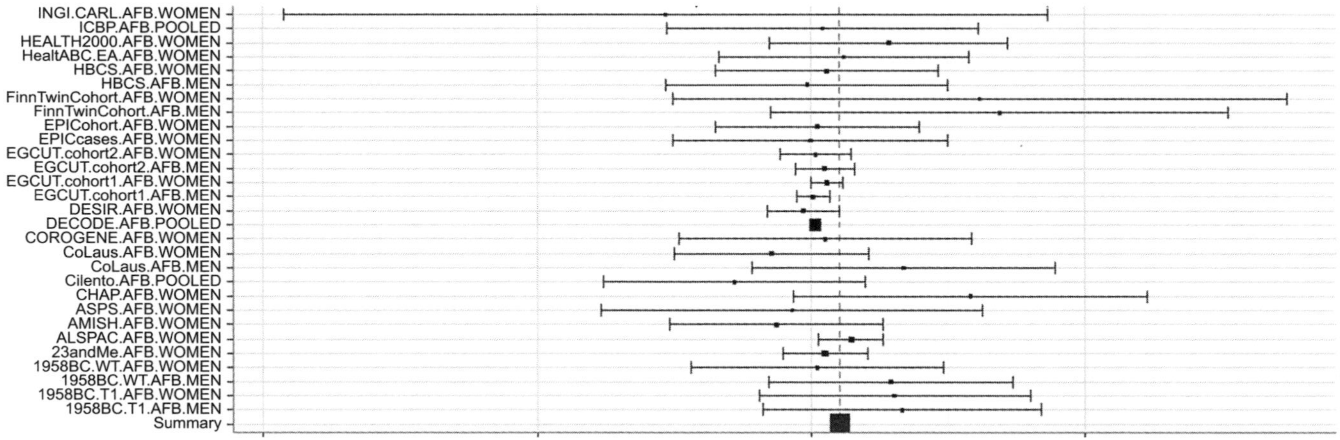

Figure 8.5
Forest plot of rs2777888.

calculating PCAs, and using an additional program called GCTA (Genome-Wide Complex Trait Analysis) [7].

Exercises

In this chapter we examined European ancestry population. Go to the following website, where were you will find other ancestry groups: http://zzz.bwh.harvard.edu/plink/res.shtml.

Download the Chinese and Japanese sample (CHB and JPT) and carry out the following analysis:

1. Recode the file into .map and .ped files.
2. Calculate allele frequencies and missing values.
3. Run a QC analysis with the following criteria:

   a. Call rate > 95%

   b. Minor Allele Frequency > 3%

   c. Hardy–Weinberg equilibrium test < 0.0001

Further reading and resources

Further reading on data QC andPLINK:

Anderson, C. A. et al. Data quality control in genetic case-control association studies. *Nat. Prot.* **5**, 1564–1573 (2010).

Chang, C. C. et al. Second-generation PLINK: Rising to the challenge of larger and richer datasets. *GigaScience* **4**(1), doi:10.1186/s13742-015-0047-8.

Laurie, C. C. et al. Quality control and quality assurance in genotypic data for genome-wide association studies. *Gen. Epid.* **34**, 591–602 (2010).

Winkler, T. W. et al. Quality control and conduct of genome-wide association meta-analyses. *Nat. Prot.* **9**, 1192–1212 (2014).

PLINK resources

Online PLINK Tutorial: http://zzz.bwh.harvard.edu/plink/tutorial.shtml.

PLINK webpage: http://www.cog-genomics.org/plink2/.

YouTube PLINK tutorial by Broad Institute: https://www.youtube.com/watch?v=ppBJqCMSBYk.

References

1. S. M. Purcell et al., PLINK: A tool set for whole-genome association and population-based linkage analyses. *Am. J. Hum. Genet.* **81**, 559–575 (2007).
2. C. C. Chang et al., Second-generation PLINK: Rising to the challenge of larger and richer datasets. *Gigascience* **4**(1) (2015), doi:10.1186/s13742-015-0047-8.
3. C. A. Anderson et al., Data quality control in genetic case-control association studies. *Nat. Protoc.* **5**, 1564–1573 (2010).

4. A. T. Marees et al., A tutorial on conducting genome-wide association studies: Quality control and statistical analysis. *Int. J. Methods Psychiatr. Res.* **27**, e1608 (2018).

5. N. Barban et al., Genome-wide analysis identifies 12 loci influencing human reproductive behavior. *Nat. Genet.* **48**, 1–7 (2016).

6. J. T. Leek et al., Tackling the widespread and critical impact of batch effects in high-throughput data. *Nat. Rev. Genet.* **11**, 733–739 (2010).

7. J. Yang, S. H. Lee, M. E. Goddard, and P. M. Visscher, GCTA: A tool for genome-wide complex trait analysis. *Am. J. Hum. Genet.* **88**, 76–82 (2011).

9

Working with Genetic Data, Part II

Association Analysis, Population Stratification, and Genetic Relatedness

Objectives

- Learn how to run *linear and logistic association* analyses
- Understand how to perform *additive, dominant,* or *recessive analysis of selected single-nucleotide polymorphisms (SNPs)*
- Be able to *run a genome-wide association study* using PLINK
- Learn how to *identify independent SNPs* through *linkage disequilibrium (LD)* pruning
- Find a *proxy genotype* of SNPs in high LD with your SNP of interest using LD
- Learn how to perform *Principal Component Analysis* in genetic data
- Calculate *genetic relatedness* using identity by state (IBS) with PLINK and Genome-wide Complex Trait Analysis (GCTA)
- Use GCTA to *estimate heritability* for different phenotypes

9.1 Introduction

9.1.1 Aim of this chapter

The previous chapter provided readers with some of the fundamentals of how to use PLINK, together with some basics of data management and quality control. The aim of this chapter is to introduce readers to the essentials of association analysis, population stratification, and genetic relatedness. Examining the relationship between SNPs and a particular trait in association analysis forms the basis of many analyses. We are then generally interested in examining whether that particular correlation between SNPs and a trait is from SNPs that are independent or redundant. For this reason, we illustrate how to isolate independent SNPs through a technique called LD pruning. This is done by finding SNPs that are in high LD with your particular SNP of interest. We know from chapter 3 that individuals from different ancestry groups differ with respect to their allele frequencies. For this reason, we demonstrate how to calculate principal components in genetic data. Another vital element to consider when working with genetic data is genetic relatedness and the bias that duplicated or related individuals can bring into your association analysis. We explain how to identify these related individuals using the measure of identity by state (IBS) in PLINK and GCTA (Genome-wide Complex Trait Analysis). We introduce readers to some of the basic commands of GCTA and provide instructions on how to calculate heritability with this program.

9.1.2   Data and computer programs used in this chapter

To actively follow along with this chapter, you will need to ensure that you have the following data installed in the appropriate directory that we will use in this chapter. Remember that data used in this book is described in appendix 2 and when relevant can be downloaded from our companion website, http://www.intro-statistical-genetics.com. With the exception of the .txt file, all have the linked (.bed, .bim, and .fam) files:

- 1kg_EU_BMI
- 1kg_EU_Overweight
- 1kg_hm3_qc
- 1kg_hm3_pruned
- 1kg_samples.txt
- BMI_pheno.txt
- 1kg_samples_EUR.txt
- hapmap-ceu

You will also need to install GCTA (https://cnsgenomics.com/software/gcta/#Overview). Instructions on how to do this are provided in appendix 1.

9.2   Association analysis

As you know from chapter 1, a primary goal of genetic analysis is to estimate the association between a genotype and a phenotype. As a simple example below, we estimate the linear association of rs9674439 alleles on body mass index (BMI). The statistical model estimates the effect of the C allele (the first allele in the .bim file) on the phenotype of interest. This type of model is by far the most common in genome-wide association studies (GWASs). Each copy of the C allele of this SNP has the same effect or, in other words, it is an additive model. A basic *linear regression with an additive effect* on a quantitative phenotype can be estimated as follows (keep in mind that commands can differ by operating system—see box 8.1):

```
./plink     --bfile 1kg _ EU _ BMI \
            --snps rs9674439 \
            --assoc \
            --linear \
            --out BMIrs9674439
```

Recall from the previous chapter that the --bfile command is the binary fileset that contains the linked .bed, .bim, and .fam files, which in this example uses the 1_kg_EU_ BMI file. The second line, --snps, specifies the SNP list, which in our case is rs9674439. Note that it is also possible to select a range of SNPs. The --assoc command provides the basic results of the association analysis. For quantitative (i.e., continuous) traits, you use the --linear command.

The result from the commands above produces the file named BMIrs9674439.assoc. linear shown below. It has the following information: chromosome number (CHR); variant identifier (SNP); base-pair position (BP); effect allele (A1); type of statistical test used (TEST), which in this case is ADD for additive; number of missing values (NMISS); regression coefficient (BETA); t-statistic (STAT); and asymptotic p-value for t-statistic (P). The following output shows that each copy of the C allele on SNP rs9674439 is associated with a reduction of 0.29 in BMI. The result, however, is not statistically significant, with a p-value of 0.20.

```
CHR  SNP         BP        A1  TEST  NMISS  BETA     STAT    P
16   rs9674439  33836510   C   ADD   379    -0.2974  -1.269  0.2052
```

When dealing with case-control studies, the regression is slightly different and the option --linear should be omitted. The statistical test calculated is a chi-square test, and the estimated coefficient is an odds ratio. A logistic regression option for binary phenotypes is also available, by substituting --logistic for --linear in the PLINK command. In the example below, we run a logistic regression on a binary trait (Overweight). Individuals with a BMI greater or equal than 25 are classified as overweight (cases), while individuals with a BMI less than 25 are coded as not overweight (controls). In PLINK, cases are coded 2, and controls 1. Here we use logistic regression to estimate the effect of rs9674439 on the probability of being overweight.

```
./plink     --bfile 1kg _ EU _ Overweight \
            --snps rs9674439 \
            --assoc \
            --logistic \
            --out Overweight _ rs9674439
```

As most readers will know, the output of a logistic regression is slightly different from a linear regression. The file `Overweight _ rs9674439.logistic`, shown below has the following columns: chromosome number (CHR), variant identifier (SNP), base pair position (BP); effect allele (A1), type of statistical test used (TEST), number of missing values (NMISS), odds ratio (OR), t-statistic (STAT) and asymptotic p-value for the t-statistic (P). As a standard output, PLINK reports the *odds ratio* estimate of a logistic regression, which in this example is the ratio between the probability of being overweight associated with each copy of the C allele, over the probability of being overweight given no copies of the C allele. In other words, it tells us how much more likely it is to be overweight if an individual has at least one copy of a particular allele (in an additive model). Odds ratios are always greater than zero. When the odds ratio is greater than 1, it indicates an increased risk; when it is less than 0 it indicates a decreased risk; and when it is equal to 1, it means that there is no association. In the example below, we see that the OR is 0.7, suggesting that the C allele is associated with decreased probability of being overweight. The associated p-value is 0.0009, meaning that the statistical association is strong. The results suggest that having a C allele on rs0674439 is protective against becoming overweight. The result is consistent with the previous model, although the level of statistical significance is different. This is attributed to the fact that we are looking at the same variable, but coded in a different way. BMI is a continuous variable, while Overweight is dichotomous. Your choice on how to code a variable is thus very important and depends on the study design and your research question.

```
CHR SNP          BP         A1  TEST  NMISS  OR       STAT   P
16   rs9674439   33836510   C   ADD   1092   0.7261  -3.32  0.0009017
```

Additive models are the most common type of genotype-phenotype association analysis, although sometimes we might be interested in different models such as dominant models or recessive models in which we study the effect of a single allele. In particular, dominant models treat the heterozygote and one of the homozygote genotypes as a single category. For example, as summarized in table 9.1, let us assume a given SNP that has alleles A and B. The three possible genotype groups are then AA, AB, and BB. If A is the effect or "risk" allele, then a dominant model will study the effect of having *at least one* copy of A, that is the effect of "AA+AB" versus "BB." On the contrary, a recessive model estimates the effect of having *two copies* of A, that is the effect of "AA" versus "AB+BB." These models can be estimated in PLINK using the options `--linear dominant` or `--linear  recessive` in case of linear models, as in the example below.

Table 9.1
Comparison of additive, dominant, and recessive regression models and their interpretation

| Model | Outcome | | Interpretation |
	Continuous	Binary	A risk allele
Additive	--linear	--logistic	AA vs AB vs BB
Dominant	--linear dominant	--logistic dominant	AA + AB vs BB
Recessive	--linear recessive	--logistic recessive	AA vs AB+BB

```
./plink    --bfile 1kg _ EU _ BMI \
           --snps rs9674439 \
           --assoc \
           --linear dominant \
           --out BMIrs9674439
```

This command writes out three files, BMIrs9674439.log, .assoc _ linear, and .qassoc. The BMIrs9674439.assoc _ linear file is show below.

```
CHR  SNP        BP        A1  TEST  NMISS  BETA     STAT    P
16   rs9674439  33836510  C   DOM   379    -0.4783  -1.462  0.1445
```

The BMIrs9674439.qassoc is:

```
CHR SNP        BP         NMISS  BETA     SE      R2        T       P
16  rs9674439  33836510   379    -0.2974  0.2343  0.004254  -1.269  0.2052
```

Association analysis in PLINK may include covariates, such as sex of the respondent, birth year, controls for population stratification, or data-specific variables. In a linear model, this can be specified by adding the --covar option followed by a tab-separated file including the variables used as covariates in the analysis. In this case, the output file will include, for each marker, a row indicating the regression estimate for each covariate included in the model.

If, instead of testing a single variant, we are interested in testing the association with all of the genetic variants included in the genotype file, this can be done by omitting the --snp command. This is a genome-wide analysis (i.e., GWAS), which was discussed extensively in chapter 4, although this type of analysis usually includes both genotyped and imputed data.

```
./plink      --bfile 1kg _ EU _ BMI \
             --assoc \
             --linear \
             --out BMIgwas
```

An excerpt of the typical output file includes the results below. Note that because this file includes all SNPs, it has millions of rows and we therefore only examine the first few lines.

```
head BMIgwas.assoc.linear
```

CHR	SNP	BP	A1	TEST	NMISS	BETA	STAT	P
1	rs1048488	760912	C	ADD	379	0.6031	2.151	0.03208
1	rs3115850	761147	T	ADD	379	0.6056	2.135	0.03343
1	rs2519031	793947	G	ADD	379	-0.9188	-1.019	0.3087
1	rs4970383	838555	A	ADD	379	-0.01473	-0.05882	0.9531
1	rs4475691	846808	T	ADD	379	-0.3347	-1.221	0.223
1	rs1806509	853954	C	ADD	379	-0.1015	-0.4786	0.6325
1	rs7537756	854250	G	ADD	379	-0.1289	-0.4769	0.6337
1	rs28576697	870645	C	ADD	379	0.1739	0.7539	0.4514
1	rs7523549	879317	T	ADD	379	0.1316	0.2271	0.8204

The output of a GWAS in PLINK is exactly the same as that of a single regression and has the following columns: chromosome number (CHR), variant identifier (SNP), base-pair position (BP), effect allele (A1), type of statistical test used (TEST), number of missing values (NMISS), regression coefficient (BETA), t-statistic (STAT), and asymptotic p-value for t-statistic (P). The regression model is repeated sequentially for each SNP included in the PLINK file. As discussed in chapter 4, we need to take *multiple testing* into account when interpreting GWAS results to avoid inflating the number of false positives. To give an indication of this, if we take the normal p-value threshold of 0.05, we expect that, under the null, 5% of variants are significantly associated by chance. This is likely acceptable in many cases when we estimate regression with only a few variables of interest. When we test 1 million variables (SNPs) at the same time, a p-value of 0.05 implies that 50,000 SNPs are false positives. To avoid this error, we adopt a much stricter p-value threshold (5×10^{-8}, that is 0.00000005). Another aspect you will notice when working with GWAS results is that SNPs that have similar positions will have a similar effect and p-value. This is because SNPs are in LD (see chapter 3, section 3.6). We will how to work with GWAS results further in chapter 12.

PLINK can perform many different types of association analysis. For instance, it is possible to run a *within-family analysis* (also called family fixed effect regression) where we examine the effect of different genotypes among family members. Within-family analyses are vital to establish a true genetic effect that is not confounded by population stratification and genetic nurture, as the differences in genotype across siblings operate as a true random experiment since alleles are transmitted randomly through meiosis (see chapter 1, section 1.2). Such analysis can be performed in PLINK with the command `--qfam`. More advanced association analyses can also be run using PLINK that we do not cover in this introductory textbook, including stratified case/control analyses, regression using dosage data, LASSO regression, and linear mixed model associations.

GWASs are usually performed with software other than PLINK, mostly attributed to the fact that the current version of PLINK is not the most suitable for using data with imputation uncertainty. Other software used for GWASs include SNPTEST, BOLT-LMM, and BGENIE. PLINK 2.0, currently in development, will efficiently manage imputed data and be suitable for large GWASs.

9.3 Linkage disequilibrium

As described in chapter 3 (section 3.6), linkage disequilibrium (LD) is the result of many factors such as selection, genetic recombination, mutation rate, genetic drift, population stratification, and genetic linkage. LD affects the way alleles are distributed across a population and creates a correlation structure across SNPs. Therefore, we are often interested in studying and detecting the correlation between SNPs or identifying SNPs that are independent. Establishing the correlation between SNPs is useful for several reasons. First, the same SNPs might not be measured in multiple studies. We want thus to isolate genetic signals even if the same SNPs are not measured across all studies or substudies. A second reason is to reduce the number of genetic variants for subsequent analysis. If, for instance, we wanted to derive ancestry information, we do not need all of the information included in the genotyped data since much of it is redundant. As we show in the next chapter, it is common to extract only independent SNPs when we calculate polygenic risk scores.

LD between SNPs is commonly measured with two measures: r^2 and D'. The r^2 measure is simply calculated as the square of the correlation coefficient of the alleles between two SNPs, and it is therefore constrained to take values between 0 and 1. It is a statistical measure of shared information between two markers and is commonly used to determine how well one SNP can act as a proxy for another. The statistic D' is a population genetics measure also scaled between 0 and 1, indicating the recombination probability between markers. A D' equal to 0 indicates complete linkage equilibrium and frequent recombination, while a D' equal to 1 implies no recombination between the two markers, implying complete LD. PLINK can also be used to check the LD between two markers. The option `--ld` inspects the relationship between a single pair of variants in more detail, which displays observed and expected (based on MAFs) frequencies of each haplotype, as well as haplotype-based r^2 and D'.

```
./plink --bfile hapmap-ceu --ld rs2883059 rs2777888--out
ld _ example
```

This produces the two files of ld _ example.log and ld _ example.hh and gives the following output:

```
--ld rs2883059 rs2777888:
R-sq = 0.715909 D' = 1
Haplotype         Frequency              Expectation under LE
CA                -0                     0.21
TA                0.45                   0.24
CG                0.466667               0.256667
TG                0.083333               0.293333
```

Here we find a r^2 of 0.715909, which suggests a fairly high correlation between rs2883059 and rs2777888. The D' is equal to 1, which means that the two SNPs are in complete LD or, in other words, co-inherited around 100% of the time. The disequilibrium value is a measure of the nonrandom association of alleles at two or more loci. If the two loci are independent (i.e., not co-inherited), then both the r^2 and D' values would be 0.0 regardless of either allele frequency. Others have noted that D' values suffer from a ceiling effect or in other words easily reaches 1. But which measure should you use? Typically the r^2 is the preferred measure if the focus of your research is on the predictability of one polymorphism given another. This is why it is often used in power studies for association designs. The D' is the measure that is used to assess recombination patterns since haplotype blocks are often defined as the basis of D', which is shown in the output above.

In some cases, we may be interested in finding a *proxy genotype* or, in other words, SNPs in high LD with the SNP of interest. The best way to find a proxy genotype is to check a reference panel such as 1000 Genomes, keeping in mind that different ancestral groups may differ substantially in their LD structure. This can be done using online databases such as LDlink (https://ldlink.nci.nih.gov) [1]. LDlink is a website containing multiple web-based applications designed to interrogate LD in population groups. Table 9.2 reports the results from LDlink on the first 10 proxy genotypes for SNP rs2777888 on chromosome 3, for the CEU population in the 1000 Genomes dataset.

LD Pruning   LD pruning is a statistical procedure used to remove redundant SNPs or, in other words, pairs of correlated SNPs. By iteratively examining all of the SNPs in the

Table 9.2

Example of proxy genotypes using LDlink

RS results	Chr	Position (GRCh37)	Alleles	MAF	Distance	D'	R^2	Correlated alleles
rs2681781	3	49898273	(A/G)	0.4646	273	1.0	1.0	A=A,G=G
rs62262093	3	49960388	(T/C)	0.4697	62388	1.0	0.9799	A=T,G=C
rs9848497	3	49951316	(T/C)	0.4697	53316	1.0	0.9799	A=T,G=C
rs7634084	3	49949834	(A/T)	0.4697	51834	1.0	0.9799	A=A,G=T
rs3733135	3	49939587	(G/A)	0.4697	41587	1.0	0.9799	A=G,G=A
rs3774758	3	49938227	(C/G)	0.4697	40227	1.0	0.9799	A=C,G=G
rs2230590	3	49936102	(T/C)	0.4697	38102	1.0	0.9799	A=T,G=C
rs9815930	3	49931343	(A/T)	0.4697	33343	1.0	0.9799	A=A,G=T
rs6795703	3	49930215	(C/T)	0.4697	32215	1.0	0.9799	A=C,G=T
rs1062633	3	49924940	(T/C)	0.4697	26940	1.0	0.9799	A=T,G=C

genotype data, LD pruning selects only one representative SNP from each LD block. In each step the SNP with the higher minor allele frequency is kept in the dataset. Two SNPs are then considered independent if their correlation (r^2) is inferior to a certain threshold or if their distance in base pairs is greater than specified. For instance, the command below tells PLINK to load the file 1kg _ hm3 _ qc and to keep SNPs with a MAF of at least 1%, with no pairs remaining with $r^2 > 0.2$. By default, variants more than 1,000 kilobases apart are considered independent. The additional parameters, here 50 and 5, affect how the computation works in windows across the genome.[1]

```
./plink    --bfile 1kg _ hm3 _ qc --maf 0.01 \
           --indep-pairwise 50 5 0.2 \
           --out 1kg _ hm3 _ qc _ pruned
```

We do not provide all of the output, but briefly, you should see output that describes the pruning of variants from each chromosome, with an excerpt of the output provided below.

```
Pruned 56392 variants from chromosome 1, leaving 13264.
Pruned 56216 variants from chromosome 2, leaving 12648.
Pruned 47468 variants from chromosome 3, leaving 11085.
..............................
Pruned 9577 variants from chromosome 21, leaving 2626.
Pruned 9323 variants from chromosome 22, leaving 2983.
Pruning complete. 680356 of 846484 variants removed.
Marker lists written to plink.prune.in and plink.prune.
out.
```

The command above produces a list of independent SNPs that can be used for further analysis. The file 1kg _ hm3 _ qc _ pruned.prune.in contains the independent SNPs, while the file 1kg _ hm3 _ qc _ pruned.prune.out lists the markers that are excluded from the pruning. In order to obtain a "pruned" dataset, we can use the --extract option in PLINK to get rid of the redundant markers.

```
./plink     --bfile 1kg _ hm3 _ qc \
            --extract 1kg _ hm3 _ qc _ pruned.prune.in \
            --make-bed \
            --out 1kg _ hm3 _ prunedf
```

LD clumping is a similar approach that selects independent SNPs based on a statistic. When calculating polygenic scores, for instance, a common choice for clumping is to use the test statistic computed from a GWAS of a given phenotype. In this way, we select the SNP with the smallest p-value for that genetic locus for each LD block. We will discuss LD clumping in more detail in chapter 10, which demonstrates how to calculate polygenic scores.

9.4   Population stratification

As discussed in chapter 3, as a consequence of human dispersal out of Africa (section 3.2), individuals from different ancestries differ substantially in their allele frequencies. Population structure and stratification was elaborated upon in detail in section 3.3. Here we noted that a SNP that is common in one population may be rare in another one or even show no variation at all. Large projects such as HapMap and 1000 Genomes have shown how genetic variations change across populations. Figure 9.1, for instance, shows the allele frequency of SNP rs2777888, a genetic marker strongly associated with human reproductive behavior in the 1000 Genomes reference panel. By using the Geography of Genetic Variants browser [2], a web-app provided by the lab of John Novembre from the University of Chicago, we can map how this SNP varies across populations. The most common allele is G in African populations, while the alternative A allele is much more common in Southeast Asia.

  As noted in chapter 3, population stratification has strong implications for genetic associations and must be carefully considered during analysis. Principal components analysis (PCA) is the most widely used approach for identifying and adjusting for ancestry differences among individuals. PCA is a statistical technique for data reduction, used to summarize multidimensional data into fewer variables. By calculating principal components from a multivariate dataset, we reduce the complexity of the data to account for the structure of the original dataset. PCA is used in genetics to explain differences in allele frequencies among a sample of individuals. Principal components are thus the "new variables" that explain part of the variability in the original data.

Chr3:49898000 G/A

Frequency scale = Proportion out of 1
The pie below represents a minor allele frequency of 0.25

Sample sizes below 30 become increasingly transparent to
represent uncertain frequencies, i.e.

Figure 9.1

Geographical distribution of SNP rs2777888 in the 1000 Genomes sample.
Source: Figure obtained using the Geography of Genetic Variants Browser (http://www.popgen.uchicago.edu/ggv).

An important property of principal components is their intrinsic ranking. The first component is always the one that has the greatest explanatory value, followed by the second and so forth. It is common to use the first 10 or 20 principal components from a genetic dataset in the analysis. As discussed in section 3.3.4 of chapter 3, PCA in genetics almost perfectly mirror the geographical variation across populations. Principal components are used to understand the ancestry of an individual. They are, moreover, used for QC, to remove individuals from the sample that are heterogenous in ancestry. Finally, PCA is one of the standard methods used in GWASs to correct for population stratification. As we discussed in detail earlier in chapter 4, GWASs currently generally focus on a single ancestry population, which is then replicated in another (unfortunately, this has meant an overrepresentation of those of European ancestry) [3]. However, this is not sufficient to account for further population stratification within the same ancestral group. Individuals from Northern Europe have, for example, allele frequencies that differ from Southern Europeans. Several software packages can be used to estimate principal components from genetic data. Other programs can be used to calculate PCs from genetic data, including EIGENSTRAT.

Here we report the commands that can be used in PLINK to estimate the first 10 principal components using the option --pca 10.

```
./plink --bfile 1kg _ hm3 _ pruned --pca 10 --out 1kg _ pca
```

The --pca command in PLINK produces two output files. In this example: 1kg _ pca. eigenval and 1kg _ pcas.eigenvec. The file with the extension .eigenvec is the list of principal components and can be used by other statistical software for further analyses. An excerpt of the 1kg _ pca.eigenval file is:

```
54.1464
40.0338
6.96377
3.375
```

An excerpt of the 1kg _ pca.eigenvec looks like:

```
0  HG00096  0.0149253  -0.0329941  0.0157409  0.00171199  0.00178966
-0.00704659  -0.00461685  -0.00735375  -0.00169564  0.0100253
0  HG00097  0.0146554  -0.0330726  0.0168457  -0.00070785
-0.000456348
-0.00860046  -0.00610165  -0.00293391  0.00189605  0.00350145
```

```
0 HG00099 0.0147324 -0.0333974 0.0160621 0.00243107
0.000503637 -0.00195932 -0.00130626 -0.00384657 0.00205159
-0.000813858
0 HG00100 0.0146498 -0.0329754 0.0158382 -0.00275797
0.00202298 0.00228241 -0.000977904 -0.00151248 0.00244192
0.00711757
0 HG00101 0.0145233 -0.0328001 0.0164791 0.000286727
-0.00121587
-0.00203153 -0.000880223 -0.00698668 0.00506124 0.0101333
```

After calculating the principal components, we can then use R to inspect how individuals differ by these dimensions. Using RStudio, we can open the PLINK output files and visually inspect the data. Figure 9.2 is a scatterplot of the two first principal components calculated from PLINK. We can recognize immediately that the data are not randomly distributed, but rather that a few groups emerge from the data (figure 9.2, upper panel).

We can then use information from the 1000 Genomes sample to identify the different groups within the data. The file 1kg _ samples.txt includes the populations of origin for all the individuals of the 1000 Genomes sample. Using RStudio, we can import this dataset and merge with the principal components calculated from PLINK. The code below produces an updated figure where we can easily distinguish each population in the data. Individuals with European ancestry are clustered in the bottom right corner of the figure with East Asians in the top right corner of the graph. Individuals with African ancestry are more dispersed across the X axis (PC 1), while American populations are more disperse along the second principal component.[2]

```
#Panel A
library(ggplot2)
columns=c("fid", "Sample.name", "pca1", "pca2", "pca3",
"pca4", "pca5", "pca6", "pca7", "pca8", "pca9", "pca10")

pca<- read.table(file="1kg _ pca.eigenvec", sep = "",
header=F, col.names=columns)[,c(2:12)]

ggplot(pca, aes(x=pca1, y=pca2))+ geom _ point()+
theme _ bw()+ xlab("PC1") + ylab("PC2")

#Panel B
geo <- read.table(file="1kg _ samples.txt",
sep = "\t",header=T)[,c(1,4,5,6,7)]
```

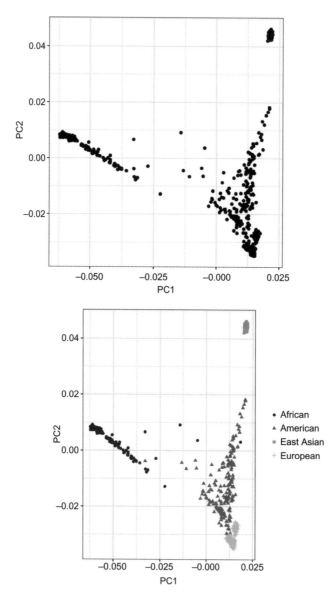

Figure 9.2
Identification of ancestry groups in data along the first two principal components.

```
data <- merge(geo, pca, by= "Sample.name")
ggplot(data, aes(x=pca1, y=pca2, col=Superpopulation.
name))+
    geom _ point()+ theme _ bw()+
    xlab("PC1") + ylab("PC2")+
    labs(col = "")
```

Figure 9.3 shows a *matrix of plots* indicating differences across the 1000 Genomes sample. Interestingly, PCA can distinguish both macro differences in the ancestral groups but also smaller details within a more homogeneous population. Restricting our analysis to an ancestry group, such as individuals with European ancestry, does not protect us from the risk of including bias in the analysis due to population stratification. As an example, we repeat the same plot, but this time only with individuals who have European ancestry (figure 9.4). Although differences across groups are less marked than in the previous figure, it is immediately apparent how different groups can be distinguished by comparing different principal components. (This textbook shows only the greyscale version of all figures, with color plots shown on the online companion website to this book.)

PCA has been shown as a very efficient method to identify individuals with different ancestry and to control for population stratification in analyses. Recently, other methods have also been introduced in GWASs to account for differences in population stratification. The most efficient method is to use a *family fixed effect model*, that is based on the analysis of DZ (dizygotic non-identical) twins or regular siblings. Genetic differences among siblings are due to randomness induced by meiosis, and, by definition, there is no population stratification among biological siblings. Running an association comparing siblings is an effective way to estimate the "real effect" of genetic variants. However, this approach comes at a cost. Since siblings share on average 50% of their genetic variations, many SNPs have exactly the same alleles among siblings. This means that a much larger sample is required if we want to use sibling differences to detect the same level of detail as with a population of unrelated individuals. Geneticists have collected many *family-based studies* in which siblings and twins are both genotyped. However, the sample size of these studies is much smaller than larger data initiatives such as the UK Biobank [4]. Family fixed effect models are often used as *robustness checks* to validate the main results obtained in a larger population. One problem, however, is that small families might be systematically excluded from these types of analyses. Another method that has increasingly been adopted in association studies is to use *linear mixed models* instead of PCA to control for population stratification. Linear mixed models are regression models that take into account the degree of genetic relatedness in the data. This class of models first estimates a matrix of genetic relatedness (see the next section) and includes this information in a regression model. The software BOLT-LMM [5] uses this approach to perform GWASs.

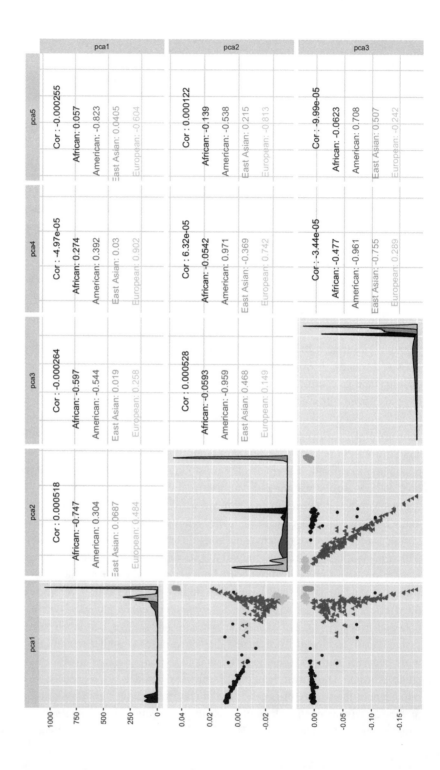

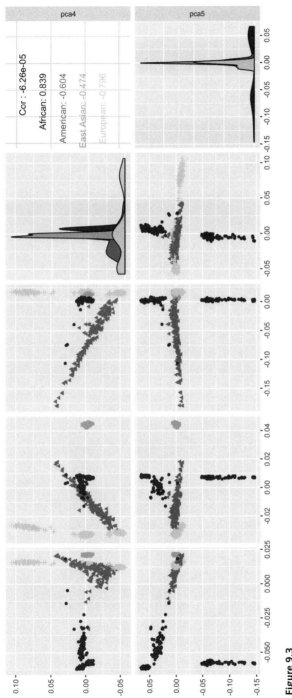

Figure 9.3
Matrix of plots of cross differences across the 1000 Genomes sample.

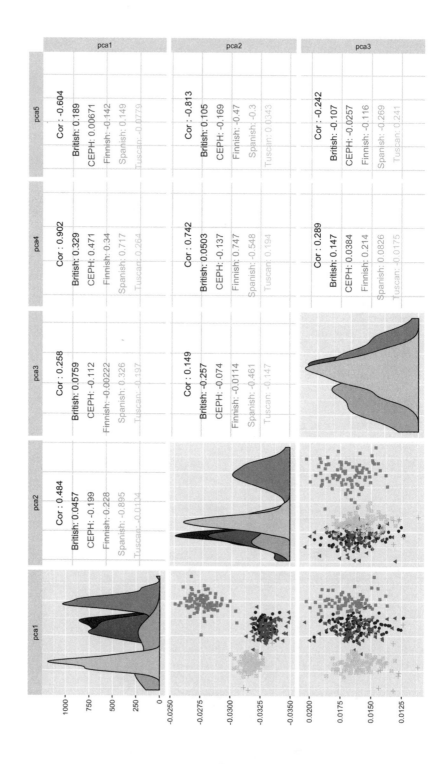

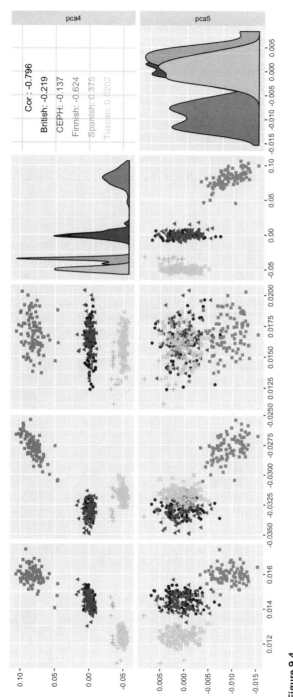

Figure 9.4
Matrix of plots of cross differences across the 1000 Genomes sample, European Ancestry Population only.

9.5   Genetic relatedness

As noted in the previous chapter, duplicates and related individuals can significantly introduce bias in association analysis. To identify related individuals in a genotypic data, a measure called *identity by state (IBS)* can be calculated for each pair of individuals based on the
average proportion of shared alleles that they have in common. IBS is usually calculated on
a set of independent genotyped SNPs. We can use PLINK to calculate IBS based on differences in genotyped SNPs across individuals using the option --distance.

```
./plink --bfile 1kg _ hm3 _ pruned --keep 1kg _ samples _
EUR.txt --distance --out ibs _ matrix
```

```
PLINK v1.90b6.5 64-bit (13 Sep 2018) www.cog-genomics.org
/plink/1.9/
(C) 2005-2018 Shaun Purcell, Christopher Chang GNU
General Public License v3
Logging to ibs _ matrix.log.
Options in effect:
--bfile 1kg _ hm3 _ pruned
--distance
--keep 1kg _ samples _ EUR.txt
--out ibs _ matrix

16384 MB RAM detected; reserving 8192 MB for main
workspace.
166128 variants loaded from .bim file.
1092 people (525 males, 567 females) loaded from .fam.
--keep: 379 people remaining.
Using up to 4 threads (change this with --threads).
Before main variant filters, 379 founders and 0 nonfound-
ers present.
Calculating allele frequencies … done.
166128 variants and 379 people pass filters and QC.
Note: No phenotypes present.
Distance matrix calculation complete.
IDs written to ibs _ matrix.dist.id .
Distances (allele counts) written to ibs _ matrix.dist.
```

The command produces a tab-delimited text file containing the Hamming distances between individuals. The file is organized as follows. The first element is the distance between genome 1 and genome 2. Distances between a genome and itself is omitted as it is of course 0. The default command produces a lower-triangular file. IBS distances are, in fact, symmetric, as a distance between genome 1 and 2 is equivalent to the distance between genome 2 and genome 1. It is possible, however, to produce a square file if we use the command `--distance square`. The distance of the first 10 genomes in the `1kg _ samples _ EUR.txt` file is shown here.

```
head ibs _ matrix.dist
```

```
74407
74208   74953
74755   74396   73099
74429   74572   73941   73346
74049   74752   74473   74218   74544
74792   74829   74704   75257   74797   74915
75499   74890   74743   74858   74740   74780   75025
74915   74522   74291   75122   74148   74892   73305   73548
74601   74286   74731   74774   74502   74734   75247   75260   74556
73911   74190   73761   74260   74132   74018   74595   74408   74186   74116
```

An alternative measure of distances is given by using the command `--make-rel`, which gives a genetic relatedness matrix using the same metric used by the software GCTA.

```
./plink --bfile 1kg _ hm3 _ pruned --keep 1kg _ samples _
EUR.txt --make-rel --out rel _ matrix
```

These results can be used to find cryptic relatedness. The expected value for MZ twins or a duplicated pair is 1; 0.5 for first-degree relatives (e.g., full-sibs or parent–offspring); 0.25 for second-degree relatives (e.g., grandparent–grandchild); and 0.125 for third-degree relatives (e.g., cousins). Please note that these are *expected values* and empirical results may differ. For instance, studies have shown that, although the average genetic distance between biological siblings is 0.5, the typical range of differences ranges from 0.4 to 0.6 [6].

By looking at the results below, we can notice, for instance, that the genetic relatedness of an individual and themselves is different from 1. This is an estimation of the inbreeding coefficient based on the relationship of haplotypes within an individual [7]. Nevertheless, it is interesting to note that some pairs of individuals have higher values of genetic relatedness. Genome 4 and genome 3 have a genetic relatedness higher than expected among unrelated individuals. Based on this value, we can conclude that these individuals are second-degree relatives. At the same time, we can also see that some cells have negative values. This is due to the fact that genetic relatedness values are normalized using the average distance among unrelated individuals. For this reason, it is very important to calculate a genetic relatedness matrix only among individuals from the same ancestry group.

```
head rel _ matrix.rel
```

Showing an excerpt of the file below:

```
0.975623
0.00309471    0.980394
0.00433624   -0.00307206    0.977987
0.00320043    0.00425093    0.0261669    0.984606
```

9.6 Relatedness matrix and heritability with GCTA

Another approach to calculate a relatedness matrix and perform additional analyses is by using the software GCTA (Genome-wide Complex Trait Analysis), developed by Jian Yang and colleagues from the University of Queensland in Australia. Instructions on how to install GCTA are included in appendix 1. GCTA is software that was originally designed to estimate heritability by using genome-wide data but was subsequently extended to perform many other analyses. In this section we briefly present some basic applications of GCTA to estimate the of relatedness of individuals and SNP-based heritability.

GCTA is a command-based software (like PLINK) that can easily work with PLINK binary files. The first step for calculating SNP-based heritability is to calculate a Genomic Relationship Matrix (GRM) using the command --make-grm. Below we provide an example on how to estimate the GRM from the binary PLINK files 1kg _ EU _ qc.bim, 1kg _ EU _ qc.bed, and 1kg _ EU _ qc.fam.

```
./gcta64    --bfile 1kg _ EU _ BMI \
            --autosome \
            --maf 0.01 \
            --make-grm \
            --out 1kg _ gcta
```

The following options are used in this command:

- `--bfile` Similar to PLINK, this part of the command specifies the name of the genotype files used in the analysis.

- `--autosome` is used to select only autosomal chromosomes in the calculation of the GRM (optional).

- `--maf 0.01` is used to select only SNPs with a minor allele frequency less than 1% (optional).

- `--make-grm` is necessary to calculate a GRM.

- `--out 1kg _ gcta` is used to specify the name of the output file.

The command produces four files:

1. `1kg _ gcta.grm.id` contains the IDs of the individuals included in the GRM.
2. `1kg _ gcta.grm.bin` is a binary file with the GRM.
3. `1kg _ gcta.grm.N.bin` is another binary file used by GCT to describe the GRM.
4. `1kg _ gcta.log` is a log file.

The second step is to remove related individuals from the sample. This can be done by removing couples of individuals from the GRM whose relationship is closer than a specified level. In the example below, we remove individuals with a relatedness greater than the cut-off 0.025, but note that one of the individuals is kept. This creates a new matrix called `1kg _ rm025`. Following this example, we removed 59 individuals from the analysis, leaving only 320 in the sample.

```
./gcta64 --grm 1kg _ gcta --grm-cutoff 0.025 --make-grm
--out 1kg _ rm025
```

The third step involves estimated SNP-heritability (h^2_{SNP}; in contrast to family- and GWAS heritability; see chapter 1). To perform the analysis, we need to provide a phenotype. We use the simulated BMI variable, also used in the previous chapter, and apply the option `--pheno` to provide an external file containing the phenotype.

```
./gcta64 --grm 1kg _ rm025 --pheno BMI _ pheno.txt --reml
--out 1kg _ BMI _ h2
```

The summary results of the analysis have been saved in a new file called `1kg _ BMI _ h2.hsq`.

If we open the file from the terminal or using any text editor, you will see the following output:

```
Source          Variance          SE
V(G)            3.363073          8.073311
V(e)            5.260860          8.059914
Vp              8.623934          0.683136
V(G)/Vp         0.389970          0.934748
logL            -506.006
logL0           -506.067
LRT             0.122
df              1
Pval            3.6369e-01
n               320
```

GCTA decomposes the variance of BMI into two parts: V(G), which is the variance explained by additive genetics, and V(e), which is the part due to environment (the nongenetic part). The estimation of heritability (h^2_{SNP}) is the proportion of V(G) over the total variance of the phenotype (Vp). The result is 0.39. We therefore estimate that almost 40% of the variance of this phenotype (BMI) is attributable to genetics.

However, the estimation of heritability comes with a degree of uncertainty. An indication of the precision of the estimation of h^2_{SNP} is given by the standard error of the estimation (SE). In this case the SE is 0.93, which is very high and more than twice the value of the point estimate. This means that the estimate is imprecise. GCTA performs a statistical test called a Log-likelihood ratio test that indicates how the h^2_{SNP} estimate is different from zero, or in other words, the null hypothesis. In this case, given the value of 0.36, we fail to reject the null hypothesis. We therefore cannot conclude from this example that the heritability of BMI is different from zero, meaning that there is no evidence of a genetic component of BMI. In reality, a larger component of BMI is attributable to genetics, (around 20% to 30%) but a larger sample size would be required to detect it using GCTA. This example is based only on our simple simulated test sample of 320 individuals, which is too small to conduct any meaningful analysis using GCTA. In this chapter we were only able to show a very basic application of GCTA. More sophisticated models can be estimated using this software including bivariate models or models including multiple matrices where the genetic variance is decomposed in multiple parts [7, 8].

9.7 Conclusion

Together with chapter 8, the current chapter provided you with a basic introduction on how to work with genomic data. These chapters provide the foundations for the more advanced analyses that we will conduct in the final part of this book. When engaging in more detailed analyses, you will see that it is important to understand how to identify independent SNPs and calculate principle components, genetic relatedness, and heritability. The next chapter now illustrates how you can create polygenic scores. These are then applied in the latter, more advanced chapters in part III of this book.

Exercises

1. Using the file `1kg _ EU _ BMI`, run three linear association models (additive, dominant, and reciessive) for the genetic variant `rs1390260`.

2. Repeat the GWAS analysis of section 9.2 using a dominant association model. Compare the results with the additive model.

3. Using `LDlink`, find possible proxy genotypes for the genetic variant `rs1390260`.

4. Using GCTA, repeat the analysis of section 9.6 to calculate the heritability of height and educational attainment. The phenotypic files are available in the textbook website at http://www.intro-statistical-genetics.com.

Further reading and resources

Chang, C. C. et al. Second-generation PLINK: Rising to the challenge of larger and richer datasets. *Giga-Science* **4**(1), doi:10.1186/s13742-015-0047-8.

Purcell, S. et al. PLINK: A tool set for whole-genome association and population-based linkage analyses. *Amer. J. of Hum. Gen.* **81**, 559–757 (2007).

Yang, J. et al., GCTA: A tool for genome-wide complex trait analysis. *Am. J. Hum. Genet.* **88**, 76–82 (2011).

Further online resources

For a more detailed coverage in PLINK on topics covered in this chapter, see the following:

association analysis: http://www.cog-genomics.org/plink/1.9/assoc.

linkage disequilibrium: http://www.cog-genomics.org/plink/1.9/ld.

population stratification: http://www.cog-genomics.org/plink/1.9/strat.

genetic relatedness: http://www.cog-genomics.org/plink/1.9/ibd and http://zzz.bwh.harvard.edu/plink/ibdibs.shtml.

For more information about GCTA, see https://cnsgenomics.com/software/gcta/#Overview.

References

1. M. J. Machiela and S. J. Chanock, LDlink: A web-based application for exploring population-specific haplotype structure and linking correlated alleles of possible functional variants: Fig. 1. *Bioinformatics* **31**, 3555–3557 (2015).

2. J. H. Marcus and J. Novembre, Visualizing the geography of genetic variants. *Bioinformatics* **15**, 594–595 (2016).

3. M. C. Mills and C. Rahal, A scientometric review of genome-wide association studies. *Commun. Biol.* **2** (2019), doi:10.1038/s42003-018-0261-x.

4.   C. Bycroft et al., The UK Biobank resource with deep phenotyping and genomic data. *Nature* **562**, 203–209 (2018).

5.   P.-R. Loh, G. Kichaev, S. Gazal, A. P. Schoech, and A. L. Price, Mixed-model association for biobank-scale datasets. *Nat. Genet.* **50**, 906–908 (2018).

6.   G. Hemani et al., Inference of the genetic architecture underlying BMI and height with the use of 20,240 sibling pairs. *Am. J. Hum. Genet.* **93**, 865–875 (2013).

7.   J. Yang, S. H. Lee, M. E. Goddard, and P. M. Visscher, GCTA: A tool for genome-wide complex trait analysis. *Am. J. Hum. Genet.* **88**, 76–82 (2011).

8.   F. C. Tropf et al., Hidden heritability due to heterogeneity across seven populations. *Nat. Hum. Behav.* **1**, 757–765 (2017).

10

An Applied Guide to Creating and Validating Polygenic Scores

Objectives

- Recall the *basics of polygenic scores (PGSs)*
- Understand how to *construct a "monogenic" score*
- Comprehend and engage in the seven steps of the *pruning and threshold method*
- Understand and *calculate a PGS using PRSice 2.0*
- Have the ability to *validate PGSs*
- Grasp how to *account for linkage disequilibrium (LD) in calculating PGSs using LDpred*
- Comprehend the *error of applying PGSs across ancestry groups*
- Understand when to consider *pruning and threshold* or an *LD weights-based PGS*

10.1 Introduction

10.1.1 Creating a polygenic score

Recall from part I of this book and the detailed discussion in chapter 5 that most of the traits that we examine are polygenic in nature. For this reason, the majority of contemporary analyses in this area of research now apply polygenic scores. A *polygenic score (PGS)* is an index that aggregates the estimated effects of individual SNPs on the trait of interest. It captures an individual's genetic predisposition to a phenotype. To reiterate from chapter 5, we define a polygenic score for an individual as the *weighted sum* of a person's genotypes at M loci. A PGS for individual i can be calculated as the sum of the allele counts a_{ij} (0, 1, or 2) for each SNP $j = 1, \ldots$ M, multiplied by a weight w_j:

$$PGS_i = \sum_{j=1}^{M} a_{ij} w_j$$

where weights w_j are transformations of genome-wide association studies (GWASs) coefficients. A polygenic score is therefore a *linear combination* of the effects of multiple SNPs on the trait of interest. The underlying model in a PGS is usually additive since we count the number of "risk alleles" for each SNP included in the score, although recessive or dominant models can be used in the construction of a PGS. As noted earlier, we adopt the terminology of score and not *risk* since many of the outcomes we study are also behavioral and would be awkwardly formulated in a risk framework. An additional assumption is the *absence of gene-gene interactions (or epistasis)* since SNP effects are assumed to be independent.

In order to create a PGS, you require summary statistics that are calculated from a GWAS of your trait of interest and the individual-level genotype data (a PLINK binary file, or other formats) in which you would like to apply your PGS. See the Further reading section of this chapter for the location of where you can download summary statistics. As discussed in chapter 4, the GWAS summary statistics should not include the same data that are used for calculating the PGS, which would introduce additional bias leading to overfitting. When constructing a PGS, there are two main decisions:

1. Which (and how many) SNPs will be used to construct the score?

2. What weights will be used?

This chapter examines these choices in more detail and provides some hands-on examples on how to compute a PGS for various traits.

There are two main methodologies used in the literature to select SNPs and their weights in a PGS. The first is called the *pruning and thresholding method*, which uses the pruning techniques described in chapter 9 (section 9.3). The software used for this method are PLINK, R, or PRSice [1] (which combines pruning and thresholding into a single procedure). The second methodology we show uses Bayesian methods and takes into account the linkage disequilibrium (LD) structure of the data to construct a score. The software used for this technique is called LDpred [2]. Although we will discuss both approaches in this chapter, we dedicate more attention to the former since it is currently the most widely used and very flexible to accommodate multiple research requirements. Other methods, in particular new methods that use machine learning approaches, are also emerging in the literature. Although this goes beyond the scope of this introductory approach, we list additional software at the end of this chapter and address some of these more advanced and emerging techniques in the concluding chapter 15 of this book. The expectation is that readers will have already read the detailed introduction of PGS in chapter 5 and preceeding chapters on how to work with genetic data and PLINK. This chapter focuses directly on the estimation of PGS with other chapters in this book outlining important discussions such as the clinically applicability of PGSs, the use of PGSs in Mendelian Randomization (MR), and gene-environment interaction research.

10.1.2   Data used in this chapter

As with the other chapters, all of the data that you will use in this chapter is available on the companion website to this book (http://www.intro-statistical-genetics.com). To summarize, you will need the following data files, some of which were used in previous chapters.

- 1kg_hm3_qc
- score_rs9930506.txt
- BMI.txt
- BMI_pheno.txt
- Obesity_pheno.txt
- 1kg_EU_qc
- pca.eigenvec
- 1kg_samples.txt
- BMI_score_MULTIANCESTRY.best
- BMI_LDpred.txt
- BMI_pheno_LDpred.txt

10.2 How to construct a score with selected variants (monogenic)

Polygenic scores are usually constructed by assembling the effects of thousands or more genetic variants across the entire genome. It is possible, however, to construct a polygenic score with far fewer variants. The most extreme case would be to use a single variant. In that case, the term polygenic is not appropriate. For this reason we term it *monogenic* where in this special case, $M = 1$ and the weights w_{ij} are not necessary. The polygenic score is in this case a list of allele counts, or in other words, a list of 0, 1, 2. In other words, this simple score is the count of a specific allele for each individual. It makes sense to count the alleles that increase the risk of a particular disease, which is why we often refer to the "risk alleles" in the construction of a polygenic score.

This simple score can be constructed in PLINK by using the command --score. This command requires an external file containing the SNP number, the allele to be counted, and a weight. We created a file called score_rs9930506.txt containing only the following elements in one line: rs9930506 G 1 (SNP id; risk allele; weight). rs9930506 is a genetic variant in the FTO gene, and the G allele has been associated with increased risk of obesity. Since we are working with a single variant, we assign the weight equal to 1. When running the syntax, keep in mind that commands can differ by operating system—see box 8.1.

```
./plink    --bfile 1kg_hm3_qc \
           --score score_rs9930506.txt 1 2 3 \
           --pheno BMI_pheno.txt \
           --out 1kg_FTOscore
```

The `--score` command in PLINK requires an external file that has the following information: variant identifier (rs number in most of the SNPs), reference allele (the allele that will be counted in the PGS), and weight (often the Beta coefficients in a GWAS). We also added a phenotype using the `--pheno` option, which is in the file BMI _ phenot.txt. The file used in the example contains only one row:

```
rs9930506 G 1
```

The `--score` command also requires the user to specify the column used in the external file to calculate the PGS. In this case the columns are 1 (SNP id), 2 (reference allele), and 3 (weight).

The result is a file called `1kg _ FTOscore.profile` as specified in the `--out` option in the previous command. It consists of 6 columns and N (number of people genotyped) rows. The first two rows (FID and IID) indicate the family and the individual IDs, the third column (PHENO) contains the phenotype, the fourth column (CNT) indicates the number of possible allele counts taking into consideration missing genotypes, while the fifth (CNT2) is the risk allele count. Finally, the last column (SCORE) indicates the ratio between the previous two columns, that is our *mono*-genic score.

```
head 1kg _ FTOscore.profile
```

FID	IID	PHENO	CNT	CNT2	SCORE
0	HG00096	25.0228	2	1	0.5
0	HG00097	24.8536	2	2	1
0	HG00099	23.6893	2	1	0.5
0	HG00100	27.0162	2	0	0
0	HG00101	21.4616	2	0	0
0	HG00102	20.6736	2	0	0
0	HG00103	25.7151	2	0	0
0	HG00104	25.2522	2	0	0
0	HG00106	22.765	2	0	0

It is important to understand that some individuals may not be genotyped on that specific SNP. In that case the correct value of the polygenic score is missing, not zero. By assigning zero to individuals with a missing genotype, we assign the same value of being homozygous on the other allele. PLINK takes this into account by counting the number of possible alleles.

10.3 Pruning and thresholding method

The so-called pruning and thresholding method is the most commonly used technique to construct a PGS and can be summarized in the following steps. We first summarize the steps and then explain each turn in detail.

1. Obtain GWAS summary statistics from a large discovery sample.
2. Obtain independent sample(s) that has genome-wide data.
3. Align and select SNPs in common between the two samples.
4. Select independent SNPs by pruning based on LD.
5. Restrict to SNPs with a p-value that is lower than a certain threshold.
6. Construct PGS by summing up risk alleles weighted by betas from the summary statistics.
7. Evaluate strength of this association by regressing the phenotype on the PGS.

The method can be used to calculate a score based on any number of genetic variants, including all SNPs. It takes into account the LD structure of the data by selecting independent SNPs to avoid oversampling of more densely genotyped SNPs. For more information on accounting for the LD structure, refer back to chapter 3 (section 3.6, "Linkage disequilibrium and haplotype blocks") and chapter 9 (section 9.3, "Linkage disequilibrium," and within that section the discussion of LD pruning).

Step 1. Obtain GWAS summary statistics from a large discovery sample   Obtaining summary statistics is essential to define PGS and set up weights and define risk alleles. We described why you need to use a large discovery sample in chapter 5 and where you can obtain these summary statistics in chapter 7 and in the "Further reading and resources" section of this chapter.

Step 2. Obtain independent sample(s) that have genome-wide data   Once you have the summary statistics, you will then need to locate genetic sample(s) that can be used to calculate the score. In chapter 7 we provided information about some of the most commonly used human genomic datasets used until now in this area of research. Also, on our GitHub site that accompanied the recent review of all GWASs [3], we produced a list of datasets that have been used in the largest GWAS collaborative samples (see also the "Further reading and resources" section).[1] Central to finding an appropriate dataset is, of course, ensuring that the data has measured the particular phenotype(s) that you are interested in studying. There are many biomedical datasets, some of which were collected to measure a particular disease. Some of the most frequently used studies are broadly phenotyped and contain many measurements. Since many journals now require you to replicate your results, most scientific publications now use more than one dataset to replicate results, particularly when the samples are small.

As we elaborated upon in detail in chapter 5 in our introduction to PGSs, when selecting your data it is also important to note that the *target sample*—which is the genotyped sample you will use—should be not too different from the *baseline sample*. The baseline sample is the sample or the collection of studies that has been used to calculate the original summary statistics of the GWAS you will use. As we have discussed at length in previous chapters, ancestry composition should not differ too much between the baseline and the target sample. If allele frequencies of the SNPs used in the score differ too much between the two samples, this will result in a very imprecise score that cannot be used for any further analysis, even for highly heritable traits. These problems were discussed previously in chapter 3 (section 3.3, "Population structure and stratification"). Using the summary statistics for the phenotype of interest that you obtained in the previous step, it is then possible to examine the genotyped SNPs and weighting of the number of alleles to obtain a PGS for each individual in the sample.

Step 3. Align and select SNPs in common between the two samples   This step is to ensure that SNPs are measured consistently in the baseline and target sample. The alleles reported in the summary statistics should thus be the same as in the genotype data. Unfortunately, this is not always the case. Different genotyping platforms (see chapter 7, section 7.2) may report different alleles for the same SNP. In the case of a mismatch, it is sometimes possible to align SNPs by "flipping" the alleles. In some other cases (see box 10.1), this is not possible, and it is recommended to drop ambiguous alleles from the calculation of the score. Another issue is that the list of SNPs available in the two samples could be different. GWAS summary data typically include both genotyped and imputed data, while this may not be the case for the available genetic data. Different QC criteria may have also filtered out different SNPs. This step, therefore, is to ensure that you select SNPs that are available in both samples.

Step 4. Select independent SNPs by pruning based on LD (optional)   To avoid the oversampling of a region of the genome that is more densely genotyped, it is recommended to select only independent SNPs (see previous discussions on LD). Although this step is optional, using independent SNPs gives more robust results. This is due to the fact that you avoid basing the PGS on specific regions of the DNA without any substantive reason other than genotype density. In other words, we want to avoid double counting causal variants.

To select independent SNPs, two main approaches can be used:

1.   Pruning is a statistical procedure that selects *one random SNP* per LD block.

2.   Clumping, instead, selects the *SNP with the lowest p-value association* in each LD block.

Both pruning and clumping can be performed in PLINK and are included in the PRSice package we describe shortly. Since we are interested in the most predictive SNPs, clumping

Box 10.1
Flipping alleles and ambiguous alleles

As you learned in the chapter 1, DNA is organized in a double-helix structure composed of four bases: adenine (A), thymine (T), cytosine (C), and guanine (G). Adenine pairs with thymine and cytosine with guanine, forming a sequence organized in two strands such as this:

Chromosome 1
Strand 1: ATCTGG**T**ACTCCAT
Strand 2: TAGACCATGAGGTA

Chromosome 2
Strand 1: ATCTGG**C**ACTCCAT
Strand 2: TAGACC**G**TGAGGTA

A SNP indicates a position in the DNA sequence where there is a variation in the population, and it is indicated by the possible alleles. In the example above, the individual genotype in Strand 1 is **TC**, but if we refer to Strand 2 it is AG (usually strands are called + and −).

Unfortunately, sometimes the strand information is not available. This may be problematic if the base and target data were generated using different genotyping chips and the chromosome strand is unknown. The problem is limited when SNPs have the following variants: A/C; A/G; C/T; T/G. For example, if a SNP in the base sample is A/C and T/G in the target, we know that they are simply measured on different strands since T is the complement of A and G the complementary base of C. If this is the case, we need to *flip* the alleles in the base sample by swapping the A with T or C with G. By doing so we ensure that the SNPs are coded consistently across the two samples. Most polygenic score software can automatically perform this flipping.

This is more difficult in the case of the so-called *ambiguous SNPS*, which are SNPs with the following alleles: A/T and C/G. In this case, it is not possible to say whether the base and target data are referring to the same allele or not. Here we can use allele frequencies to infer the matching alleles. For instance, if in our baseline sample we have an A as a minor allele with a frequency of 5% and the target sample has the same allele frequency for the T allele, we can flip the alleles. However, allele frequencies provided in baseline GWAS results are often those from resources such as the 1000 Genomes project. This means that aligning alleles according to their frequency could lead to systematic biases in PGS analyses. A common solution adopted in the construction of genome-wide PGSs is therefore to remove completely ambiguous SNPs.

is preferred since it selects the most statistically significant variant in the locus. For both procedures, we need to provide some parameters indicating the LD window and the r^2 threshold. The recommended clumping r^2 is between 0.5 and 0.7. An r^2 window that is too small would drop potential causal SNPs, while one that is too large would allow too much "double counting." The LD window defines the distance after which we suppose that variants are statistically independent. A common choice for the LD window is 1Mb (see chapter 1, section 1.1.3: a megabase [Mb] is a measure of the length of a genome segment).

Step 5. Restrict to SNPs with a *p*-value lower than a certain threshold   A frequent and vital question is: how many SNPs do we need to use to construct a PGS? The answer depends on multiple factors. A common approach is to select SNPs based on the *p*-value of the association within the summary statistics. We can decide to select only GWAS

significant SNPs (p-value $< 5 \times 10^{-8}$) or at the other extreme, select all SNPs (p-value ≤ 1). The choice depends on the phenotype and the type of application you will conduct. A guideline—which is certainly not exhaustive—is to consider the choice in terms of the level of polygenicity of the trait and the goal of your research. You may recollect from figure 1.3 that there is a spectrum of genetic contributions to a phenotype. Generally speaking, stricter p-value thresholds are more suitable for traits that are not polygenic while more lenient thresholds perform the best for polygenic traits. The aim of your research will also shape your decision. If the goal is to maximize prediction, having more SNPs would be the better choice. However, the more variants that are included in the calculation, the greater the risk that you include unnecessary "noise" in the PGS. In other words, you are likely to include many variants that are not causal. As we describe in box 10.2, most phenotypes are highly polygenic. In general, you will have more predictive results when you include all of the SNPs in the calculation of PGSs for highly polygenic traits.

As we discussed in chapter 6, the use of these scores in gene-environment (G×E) interaction studies is even more complex. In G×E studies we are interested in genetic variants that have a differential effect across different environments. As we explored in previous chapters, some have opted to select a single variant in the FTO gene to demonstrate how birth cohort interacts with genetic predisposition in relation to obesity [6]. Others have opted to use all alleles in their prediction. Barcellos and colleagues [5], for instance, did not impose any p-value threshold in their recent publication. They used the PGS for years of education [7] in interaction with a one-year increase in compulsory education via an econometric technique called regression discontinuity. By comparing individuals before and after the educational reform, they found that differences in health outcomes due to

Box 10.2
Prediction and inclusion of SNPs for highly polygenic traits

As we have argued throughout this book, many of the traits under study and particularly behavioral traits are in fact highly polygenic. PGSs allow us to combine millions of tiny effects that are scattered across the genome. A classic example of a highly polygenic trait is human height. Height, particularly in many Western and European countries, has been shown to be highly heritable at around 80%, with 90% of the variance attributed to additive genetic effects [8]. In 2007, Peter Visscher and colleagues [9] showed that there is a linear relationship between chromosome length and heritability decomposed by chromosome (CIT). In other words, the longer the chromosome, the more variation it explains in adult height. They found that additive genetic variance was spread across multiple chromosomes and that around six chromosomes (3, 4, 8, 15, 17, 18) were responsible for the observed variation. Subsequent studies have confirmed that height is the product of many genes across the entire genome (see also box 1.2 for the discussion of the PGSs in the case of 7′6″ former NBA star Shawn Bradley). For these highly polygenic traits it is almost impossible to select a small number of SNPs to create a polygenic score.

genetic propensity were reduced among the more educated. They concluded that education reduced health risks later in life.

Step 6. Construct PGS by summing up risk alleles weighted by betas from summary statistics   In this sixth step, the PGSs are calculated. This can be constructed in PLINK, or as we will see in the next section, in PRSice.

Step 7. Evaluate strength of this association by regressing phenotype on PGS   A final and important step is to evaluate the *explanatory or "predictive" power* of a PGS. We note that prediction is a misleading term here since we are actually usually interested in understanding how much variability can be explained by including a PGS in a model. In other words, our aim is generally to try to describe the trait in the best way that we can. In this sense, we therefore want to understand the additional gain of including a PGS in a statistical model. A standard procedure is to include population stratification variables (for example the first 10 or 20 principal components) and other covariates in a model. The most frequent manner to describe how much variance is explained by a PGS is to run a regression model, with the phenotype as dependent variable and the PGS as independent variable, and calculate the R^2. We will show you how to do this in the next chapter.

   The interpretation of R^2 is that it is the proportion of variance explained by the regression model. When using covariates, it is common to estimate the gain in R^2 in two steps. First, we estimate a regression model with covariates but without a PGS. In the second step, we add the PGS to the models and estimate the differences in the R^2 from the two models. Since the statistical distribution of the R^2 is not a standard one, it is not possible to estimate a confidence interval unless we use nonparametric statistical techniques such as bootstraping.[2] For binary traits, the approach is very similar, but instead of estimating a series of linear regression models, we estimate logistic regression models and report the gain in pseudo-R^2. Alternatively, it is possible for binary traits to estimate the area under the curve as a measure of accuracy of the PGS to explain the phenotype.

10.4   How to calculate a polygenic score using PRSice 2.0

In this section we provide a hands-on exercise on how to calculate a PGS using PRSice 2.0 (pronounced as "precise"), software based on R and PLINK for calculating, applying, evaluating, and plotting the results of PGS analyses. PRSice requires two sets of information: a list of summary statistics called *base data*, and a genotype file in PLINK format called *target data*. If you would like to actively follow this tutorial, you will first need to install PRSice on your computer. (See appendix 1 for more information on software installation.) Once it is installed, go to the directory where the software is installed and type: ./PRSice to view all of the available parameters. Note that depending on your operating system the downloaded files might have different extensions (i.e., PRSice _ linux, PRSice _ mac, PRSice _ win32.exe, or PRSice _ win64.exe). Please adapt the syntax accordingly or rename the respective file, deleting the _OS extension to execute the PRSice commands in this book.

A genome-wide polygenic score for continuous phenotype

The code below is used to calculate a polygenic score for BMI with no *p*-value thresholds or, in other words, using all available SNPs. The software automatically aligns the *target* dataset to the *base* and performs clumping to avoid double counting of SNPs in LD. Finally, it then calculates a PGS for all the individuals in the target dataset.

```
Rscript PRSice.R --dir . \
    --prsice ./PRSice \
    --base BMI.txt \
    --target 1kg _ hm3 _ qc \
    --snp MarkerName \
    --A1 A1 \
    --A2 A2 \
    --stat Beta \
    --pvalue Pval \
    --pheno-file BMI _ pheno.txt \
    --bar-levels 1 \
    --fastscore \
    --binary-target F \
    --out BMI _ score _ all
```

To explain each of the commands above, we decompose it into various sections.

Setting up the directory and the commands This chunk of code informs your computer about the location of the working directory and the PRSice files. We remind readers of the potential differences in commands that may arise operating systems (recall box 8.1).

```
Rscript PRSice.R --dir . \
    --prsice ./PRSice \
```

Setting up the base and target files The next lines are used to set up the base and target files. As with the previous exercises using PLINK, it easiest to have all of the files in the same directory. These two files thus need to be in the same working directory where you are also executing PRSice.

```
--base BMI.txt \
--target 1kg _ hm3 _ qc \
```

Input file columns   Here we specify the column names of the base file. Alternatively, it is possible to specify the column's numbers if the base file has no header.

```
--snp MarkerName \
--A1 A1 \
--A2 A2 \
--stat Beta \
--pvalue Pval \
```

***p*-value thresholds and phenotype setup**   The last part of the code informs PRSice about which file is the phenotype (--pheno-file) and what *p*-value thresholds (--bar-levels 1) are used in the calculation of the PGS. In this case, we want to calculate only one score with all SNPs thereby specifying (*p*-value ≤ 1). The option --fastcore is used to calculate the score only at the *p*-value thresholds that were specified above, while the option --binary-target F is used to indicate that the phenotype is not binary. Finally, the standard option --out gives the prefix to the output files

```
--pheno-file BMI _ pheno.txt \
--bar-levels 1 \
--fastscore \
--binary-target F \
--out BMI _ score _ all
```

Output
PRSice produces several outputs. In this example, the output files are as follows:

1. A summary file BMIscore _ all1.summary. This file includes information on the number of SNPs used in the score and the R^2 of the PGS. For the PGS that we just calculated, it is based on 118,033 SNPs (after clumping) and explains 13.78% of the variance in BMI.

2. A log file BMIscore _ all1.log contains a log of the operations performed by PRSice.

3. A ".prsice" file contains a summary for each of the *p*-value thresholds specified in the command.

4. A ".best" file contains the most accurate PGS. In this case, as we calculated a single score, this is equivalent to a value of the PGS for each individual in the target sample.

5. A ".png" file with a barplot of R^2 for the different *p*-value thresholds.

PRSice has many options, and we are unable to examine all of them here. Below we have chosen some of the most relevant ones.

Input files

Base dataset As you know, to run PRSice, we need to provide a base dataset. The dataset needs to be whitespace delimited and in a compressed format. The required information for the base file is: effective allele (`--A1`), effect size estimates (`--stat`), *p*-value for association (`--pvalue`), and the SNP ID (`--snp`). We can specify which column of the base file is used in the calculation of the score by specifying them in the PRSice command. In the situation where the input file does not contain a header, columns can be specified using their positions and the `--index` option. For example, the parameters for a base file with the following columns is `--snp SNP --chr CHR --bp BP --A1 A1 --A2 A2 --stat OR --se SE --pvalue P`.

SNP	CHR	BP	A1	A2	OR	SE	P
rs3094315	1	752566	A	G	0.9912	0.0229	0.7009
rs3131972	1	752721	A	G	1.007	0.0228	0.769
rs3131971	1	752894	T	C	1.003	0.0232	0.8962

Target dataset Two different target file formats are supported by PRSice:

1. PLINK Binary files (.bed, .bim., .fam files). Note that the **.bed** and **.bim** file must have the same prefix. If the binary file is separated into individual chromosomes, then a # can be used to specify the location of the chromosome number in the file name. PRSice will automatically substitute # with 1–22.

2. BGEN files contain information on imputation uncertainty. PRSice currently supports BGEN v1.1 and v1.2. To specify type of file you need to specify the option `--type bgen` or `--ld-type bgen`. If this is not specified, then PRSice will use dosage information to compute the score.

Phenotype If a phenotype is included in the .fam file (it is possible to calculate the polygenic score without providing a phenotype), the option `--pheno-file` can be used to specify the name of the phenotype file. The file must include in the first two columns of the family ID (FID) and the individual ID (IID). The third column contains the actual phenotype. Missing values are specified with NA or –9 (only for binary phenotypes).

Clumping

By default, PRSice will perform clumping to remove SNPs that are in LD with each other. Clumping parameters can be changed by using the --clump-kb, --clump-r2, and --clump-p options. Clumping can likewise be disabled by using --no-clump. When the target sample is small (e.g., < 500 samples), an external reference panel can be used to improve the LD estimation for clumping using the option --ld. The default values for clumping are: --clump-kb 250 (PRSice will clump any SNPs that are within 250 kb to both ends of the index SNP, that is a 500 kb window); --clump-r2 0.1 and --clump-p 1. Refer to the previous chapter for a more detailed discussion of LD pruning. By default, PRSice will perform clumping to remove SNPs that are in LD with each other. Clumping parameters can be changed by using the --clump-kb, --clump-r2, and --clump-p options. Clumping can likewise be disabled by using --no-clump.

p-value thresholding

If you do not specify it, PRSice will automatically calculate the PGS for different *p*-value thresholds. It will then also perform a regression to test the level of association of the PGS with your phenotype. If you desire to run an iterative model that specifies the testing of different *p*-value thresholds, you need to specify the interval steps using the options --interval to select the step size of the threshold, and --lower to set the starting *p*-value threshold. If we are interested in specific *p*-value thresholds, it is possible to specify them using the option --bar-levels together with --fastscore. The following options, for example, can be used to calculate the PGS at the *p*-value thresholds: 5×10^{-08}, 5×10^{-07}, 5×10^{-06}, 5×10^{-05}, 5×10^{-04}, 5×10^{-03}, 0.05, 0.5.

```
--bar-levels 5e-08,5e-07,5e-06,5e-05,5e-04,5e-03,5e-02,5e-01 \
--fastscore \
```

Graphical outputs

PRSice will also produce a barplot showing the R^2 of the fitted model, shown in figure 10.1, and the *p*-value threshold that we indicated by the command --bar-levels. If you prefer not to produce this plot, you can specify the command --fastscore.

Additional options

The option --model can be used to select the genetic model used for regression. Other available models include additive, recessive, dominant, and heterozygous-only model. The option --cov-col can be used to include covariates in the calculation of the polygenic score.

Examples

We now provide some examples of how to calculate a PGS using PRSice.

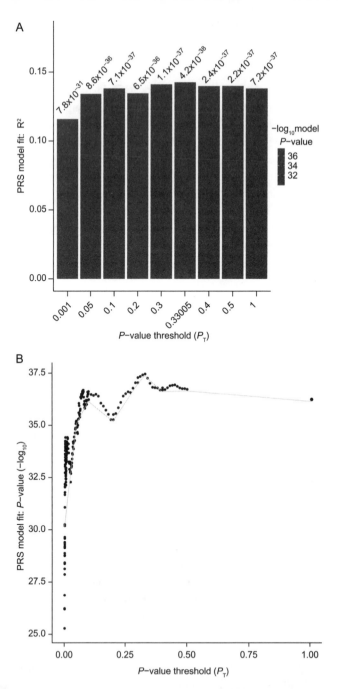

Figure 10.1
Example of PRSice graphical outputs indicating model fit of PGS using different p-value thresholds.
Note: Panel A shows the R^2 associated with different p-value thresholds. The figures on top of the bars in the left panel report the p-value of a statistical test where the null hypothesis is no association between the PGS and the phenotype. Small values indicate that the PGS is statistically significant. Panel B is a high-resolution plot that shows how predictive the different PGSs are, calculated using multiple thresholds. Higher values in the y-axis indicated higher prediction of the associated models. The model with the best prediction is the one using a p-value threshold of 0.33005.

Example 1. Polygenic scores using specific *p*-value thresholds; no phenotype specified

```
Rscript PRSice.R --dir . \
    --prsice ./PRSice \
    --base BMI.txt \
    --target 1kg _ hm3 _ qc \
    --thread 1 \
    --snp MarkerName \
    --A1 A1 \
    --A2 A2 \
    --stat Beta \
    --pvalue Pval \
    --bar-levels 5e-08,5e-07,5e-06,5e-05,5e-04,5e-03,
      5e-02,5e-01 \
    --fastscore \
    --all-score \
    --no-regress \
    --binary-target F \
    --out BMIscore _ thresholds
```

Since we specified the option `--no-regress` above and did not include any phenotype, PRSice will not perform any regression nor will it calculate an R^2. The option `--all-score` produces a new output file called `BMIscore _ thresholds.all.score` with a score calculated at each *p*-value threshold (see below the first 10 rows of this output file).

```
head BMIscore _ thresholds.all.score
```

```
FID IID 5e-08 5e-07 5e-06 5e-05 0.0005 0.005 0.05 0.5 1
0 HG00096    -0.000479696398    -0.000463195405
-0.000355091714    -0.000250196117    -0.000169979944
-0.000102162973    -6.57848378e-05    -3.28632122e-05
-2.4984115e-05
0 HG00097    -0.000661353581    -0.000657264373
-0.000460262493    -0.000359052436    -0.000245826021
```

```
-0.000158377441     -0.000103819954      -5.01441736e-05
-3.75733909e-05
0 HG00099    -0.000609108165    -0.000603264372
-0.00044481341     -0.000327043764     -0.000196803709
-0.00012038489    -8.2112866e-05     -4.14073939e-05
-3.13310687e-05
0 HG00100    -0.000396331442    -0.00044547127
-0.000338678054    -0.000258639555    -0.000165711959
-0.000101754557    -7.16410241e-05    -4.21387148e-05
-3.09468542e-05
0 HG00101    -0.000692441503    -0.000715379318
-0.000546631881    -0.000372120148    -0.000243275257
-0.000140449583    -9.25905804e-05    -4.83711745e-05
-3.64876774e-05
```

Example 2. Polygenic scores using interval *p*-value thresholds; quantile graph included

```
Rscript PRSice.R --dir . \
    --prsice ./PRSice \
    --base BMI.txt \
    --target 1kg _ hm3 _ qc \
    --thread 1 \
    --snp MarkerName \
    --A1 A1 \
    --A2 A2 \
    --stat Beta \
    --pvalue Pval \
    --pheno-file BMI _ pheno.txt \
    --interval 0.00005 \
    --lower 0.0001 \
    --quantile 5 \
    --all-score \
    --binary-target F \
    --out BMIscore _ graphics
```

In this example, we calculated different scores on an interval of *p*-value thresholds. We also include an additional graph that is a quantile plot showing the changes in the phenotype associated with different levels of PGS (see figure 10.2).

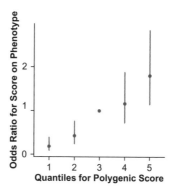

Figure 10.2
Quantile plot from PRSice.
Note: The plot shows changes in phenotype related to different quantiles (in this case 5 quantiles) of the calculated polygenic scores. Bars represent 95% confidence intervals. Higher quantiles are associated with positive changes in the phenotype.

Example 3. Polygenic scores on a binary outcome using interval *p*-value thresholds; no clumping

We now calculate a PGS for a binary phenotype. In the example below, we reclassified the BMI phenotype into binary categories (0 if BMI ≤ 30; 1 if BMI > 30) to mirror the definition of clinical obesity. Note that the quantile plot now reports an odds ratio on the y-axis, since the phenotype is a binary variable.

```
Rscript PRSice.R --dir . \
    --prsice ./PRSice \
    --base BMI.txt \
    --target 1kg _ hm3 _ qc \
    --thread 1 \
    --snp MarkerName \
    --A1 A1 \
    --A2 A2 \
    --stat Beta \
    --pvalue Pval \
    --no-clump F \
    --pheno-file Obesity _ pheno.txt \
    --interval 0.00005 \
    --lower 0.0001 \
    --quantile 5 \
    --all-score \
    --binary-target T \
    --out Obesity _ score _ graphics
```

10.5   Validating the PGS

PRSice may also be used to validate your PGS. In many cases, however, it may be useful to import the scores into another software (for instance R) and run additional checks or to run different models, for instance, gene-environment interaction models (see also chapter 11 on applications of PGS). For this reason, we now import the files and continue further in RStudio.

Importing scores into R and standardizing scores

We can use the scores calculated in the file .all.score or .best to import and examine them further using RStudio. Since the scale of the PGS highly depends on the number of SNPs used in the score (its weighted allele count), PGSs are usually not directly comparable. It is therefore common to *standardize the PGS* by subtracting the mean and dividing the score by the standard deviation. Since it is the sum of independent random variables, the score will be approximately normal (see box 5.1). We can then merge the scores with the phenotypes and the covariates such as principal components. Note that lines that start with the symbol # are not commands but rather notes to explain what each command is doing.

```
In RStudio

# Import external data
data<-read.table("BMIscore _ thresholds.all.score",
header=T)

# Show first rows of data
head(data)

# Calculate new standardized variable
data$PGS=(data$X1-mean(data$X1))/sd(data$X1, na.rm=T)

# Plot histogram of the polygenic score
hist(data$PGS)

# Import external data with phenotype
pheno _ BMI<-read.table("BMI _ pheno.txt", header=T)

# Merge the two datasets
data.with.pheno<-merge(data,pheno _ BMI, by="IID")
```

Run a regression model with covariates

Once you have imported the scores and merged them with the phenotype and covariates, it is possible to run regression models in RStudio and calculate how much of the variability of BMI is explained by including a PGS.

```
In RStudio

# Run a linear regression model
mod1<-(lm(BMI~PGS, data=data.with.pheno))
summary(mod1)
```

Which results in the following output:

```
Call:
lm(formula = BMI ~ PGS, data = data.with.pheno)

Residuals:
Min             1Q              Median          3Q          Max
-7.9255         -1.8779         0.0327          1.8886      9.7165

Coefficients:
                Estimate        Std.Error       tvalue      Pr(>|t|)
(Intercept)     25.00000        0.08432         296.50      <2e-16***
PGS             1.11510         0.08436         13.22       <2e-16 ***

Signif. codes:  0 '***' 0.001 '**' 0.01 '*' 0.05 '.' 0.1 ' ' 1

Residual standard error: 2.786 on 1090 degrees of freedom
Multiple R-squared:  0.1382, Adjusted R-squared:  0.1374
F-statistic: 174.7 on 1 and 1090 DF,  p-value: <2.2e-16
```

Retrieve the model R^2

```
summary(mod1)$r.square
[1] 0.1381616
```

Results show that one standard deviation of the PGS (remember that the score was standardized) is associated with a 1.11 unit increase in BMI. The regression coefficient is statistically significant with a p-value lower than 2×10^{-16}. The R^2 of this regression is

0.13, meaning that more than 13% of the variability of BMI can be explained by regressing the phenotype on the PGS.

```
In RStudio

# Create a vector with column names
columns=c("FID", "IID", "pca1", "pca2", "pca3", "pca4",
"pca5", "pca6", "pca7", "pca8", "pca9", "pca10")

# Read external data with PCAs
pca<- read.table(file= "pca.eigenvec", sep = "" , header=F,
col.names=columns)[,c(2:12)]

# Merge file with covariates with the rest of the file
data.with.covars<-merge(data.with.pheno,pca, by="IID")

# Estimate new linear regression model
mod2<-lm(BMI~PGS+pca1+pca2+pca3+pca3+pca5+pca6+pca7+pca8+
pca9+pca10, data=data.with.covars)

# Calculate R-squared
summary(mod2)$r.square
[1]  0.1443324
```

As described throughout the book, *population stratification* may have a strong effect on many traits and needs to be accounted for. We therefore merge the dataset with the principal components calculated in the previous chapter and re-estimate the model. The result for this new model shows that the R^2 is now larger (14% variance explained), meaning that population stratification matters in explaining part of the variance in BMI—even in an homogenous population with European ancestry. We are therefore interested in calculating how much of the variance is explained by the PGS, net of the first 10 principal components. To do this, we can calculate the differential in the R^2 in the following way:

```
In RStudio

# Estimate linear regression model
mod2.no.pgs<-update(mod2, ~ . - PGS)

# Estimate difference in R-squared
print(summary(mod2)$r.square-summary(mod2.no.pgs)$r.square)
[1]  0.08409551
```

The differential R^2 is 0.08, meaning that, net of population stratification (first 10 components), the additional proportion of variance explained by the PGS is about 8%. We note that since this is a simulated dataset; this would be a relatively extreme example.

It is important, however, to also produce *confidence intervals* on this differential R^2. Because the statistical distribution is not known, it is necessary to use nonparametric statistical techniques such as bootstrap. Below we build a small function in R that calculates confidence intervals for the differential R^2.

```
In RStudio

# Install new package in R to perform bootstrap

install.packages("boot")
library(boot)
set.seed(12345)

# Define new function to calculate differential
R-squared.
# The following commands define the new function
rsq <- function(formula, PGS=PGS, data, indices) {
  d <- data[indices,]
  fit1 <- lm(formula, data=d)
  fit2 <- update(fit1, ~ . + PGS)

return(summary(fit2)$r.square-summary(fit1)$r.square)
}
```

Now that we have created this new R function, it can be used to calculate the *incremental R^2*.

```
# bootstrapping with 1000 replications
results <- boot(data=data.with.covars, statistic=rsq,
     R=1000, formula=BMI~pca1+pca2+pca3+pca3+pca5+pca6+pca7
+pca8+pca9+pca10, PGS=PGS)
boot.ci(results, type="norm")
```

The results of the incremental R^2 are as follows: A 95% confidence interval based on 1,000 bootstrap replications is (0.0560, 0.1133), meaning that the PGS, net of population stratification, explains between 5% and 11% of the variance in BMI.

```
BOOTSTRAP CONFIDENCE INTERVAL CALCULATIONS
Based on 1000 bootstrap replicates

CALL:
boot.ci(boot.out = results, type = "norm")

Intervals:
Level Normal
95% (0.0564, 0.1124)
Calculations and Intervals on Original Scale
```

Why polygenic scores cannot be applied across different ancestry groups: An example
In chapter 3, when we discussed human dispersal out of Africa and population structure
and stratification, we clarified why we cannot apply PGS that have been derived from one
ancestry group directly to another ancestry group (see section 3.3.3 and box 3.2). To clarify why this is problematic, we now estimate an example. Instead of calculating the PGS
on the European-ancestry sample only, from which it was derived, we now (incorrectly)
calculate it on the entire 1000 Genomes sample with PRSice. We then merge the data with
the geographical information from 1000 Genomes and estimate a regression model for the
African subsample.

```
In RStudio

# Import external data with multi-ancestry information

data<-read.table("BMIscore _ MULTIANCESTRY.best", header=T)
head(data)

# Calculate standardized PGS
data$PGS=(data$PRS-mean(data$PRS))/sd(data$PRS, na.rm=T)

data.with.pheno<-merge(data,pheno _ BMI, by= "IID")

# Import dataset with geographical information (ancestral
populations)
geo<-read.table(file= "1kg _ samples.txt", sep = "\t",
header=T)[, c(1,4,5,6,7)]
names(geo)[1]<- "IID "

# Create a vector with new column names
columns=c("FID", "IID", "pca1", "pca2", "pca3", "pca4",
"pca5", "pca6", "pca7", "pca8", "pca9", "pca10")
```

```
# Import exteral data with PCAs
pca<- read.table(file= "pca.eigenvec", sep = "" , header=F,
col.names=columns)[,c(2:12)]

# Merge information with covariates and ancestral
information
data.with.covars<-merge(data.with.pheno,pca, by= "IID")
data.with.geo<-merge(data.with.covars,geo, by= "IID")

# Fit a linear regression model
mod.AFR<-lm(BMI~PGS, data=subset(data.with.geo,
Superpopulation.name== "African"))

summary(mod.AFR)$r.square
[1] 0.03478307
```

The model explains only 3% of the variance for individuals of African ancestry (compared to 12% for European ancestry). If we control for population stratification, the additional R^2 is 0.022 for individuals of African ancestry, with 95% confidence intervals that are overlapping zero, meaning that they are not significantly different from the null hypothesis. In other words, the PGS does not explain the variance of BMI among individuals with African ancestry at all.

```
In RStudio

# Run bootstrap function to estimate increased R-squared
results <- boot(data=subset(data.with.geo, Superpopulation.
name== "African"), statistic=rsq,
    R=1000, formula=BMI~pca1+pca2+pca3+pca3+pca5+pca6+pca7
+pca8+pca9+pca10, PGS=PGS)
boot.ci(results, type= "norm")
```

```
BOOTSTRAP CONFIDENCE INTERVAL CALCULATIONS
Based on 1000 bootstrap replicates

CALL:
boot.ci(boot.out = results, type = "norm")

Intervals:
Level Normal
95% (-0.0118, 0.0528)
Calculations and Intervals on Original Scale
```

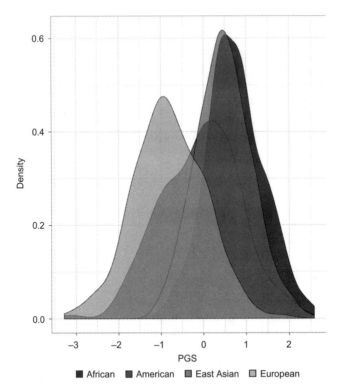

Figure 10.3
Distribution of PGS for BMI for different continental ancestry groups.

Note that our simulated data based on GWAS results from the GIANT consortium (see appendix 2) have a very small sample size for an actual genetic study. Nevertheless, they teach an important lesson. Due to the factors discussed previously, such as differences in genetic diversity (section 3.3, box 3.1) and different LD structure and underlying distribution of scores, it is not correct to simply estimate cross-ancestry PGSs. In figure 10.3 we show the statistical distribution of a PGS for BMI across different populations based on the 1000 Genomes data.

The scores are far from being homogeneous across groups and differ substantially. This has strong implications in the use of polygenic scores as a diagnostic tool in health and other research, for instance, as described in the recent work by Martin and colleagues [10]. As noted elsewhere (see, for example box 3.2 and 14.1), those who do not warrant a citation have also attempted to falsely apply PGSs derived from European ancestry to traits such as intelligence and cognitive ability across different ancestry groups. They then falsely claim that the lower predictive value of the PGS has an inherent value that can be interpreted, which is patently incorrect.

10.6 LDpred: Accounting for LD in polygenic score calculations

10.6.1 Introduction and three steps

Another popular approach to calculate PGS is to use the package LDpred [2]. LDpred is a Python-based software package that adjusts GWAS summary statistics for the effects of LD (see https://github.com/bvilhjal/ldpred and appendix 1). Rather than selecting independent SNPs with clumping (or pruning), this approach uses the entire correlation structure of the data. In practice, LDpred is a different way to estimate the weights w_j—using a **Bayesian approach**—that was shown in the equation at the start of this chapter. Although we do not delve into any detail in this introductory book, the method assumes a point-normal mixture prior for the distribution of effect sizes[3] and takes into account the correlation structure of SNPs by estimating the LD patterns from a reference sample of unrelated individuals. The weight for each variant is set to be equal to the mean of the posterior distribution (approximated via Monte Carlo Markov Chain MCMC[4] simulation) after accounting for LD. LDpred requires an assumption about the fraction of SNPs that are truly associated with the outcome. A common choice for polygenic traits, followed in this book and many studies, is to assume that all of the SNPs are associated with the outcome of interest. Calculation of PGSs in LDpred is conducted in three steps:

1. *Coordinate the base file and the target file*, making sure that the same SNPs are selected and the alleles are aligned.

2. *Calculate the posterior distribution* or, in other words, recalculation of the weights using simulation methods.

3. An optional step is *the calculation of the score using the new weights*. Note that this step is not strictly necessary since the score could be calculated in PLINK or PRSice (excluding clumping).

Since LDpred is software that is written in Python, you will need to install Python on your computer (see appendix 1) in addition to three packages as well as installing the master files for LDpred. Detailed instructions are included in appendix 1. To run the code presented in the last part of this chapter, we assume that the LDpred software is installed in a folder called `ldpred` and that Python 3 is installed in your system and can be invoked by typing `python --version` in your terminal. If this is the case, you should see the following output on your computer screen.

```
python --version
Python 3.6.7:: Anaconda, Inc.
```

If Python 2 is active in your system, please refer to the instructions in appendix 1 on how to switch between different versions of Python.

In the previous section describing PRSice, we used a small simulated dataset. For this section we now apply LDpred to a more realistic example. Here we use the summary statistics for BMI and calculate a polygenic score on the 1000 Genomes sample.

Step 1. Coordinate files   The first step of LDpred coordinates the base and the target file. The base file contains information on summary statistics that are used to define the weights used in the PGS calculation. We created a test file (`BMI _ LDpred.txt`) that has the following columns: Chromosome number (chr), base position (pos), reference allele (ref), alternative allele (alt), reference allele frequency (ref), imputation quality (info), marker name (rs), pvalue (pval), and beta (effalt).

```
head BMI _ LDpred.txt
```

```
chr    pos         ref alt reffrq     info rs         pval        effalt
chr12  126890980   A   G   .2401      1    rs1000000  .95999998   -.000099999997
chr4   21618674    C   T   .49340001  1    rs10000010 .94          .000099999997
chr3   183635768   T   C   .4763      1    rs1000002  .0013        .0055
chr4   95733906    G   T   .38519999  1    rs10000023 .0071999999 .0046999999
chr3   98342907    G   A   .1794      1    rs1000003  .23         -.0029
chr4   139599898   C   T   .47889999  1    rs10000033 .048999999   .0034
chr4   38924330    A   G   .2296      1    rs10000037 .75         -.00060000003
chr4   165621955   G   T   .1478      1    rs10000041 .81         -.00060000003
chr4   17348363    C   T   .26910001  1    rs10000081 .13          .003
```

LDpred needs slightly more information than PRSice. The various types of formats are listed in the accompanying software webpage (https://github.com/bvilhjal/ldpred). There is a good chance, however, that your data are not in the format that may be required. If this is the case, you will first need to rearrange and reformate the data using other software such as R or Stata. The target file is a PLINK binary set of files (see chapter 7).

First, ensure that you are in the directory where LDpred is installed. Then type the following command, which will present all of the options that are available to coordinate the base file and phenotype:

```
python ldpred/LDpred.py coord --help
```

The command shows all of the options that can be used in LDpred. This step requires at least one genotype file that will be used as the LD reference. The software developers recommend that you use at least 1,000 unrelated individuals with the same ancestry composition as the individuals for which summary statistics datasets are obtained from. It is also possible to use another genotype to validate the predictions using a separate set of genotypes.

The most important information required by the command are: `--gf` followed by the prefix of a target genotype file; `--ssf` followed by a file containing summary statistics; `--N` indicating the sample size used in the summary statistics; `--out` followed by the prefix of the coordinated output file; and `--ssf-format=STANDARD`, which indicates that the summary statistics are in the standard file format. This command will produce a file called BMI _ ldpred _ coord that will be used later in the second step of the analysis.

```
python ldpred/LDpred.py coord \
    --gf=1kg _ EU _ qc \
    --ssf=BMI _ LDpred.txt \
    --ssf-format=STANDARD \
    --N 500000 \
    --out BMI _ ldpred _ coord
```

Step 2. Calculate LDpred weights   Once you have coordinated the base and the target file, it is then necessary to calculate the weights that will be used in the PGS calculation. At this time the LD structure is also taken into account in the calculation of the weights. There are two ways to account for LD structure in LDpred. First, by providing a reference file with LD scores, or second, by calculating it from the target dataset. In this example we calculate the LD scores from the target dataset.

Warning! This command can take up to several hours, depending on LD radius and the number of SNPs. The following command can be typed to explore all the options available in this step:

```
python ldpred/LDpred.py gibbs --help
```

The following options then need to be specified next in this step:

`--cf` followed by the coordinated file calculated in step 1;

`--ldr` with the LD radius used to calculated the weights (the recommended number is the number of SNPs used in the analysis divided by 3,000;

`--f` with fraction of causal SNPs;

`--ldf` indicating the prefix of the newly generated LD file based on the LD radius;

`--out` with the prefix of the output file.

In this example, we specified the fraction of causal SNPs equal to 1, meaning that we think that all SNPs have a contribution in BMI. This is used by LDpred to set up the prior distribution. It is possible to provide several values for --f, and several models will be used in the calculation.

```
python ldpred/LDpred.py gibbs \
    --cf BMI _ ldpred _ coord \
    --ldr 240 \
    --ldf 1kg _ bmi \
    --f 1 \
    --N=500000 \
    --out BMI _ weights _ LD
```

This example produces three output files: 1kg _ BMI _ ldradius240.pickled.gz with the LD scores and two files containing LD weights: BMI _ weights _ LD _ LDpred _ p1.0000e+00.txt and BMI _ weights _ LD _ LDpred-inf.txt. LDpred produces a weight file for each value of --f specified (in this case only one), in addition to a baseline file based on an *infinitesimal model*,[5] where all SNPs are assumed to have an infinitesimal contribution to the calculation of the score.

Attention! It can take several hours to compute LD weights, and the resulting file can be very large. The output file 1kg _ BMI _ ldradius240.pickled.gz is 1.77 GB, and larger LD radius will produce even larger files. For this reason you may not want to run this command, and we provide a version of the prepared LD weights file on the companion website to this book (http://www.intro-statistical-genetics.com).

The content of the output files is shown below by examining the first few lines. The columns are: chrom, pos, and sid, indicating chromosome number, base position, and marker name, respectively; nt1 and nt2 indicating reference and alternative alleles; raw _ beta reporting the original association statistics; and finally ldpred _ beta with the calculated LDpred weights.

```
head BMI _ weights _ LD _ LDpred _ p1.0000e+00.txt
```

chrom	pos	sid	nt1	nt2	raw _ beta	ldpred _ beta
chrom _ 1	1005806	rs3934834	T	C	2.3262e-03	6.2367e-04
chrom _ 1	1017587	rs3766191	T	C	3.1698e-03	9.9807e-04
chrom _ 1	1018704	rs9442372	A	G	-4.1348e-04	-4.0554e-05

```
chrom_1   1021346   rs10907177   G   A   -3.2391e-03   -1.3566e-03
chrom_1   1021415   rs3737728    A   G   -3.1074e-03   -6.5916e-04
chrom_1   1021695   rs9442398    A   G   -3.1921e-03   -7.5797e-04
chrom_1   1048955   rs4970405    G   A   -7.0124e-04   -3.0608e-04
chrom_1   1053452   rs4970409    A   G    1.0217e-03    4.0224e-04
chrom_1   1060608   rs17160824   A   G    1.0449e-03    4.4812e-04
```

Step 3. Calculate the LD scores   In the last step, the PGS is then calculated using the option `score`. As in the previous steps, we can list all of the options by adding `--help` at the end of the command:

```
python ldpred/LDpred.py score --help
```

The necessary options are: `--gf`, which specifies the prefix of the target genotype data files; `--rf`, indicating the prefix of the SNP weights file; `--pf`, which directs to an external file with individual IDs and phenotypes; and `--out`, which can be used to specify the name of the output file. The following example is used to calculate the scores:

```
python ldpred/LDpred.py score \
    --gf 1kg_EU_qc \
    --rf BMI_weights_LD \
    --pf BMI_pheno_LDpred.txt \
    --rf-format LDPRED \
    --out BMI_score_LD
```

The command produces a series of outputs with the scores calculated according to the various model specifications. Below is the output file for the model where we assumed that the proportion of causal SNPs in the data is equal to 1. The columns are: `IID`, with the person identifier; `true_phens`, with the original phenotype; `raw_effects_prs`, with the polygenic score using the original weights; and `pval_derived_effects_prs`, the PGS calculated using the LD weights.

```
head BMI_score_LD_LDpred_p1.0000e+00.txt
```

```
IID,          true phens,      raw effects prs,    pva derived effects prs
HG00096,     2.502283e+01,     -9.098944e-01,       1.979399e+00
HG00097,     2.485364e+01,     -2.628862e+00,       1.076786e+00
HG00099,     2.368930e+01,     -1.494926e+00,       1.853949e+00
HG00100,     2.701620e+01,     -9.014597e-01,       1.507434e+00
HG00101,     2.146162e+01,     1.456715e-01,        2.083895e+00
HG00102,     2.067364e+01,     -5.346663e-01,       2.053100e+00
HG00103,     2.571508e+01,     -4.152090e-01,       1.790609e+00
HG00104,     2.525224e+01,     -3.725402e-01,       1.642239e+00
HG00106,     2.276505e+01,     -6.815518e-01,       1.639680e+00
```

An alternative way to calculate polygenic scores from LD-based weights is to apply the PLINK command used in section 10.1 to calculate the score starting from the output files obtained in step 2. This is particularly useful when you do not have the phenotype files. The scores can be calculated as follows (3 4 7 indicate the columns for marker name, reference allele and LD weight; header uses the first raw of the file as file header):

```
./plink --bfile 1kg _ EU _ qc \
    --score BMI _ weights _ LD _ LDpred _ p1.0000e+00.txt 3 4
7 header \
    --out scores _ BMI _ LD
```

10.7 Conclusion

The intention of this chapter was to provide practical guidance on how to construct and examine polygenic scores. We reviewed the basic structure of a polygenic score, starting from the selection of variants and then extending it to the use of summary statistics of genome-wide polygenic scores. A secondary intention of this chapter was to also include a collection of good practices for the estimation of polygenic scores. There are several ways to construct and use polygenic scores (e.g., prediction, G×E model, causal analyses), and it is important for the researcher to choose which method is the most suitable for a particular research question.

After showing the reader two different ways—PRSice and LDpred—to create PGS, your obvious question is: which method should we use? The two methods are complementary and based on slightly different methodology. PRSice is a pruning and thresholding method that selects independent SNPs and calculates the score based on association statistics. LDpred is software based on a Bayesian approach that calculates LD-based weights that can be used to construct PGS. LDpred can give more precise estimates of PGS and increase prediction accuracy, but only under certain circumstances.

For the general calculation of PGS, we tend to recommend PRSice over LDpred in many instances for the beginning researcher for the following reasons. First, LDpred requires a very careful QC of the reference file and small genotyping issues may affect substantially the calculation of LD scores. Genetic ancestry in summary data needs to be consistent with the target file in LDpred, otherwise biases are introduced in the calculation of the PGS. Second, it is recommended to use large genotyping files (at least 1,000 individuals). In our example, we used data from the CEU ancestry (Utah residents with Northern and Western European ancestry) of 1000 Genome consisting of only 379 individuals. Third, we need to make some assumptions on the *prior distribution* that are not always straightforward. This information relates to the proportion of causal SNPs included in the sample. A possible solution (recommended by the software developers) is to estimate different models with different priors and select the model with the best performance. Lastly, the calculation of LD weights is computationally intensive and requires adequate computer power. However, it is vital to note that LDpred has superior prediction performances when using summary statistics from large summary statistics when applied to large datasets.

Exercises

1. Use PRSice to replicate examples 1, 2, and 3 for the following phenotypes: height, age at menarche, and educational attainment. The phenotypic files are available in the textbook website: http://intro-statistical-genetics.com/data.

2. Use RStudio to calculate the R^2 of the polygenic scores calculated in the previous exercise.

3. Use RStudio to reproduce figure 10.3.

Further reading and resources

Euesden, J., C. M. Lewis, and P. F. O'Reilly. PRSice: Polygenic risk score software. *Bioinformatics* **31**(9), 1466–1468 (2014).

Maier, R. M. et al. Embracing polygenicity: A review of methods and tolls for psychiatric genetics research. *Psych. Med.* **48**(7), 1055–1067 (2018).

Maier, R. M. et al. Improving genetic prediction by leveraging genetic correlations among human diseases and traits. *Nat. Comm.* **9**, 989 (2018).

Vilhjálmsson, Bjarni J. et al. Modeling linkage disequilibrium increases accuracy of polygenic risk scores. *Amer. J. of Hum. Gen.* **97**(4), 576–592 (2015).

Web references

LDpred: https://github.com/bvilhjal/ldpred.

List of top genetic data used on GitHub site accompanying Mills and Rahal (2019): https://github.com/crahal/GWASReview/blob/master/tables/Manually_Curated_Cohorts.csv.

Polygenic scores from the Health & Retirement Study: http://hrsonline.isr.umich.edu/index.php?p=shoavail&iyear=ZA.

Polygenic scores from the Wisconsin Longitudinal Study: https://www.ssc.wisc.edu/wlsresearch/documentation/GWAS/.

PRSice-2: https://choishingwan.github.io/PRSice/.

1000 Genomes project data: https://www.cog-genomics.org/plink/1.9/resources.

Video Why are polygenic scores important? Broad Institute, https://www.broadinstitute.org/videos/why-are-polygenic-risk-scores-important.

GWAS summary statistics

The Atlas of GWAS summary statistics includes a database of publically available summary statistics at: https://atlas.ctglab.nl/.

The NHGRI-EBI Catalog (https://www.ebi.ac.uk/gwas/summary-statistics) also contains summary statistics.

References

1.   J. Euesden, C. M. Lewis, and P. F. O'Reilly, PRSice: Polygenic risk score software. *Bioinformatics* **31**(9), 1466–1468 (2014).

2.   B. J. Vilhjálmsson et al., Modeling linkage disequilibrium increases accuracy of polygenic risk scores. *Am. J. Hum. Genet.* **97**, 576–592 (2015).

3.   M. C. Mills and C. Rahal, A scientometric review of genome-wide association studies. *Commun. Biol.* **2** (2019), doi:10.1038/s42003-018-0261-x.

4.   K. Watanabe et al., A global view of pleiotropy and genetic architecture in complex traits. *bioRxiv* (2019), doi:10.1101/500090.

5.   S. H. Barcellos, L. S. Carvalho, and P. Turley, Education can reduce health differences related to genetic risk of obesity. *Proc. Natl. Acad. Sci.* **115**, E9765–E9772 (2018).

6.   J. N. Rosenquist et al., Cohort of birth modifies the association between FTO genotype and BMI. *Proc. Natl. Acad. Sci. USA* **112**, 354–359 (2015).

7.   J. J. Lee et al., Gene discovery and polygenic prediction from a genome-wide association study of educational attainment in 1.1 million individuals. *Nat. Genet.* **50**, 1112–1121 (2018).

8.   A. R. Wood et al., Defining the role of common variation in the genomic and biological architecture of adult human height. *Nat. Genet.* **46**, 1173–1186 (2014).

9.   P. M. Visscher et al., Genome partitioning of genetic variation for height from 11,214 sibling pairs. *Am. J. Hum. Genet.* **81**, 1104–1110 (2007).

10. A. R. Martin et al., Human demographic history impacts genetic risk prediction across diverse populations. *Am. J. Hum. Genet.* **100**, 635–649 (2017).

Applications and Advanced Topics

11

Polygenic Score and Gene-Environment Interaction (G×E) Applications

Objectives

- Understand and be able to conduct *out-of-sample prediction* using polygenic scores (PGSs)

- Perform *cross-trait prediction* and estimate *genetic covariation* using PGSs

- Understand and estimate *genetic confounding* using PGSs

- Engage in an *applied analysis of gene-environment interaction* of BMI×Birth Cohort using an interaction model

- Understand and correct for *interaction dependence on scaling metric* and show *interaction effects by subgroups*

- Grasp the central methodological and statistical concerns of these models including measuring the environment, the importance of main effects, environmental confounding, and lack of diversity of samples

11.1 Introduction

Until now, we have introduced readers to the basic theory and concepts underpinning applied statistical genetic research; provided a background of genetic discovery, data types, and management; and explained how to construct a polygenic score (PGS). Armed with this basic knowledge, it is now possible to learn how to apply these skills. The aim of this chapter is to provide hands-on experience on how to conduct statistical analyses using PGSs, model gene-environment interaction and understand challenges within this area of research. Building on the previous chapters (specifically chapters 5, on how to construct PGSs, and 6, on G×E), we will first provide examples of some basic PGS analyses including out-of-sample prediction, cross-trait prediction, genetic covariation, and genetic confounding. We then focus on gene-environment (G×E) interaction models, with an applied example of body mass index (BMI)×Birth Cohort, and explore the basic conceptual and statistical concerns. The final section expands our discussion from chapter 6 and describes some of the methodological challenges in G×E research and potential solutions.

In part II of this book we used small and simulated datasets to enable you to focus on the commands and analyze data quickly. To ensure that the examples in this textbook are as realistic as possible to allow you to independently carry out your own research, we opted to use a publicly available dataset, which is the U.S. Health and Retirement Study (HRS) (for a description, see chapter 7, box 7.1). A detailed description of how to obtain this data is provided in appendix 2 at the end of this book. The HRS is an excellent and widely used data source that contains both genetic information but also a rich variety of social and phenotypic data. It is also a large sample of more than 37,000 individuals from around 23,000 households and representative for the U.S. population aged 50+ and household members/spouses. Of those, PGS information is available for around 12,090 individuals. Attractive features of this data include that it is longitudinal—individuals have been interviewed at multiple points in time—and it provides prepared PGSs for several traits together with the publicly available phenotypic data. While it is still very important to learn how to generate a polygenic score (as discussed in chapter 10), the inclusion of expertly prepared PGSs by the HRS staff ensures that you can confidentially use them for your analyses. Ensure that you follow the instructions in appendix 2 and use the datafile named hrs_GePhen_uni, which contains 8,451 individuals and more than 10,000 variables (which may change over time with each new data update), before embarking upon this analysis.

11.2   Polygenic score applications: (Cross-trait) prediction and confounding

In chapter 10 we showed how to construct a PGS using different techniques, such as the LDPred and PRSice software [2]. By default, PRSice produces prediction results for the PGSs as well as goodness-of-fit statistics in addition to other outputs. You also learned how to read the generated scores into RStudio. Equipped with this knowledge, we will now show you how to flexibly integrate PGSs into your (causal) modeling. While chapter 10 was based on data from the 1000 Genomes Project and simulated phenotypes, we now use real data (HRS) in order to reproduce or replicate a selected number of findings in the literature. Please note that results may at times differ from the published studies that we replicate. This is due to the fact that we might engage in slightly different—and simplified—analytical strategies as well as the fact that the HRS data is regularly updated and thus may have slightly altered from the original study.

11.2.1   Out-of-sample prediction

In this first exercise, we will show you how to perform out of sample prediction of a phenotype using the HRS data. *Out-of-sample prediction* is based upon selection of single-nucleotide polymorphisms (SNPs) obtained via a genome-wide association study (GWAS) via validation (prediction) on an independent sample that also has the same phenotype [3]. We focus on educational attainment (years of education) because there are two different scores currently available in the HRS data at the time of writing this book. The first is from the 2016 GWAS by Okbay et al. [4] and the second is from the 2018 GWAS by Lee et al. [5]. The main difference between the two studies is the sample sizes, which was around 350,000 in the 2016 study and increased to over 1.1 million individuals in 2018. As shown in chapter 4 (figure 4.4), there was a sharp growth in published GWAS sample sizes, with this 2018 study being one of the largest to date as of the end of 2018. The increased sample size meant substantial increases in both the discovery rate of SNPs as well as the predictive power of the PGS.

We first predict the phenotype years of education based on both PGSs from the 2016 and 2018 study. These are sometimes referred to in the literature as EA2 (i.e., educational attainment 2) for the 2016 score and EA3 for the 2018 score because these were the second and third EA GWASs that were conducted. As we noted in our introduction to PGSs in chapter 5, if you plan to engage in prediction, it is essential that the data that you are using is an *independent sample* or, in other words, was not included in the original GWAS [3]. In our example, although HRS did participate in both GWASs, the score we use—and all HRS PGSs—exclude HRS from the GWAS meta-analysis, which the score construction is based on. Using these scores also provides us with the opportunity to validate the findings from the GWAS and evaluate the missing heritability issue for educational attainment (see chapter 1, section 1.6.4).

In the data set we created, we have 8,451 observations and more than 10,000 variables (see appendix 2). Since we will not use a majority of the variables in this chapter, we first generate a subset of variables to use in our prediction model, with the R code provided below. The model we are interested in can be expressed as follows:

$$EduYears = \mu + \beta_{PGS\_edu}PGS_{edu} + \gamma Z + \varepsilon$$

Where μ is an intercept, β_{PGS_edu} estimates the effect of the PGS for education PGS_{edu} on the observed phenotype years of education (*EduYears*), γ represents a vector of estimates for a vector of control variables Z, which include in our case sex, birth year, and the first five to 10 principal components. The predictive power of the PGS can be evaluated as the incremental R^2 of the PGS, which we obtain if we compare the R^2 from this equation with the one from a model without the score, or in other words:

$$EduYears = \mu + \gamma Z + \varepsilon$$

For a better overview, we first extract the variables that we actually need from the larger file. It is perhaps important to note the naming conventions of the variables in HRS. In general, the first character indicates whether the variable refers to the reference person ("*r*"), spouse ("*s*"), or household ("*h*"). The second character indicates the wave to which the variable pertains: "1"–"10" or "*A*" for all waves. We, for example, need education measured in years (*edyrs*), from the reference person (*r*), and this is the same across all waves (*a*) since the sample is of older individuals, so the variable is named: raedyrs.

Also note that if you read the data in from alternative software packages, the letters in the variable names might appear as capitalized. We also extract two PGSs from the data, which as noted earlier, is from Okbay et al. 2016 [4] (ea_pgs3_edu2_ssgac16) and Lee et al. 2018 [5] (ea_pgs3_edu2_ssgac16) as well as birth year (rabyear), gender (ragender), and the first 10 principal components (pc1_a-e and pc6_a-e).

```
#### Generate a subset of variables for the prediction
exercise
# Define the variables for subset

vars _ prediction <- c("raedyrs","ea _ pgs3 _ edu2 _ ssgac16",
"ea _ pgs3 _ edu3 _ ssgac18","rabyear","ragender","pc1 _ 5a",
"pc1 _ 5b","pc1 _ 5c","pc1 _ 5d","pc1 _ 5e","pc6 _ 10a", "pc6 _ 10b",
"pc6 _ 10c","pc6 _ 10d","pc6 _ 10e")

### Create subset of data
### Follow the instructions in Appendix 2 to create the
dataset hrs _ GePhen _ uni
data _ prediction <- hrs _ GePhen _ uni[vars _ prediction]
```

Our new dataset is thus called: data_prediction. Whenever you generate a new dataset, it is always good to first examine the data to get a grasp of the value ranges, mean level, and distributions. It is also important to check whether there are errors or anomalies, which can happen in even the most expertly curated data sources. For example, it is useful to produce some basic summary statistics and simple density plots. As with the previous chapters, all commands are listed first and all output is shown in as shaded in gray.

```
#### We first generate some descriptive statistics for
the data
# output is shown as shaded in gray

# First, generate some summary statistics for (in ####)

summary(data _ prediction$raedyrs)      ### education years
```

Min.	1st Qu.	Median	Mean	3rd Qu.	Max.	NA's
0.00	12.00	13.00	13.22	16.00	17.00	21

```
summary(data _ prediction$rabyear)        ### birth year
```

Min.	1st Qu.	Median	Mean	3rd Qu.	Max.
1905	1932	1939	1940	1950	1980

```
summary(data _ prediction$ragender)       ### gender
```

Min.	1st Qu.	Median	Mean	3rd Qu.	Max.
1.000	1.000	2.000	1.552	2.000	2.000

```
summary(data _ prediction$ea _ pgs3 _     ### PGS education 2
  edu2 _ ssgac16)
```

Min.	1st Qu.	Median	Mean	3rd Qu.	Max.
-3.688239	-0.679136	-0.001499	-0.005816	0.667879	3.809949

```
summary(data _ prediction$ea _            ### PGS education 3
  pgs3 _ edu3 _ ssgac18)
```

Min.	1st Qu.	Median	Mean	3rd Qu.	Max.
-3.381941	-0.675275	-0.014965	-0.006925	0.663537	3.369625

```
# Now look at the distributions of education years the PGSs

d _ PGS2
<- density(data _ prediction$ea _ pgs3 _ edu2 _ ssgac16)
plot(d _ PGS2, main= "Kernel Density PGS2")
```

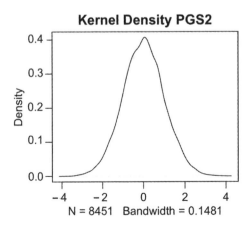

```
d _ PGS3
<- density(data _ prediction$ea _ pgs3 _ edu3 _ ssgac18)
plot(d _ PGS3, main= "Kernel Density PGS3")
```

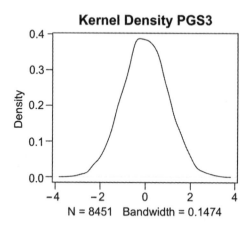

```
d _ EduYears <- density(data _ prediction$raedyrs, na.rm=T)
plot(d _ EduYears, main= "Kernel Density EduYears")
```

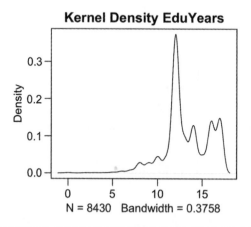

We see that people stayed on average for 13.22 years in education, with a minimum of 0 and a maximum of 17 years. Importantly, 21 observations have missing values here (NAs), so we should not be surprised if they are removed from the analysis. It is also important to consider these missing values in subsequent R commands. Individuals in our HRS sample are born between 1905 and 1980, with a median birth year of 1939 and a mean of 1940.

Gender takes values of 1 and 2, with men coded as 1. The mean of this variable is 1.55, so around 55% of the sample consists of women.

Another important aspect to note is that the mean value of the education PGSs is close to 0. This is attributed to the fact that the PGSs are standardized (as discussed in chapter 10). This means that for each individual's value, the variable mean has been subtracted and the results were divided by the standard deviation. If you ask to view the standard deviations of the scores, you will see that they are close to 1 (type, for example: `sd(data _ prediction$ea _ pgs3 _ edu3 _ ssgac18, na.rm=TRUE`) and the result will be = 0.999).

You will also see that, as we discussed in chapter 5, the PGSs are almost perfectly normally distributed (see box 5.1 for an explanation). The distribution of educational years looks slightly less elegant and, in fact, somewhat left skewed. There are several ways that we could adjust various distributions of the dependent variable, but since this is not the focus of this book and not essential to our exercise, we leave the distribution intact.

The final step is to engage in the actual estimation of the out-of-sample prediction. Below we show how we estimate three linear regression models with all covariates and the PGS for education from the 2016 GWAS, all covariates and the PGS from the 2018 GWAS, and one with only covariates.

```
#### Estimation of statistical models

# Model 1: Edu = PGS2 + Cov

predic _ control _ PGS2 <- lm(raedyrs~ ea _ pgs3 _ edu2 _
ssgac16+rabyear+ragender +pc1 _ 5a+pc1 _ 5b+pc1 _ 5c+pc1 _ 5d
+pc1 _ 5e++pc6 _ 10a+pc6 _ 10b+pc6 _ 10c+pc6 _ 10d+pc6 _ 10e,
data = data _ prediction)

# Model 2: Edu = PGS3 + Cov

predic _ control _ PGS3 <- lm(raedyrs~ ea _ pgs3 _ edu3 _
ssgac18+rabyear+ragender +pc1 _ 5a+pc1 _ 5b+pc1 _ 5c+pc1 _ 5d
+pc1 _ 5e++pc6 _ 10a+pc6 _ 10b+pc6 _ 10c+pc6 _ 10d+pc6 _ 10e,
data = data _ prediction)

# Model 3: Edu = Cov

predic _ control <- lm(raedyrs~ rabyear+ragender +pc1 _ 5a+
pc1 _ 5b+pc1 _ 5c+pc1 _ 5d+pc1 _ 5e++pc6 _ 10a+pc6 _ 10b+pc6 _ 1
0c+pc6 _ 10d+pc6 _ 10e,data=data _ prediction)
```

We use the the the arrow <- sign to save the results of the regression models in the objects predic _ control _ PGS2, predic _ control _ PGS3, and predic _ control, which we can later inspect. To examine the results, use the summary() command.

```
#### Results for Model 1: Edu = PGS2 + Cov

summary(predic _ control _ PGS2)
```

```
Call:
lm(formula = raedyrs ~ ea _ pgs3 _ edu2 _ ssgac16 + rabyear + ragender +
    pc1 _ 5a + pc1 _ 5b + pc1 _ 5c + pc1 _ 5d + pc1 _ 5e + pc6 _ 10a + pc6 _ 10b +
    pc6 _ 10c + pc6 _ 10d + pc6 _ 10e, data = data _ prediction)

Residuals:
Min          1Q        Median      3Q          Max
-14.1182    -1.4157    -0.2347     1.8519      6.3742

Coefficients:
                              Estimate     Std. Error    t value     Pr(>|t|)
(Intercept)                  -66.053976    4.260222      -15.505     < 2e-16  ***
ea _ pgs3 _ edu2 _ ssgac16     0.608594    0.027135       22.429     < 2e-16  ***
rabyear                        0.041196    0.002193       18.783     < 2e-16  ***
ragender                      -0.409135    0.052715       -7.761     9.40e-15 ***
pc1 _ 5a                      24.217770    2.827632        8.565     < 2e-16  ***
pc1 _ 5b                     -11.578142    2.854081       -4.057     5.02e-05 ***
pc1 _ 5c                      -6.268370    2.877701       -2.178     0.0294   *
pc1 _ 5d                      -5.975963    2.880325       -2.075     0.0380   *
pc1 _ 5e                      -1.527099    2.869198       -0.532     0.5946
pc6 _ 10a                     -1.832483    2.900183       -0.632     0.5275
pc6 _ 10b                      2.883569    2.858018        1.009     0.3130
pc6 _ 10c                      4.604652    2.869050        1.605     0.1085
pc6 _ 10d                     -3.448237    2.888447       -1.194     0.2326
pc6 _ 10e                      2.365389    2.852679        0.829     0.4070
---

    Signif. codes:  0 '***' 0.001 '**' 0.01 '*' 0.05 '.' 0.1 ' ' 1

    Residual standard error: 2.399 on 8416 degrees of freedom
       (21 observations deleted due to missingness)
    Multiple R-squared:  0.1139,  Adjusted R-squared:  0.1125
    F-statistic: 83.22 on 13 and 8416 DF,  p-value: < 2.2e-16
```

Examining the regression coefficients first, we see that the PGS for the 2016 educational attainment study is a highly significant predictor of years of education (p-value $< 2e–16$). The regression coefficient is 0.608594 meaning that for an increase of 1 standard deviation in the educational attainment PGS score, schooling increases more than 7 (0.6×12) months. We know from the educational literature that more recent birth cohorts have substantially higher years of education (p-value $< 2e–16$) [6]. In this dataset, men stay significantly longer in education than women (p-value $< 9.40e–15$; coding: men = 1, women = 2). The first four principal components also show significant effects. This implies that within the European ancestral population in the United States, ancestry-based allele frequencies are associated with educational attainment differences, or that population stratification is important for education in this data, respectively. This is most likely attributed to the fact that these allele differences are correlated with environmental differences, and for this reason we should control for PCs in models using PGSs. Refer to chapter 3, section 3.3 for a more detailed explanation about these PCs and how they related to population structure and geography.

Central to this prediction exercise is the R^2, which adjusted for covariates is 0.1125. This means that the full model, including the 2016 PGS for education attainment, explains around 11% of the total variance (see chapter 3) in observed years of education. We now present the results for the 2018 PGS for educational attainment and for the model without PGSs. For parsimoniousness, we omit the estimates of the covariates although note that they will be displayed when you execute the summary commands yourself.

```
#### Results for Model 2: Edu = PGS3 + Cov

summary(predic _ control _ PGS3)
```

```
Call:
lm(formula = raedyrs ~ ea _ pgs3 _ edu3 _ ssgac18 + rabyear +
    ragender + pc1 _ 5a + pc1 _ 5b + pc1 _ 5c + pc1 _ 5d + pc1 _ 5e
    + pc6 _ 10a + pc6 _ 10b + pc6 _ 10c + pc6 _ 10d + pc6 _ 10e, data
    = data _ prediction)

Residuals:
Min            1Q          Median      3Q          Max
-14.1360       -1.4072     -0.1918     1.8007      6.7903

Coefficients:
                        Estimate    Std. Error  t value  Pr(>|t|)
(Intercept)             -68.055982  4.216572    -16.140  < 2e-16 ***
ea _ pgs3 _ edu3 _ ssgac18  0.711476    0.026899    26.450   < 2e-16 ***
```

```
                          [control variables omitted]
---
Signif. codes:   0 '***' 0.001 '**' 0.01 '*' 0.05 '.' 0.1 ' ' 1

Residual standard error: 2.373 on 8416 degrees of freedom
  (21 observations deleted due to missingness)
Multiple R-squared:  0.133,      Adjusted R-squared:  0.1317
F-statistic: 99.32 on 13 and 8416 DF,  p-value: < 2.2e-16
```

```
#### Results for Model 3: Edu = Cov

summary(predic _ control)
```

```
Call:
lm(formula = raedyrs ~ rabyear + ragender + pc1 _ 5a + pc1 _ 5b +
    pc1 _ 5c + pc1 _ 5d + pc1 _ 5e + pc6 _ 10a + pc6 _ 10b + pc6 _ 10c
+
    pc6 _ 10d + pc6 _ 10e, data = data _ prediction)

Residuals:
Min            1Q         Median      3Q            Max
-13.4218       -1.4249    -0.3964     2.0515        5.3466


---
Signif. codes:   0 '***' 0.001 '**' 0.01 '*' 0.05 '.' 0.1 ' ' 1

Residual standard error: 2.469 on 8417 degrees of freedom
  (21 observations deleted due to missingness)
Multiple R-squared:  0.06094,    Adjusted R-squared:  0.0596
F-statistic: 45.52 on 12 and 8417 DF,  p-value: < 2.2e-16
```

Now we see that the adjusted R^2 of the model with the 2018 PGS is higher than the 2016 score (= 0.1317). This means that the 2018 PGS for years of education is a better predictor of educational attainment than the 2016 PGS. This is logical, since as we discussed in our previous chapter on polygenic scores and GWA studies, increasingly larger samples and advances in techniques have resulted in improvements for multiple phenotypes. To assess the h^2_{GWAS} (i.e., the GWAS heritability, see chapter 1) for each PGS, we need to subtract the adjusted R^2 of the model with controls only from the adjusted R^2 of the model including the respective score.

```
#### Calculating explained variance explained by different
models
# PGS 2

summary(predic _ control _ PGS2)$r.square-summary(predic _
control)$r.square
```

0.052964

```
# PGS 3
summary(predic _ control _ PGS3)$r.square-summary(predic _
control)$r.square
```

0.0720726

From this analysis we can draw three central conclusions. First, both the 2016 and later improved 2018 PGSs for years of education are significant predictors of observed educational attainment. This suggests that they can be useful in regression models to model the genetic predisposition for the phenotype. As figure 11.1 shows, in comparison to the standard nongenetic predictors of educational attainment, such as parental educational, the PGS performs well. Results validate the claim from discovery studies to provide us with genetic association results informative for educational attainment.

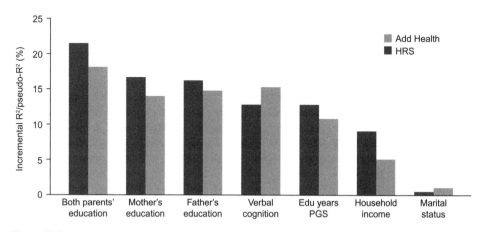

Figure 11.1
Incremental R^2 of years of education (EduYears) PGS compared to other typical variables that predict educational attainment.
Source: Adapted from Lee, Wedow, and Okbay (2018), figure 4 [5]. Add Health refers to the U.S. survey, The National Longitudinal Study of Adolescent to Adult Health.

Second, the PGS of the last 2018 GWAS is a better predictor than the 2016 score, principally attributed to a larger sample size used in the discovery. Note that h^2_{GWAS} for the 2018 PGS we report here is around 0.07 and smaller than reported in the original study (> 0.10, see [4]). The main reason for this is likely that the base dataset (GWAS summary results; see chapter 10) provided for the score construction in HRS excluded not only HRS but also the 23andMe data, which is a large cohort. This therefore makes the discovery sample size much smaller than 1.1 million. In chapter 14, on ethics, and chapter 4, on GWAS [7], we discuss the selective release of summary statistics by direct-to-consumer companies such as 23andMe.

Third and finally, as noted in detail already in chapter 1 (section 1.6.4) in relation to missing and hidden heritability, twin studies suggest that we might be able to explain 40–50% of the variance in education by additive genetic variance [8, 9]. SNP-based studies suggest this might be between 15% and 25% depending on the level of heterogeneity across studies [10, 11]. Whereas here we find that around 7% of the variance can be explained by h^2_{GWAS}, suggesting a substantial part is still missing.

We suggest some basic "data analysis hygiene" to keep your work overviewable. Therefore, after each exercise we will suggest that you clean the working space in RStudio using the following commands:

```
### Workspace hygiene

rm(vars _ prediction,predic _ control,predic _ control _
PGS2,predic _ control _ PGS3,data _ prediction,d _ EduYears,
d _ PGS2,d _ PGS3)
```

11.2.2 Cross-trait prediction and genetic covariation

Perhaps the most rudimentary but also interesting question is whether the same genetic variants are associated with different phenotypes. This question is often key in the study of the etiology of diseases [12, 13]. It is also central in the examination of natural selection and whether genes for a specific phenotype are also associated with lifetime reproductive success [14–16].

Traditionally, twin models have been used to estimate whether the heritability between traits overlaps. In these models, genetic correlations are estimated that examine whether, for instance, phenotype one of twin one predicts phenotype two of twin two better than if they are genetically identical, compared to if they were not identical. As noted earlier, in current research, researchers often use GWAS summary statistics instead of twin models to quantify the genetic overlap between phenotypes (see, for example, [14]). PGSs, however, can also be used to investigate whether the genetic variants that are associated with a specific trait predict another one. The main advantages of using PGSs are again the flexible implementation in regression models and that we can examine these relationships within specific contexts. Remember that when we use the HRS data, we are looking at the older population in the United States.

We now estimate a series of simple models that correlate four phenotypes with the PGS for years of education, for which a phenotype association is well-established. Those phenotypes are: BMI (r*bmi) [18], height (r*height) [19], number of children ever born (r*evbrn) [20], and health [21]. Here we use one indicator in HRS, which is the count of ever reported health problem across the domains of blood pressure, blood sugar, tumors, lung diseases, coronary issues, stroke, emotional problems, and rheumatism (r*conde). The basic model can be written as:

$$Y = \mu + \beta_{PGS\_edu}PGS_{edu} + \gamma Z + \varepsilon$$

Where the notation parallels the one previously discussed and Y is the respective phenotype under study. Since BMI, height, and health problems were measured across multiple waves, for the purpose of this exercise, we first average those variables as well as the respondent's age across the waves.

```
### Average BMI across waves: BIM _ AV

bmi <-c("r1bmi", "r2bmi", "r3bmi", "r4bmi", "r5bmi", "r6bmi",
"r7bmi", "r8bmi", "r9bmi", "r10bmi", "r11bmi", "r12bmi")
hrs _ GePhen _ uni$BMI _ AV = rowMeans(hrs _ GePhen _
uni[,bmi], na.rm = TRUE) summary(hrs _ GePhen _ uni$BMI _ AV)
```

Min.	1st Qu.	Median	Mean	3rd Qu.	Max.	NA's
15.33	24.08	26.87	27.75	30.50	61.78	3

```
### Average height across waves: Height _ AV

height <-c("r1height", "r2height", "r3height", "r4height",
"r5height", "r6height", "r7height", "r8height", "r9height",
"r10height", "r11height", "r12height")
hrs _ GePhen _ uni$Height _ AV = rowMeans(hrs _ GePhen _
uni[,height], na.rm = TRUE)
summary(hrs _ GePhen _ uni$Height _ AV)
```

Min.	1st Qu.	Median	Mean	3rd Qu.	Max.
1.372	1.619	1.689	1.696	1.778	2.053

```
### Average age across waves: AGE_AV

age <-c("r1agey_b", "r2agey_b","r3agey_b","r4agey_b","r5
agey_b","r6agey_b","r7agey_b","r8agey_b","r9agey_b","r1
0agey_b","r11agey_b", "r12agey_b")
hrs_GePhen_uni$Age_AV = rowMeans(hrs_GePhen_
uni[,age], na.rm = TRUE)
summary(hrs_GePhen_uni$Age_AV)
```

Min.	1st Qu.	Median	Mean	3rd Qu.	Max.
32.00	58.00	63.17	65.12	70.92	96.00

```
### Average health across waves: Health_AV

health <- c("r1conde", "r2conde","r3conde", "r4conde",
"r5conde", "r6conde", "r7conde", "r8conde", "r9conde",
"r10conde", "r11conde", "r12conde")
hrs_GePhen_uni$Health_AV = rowMeans(hrs_GePhen_
uni[,health], na.rm = TRUE)
summary(hrs_GePhen_uni$Health_AV)
```

Min.	1st Qu.	Median	Mean	3rd Qu.	Max.
0.000	1.000	1.667	1.760	2.500	7.000

Second, we select the variables that we want to include in our study along with the covariates and generate a subset of data.

```
#### Select variables we would like to use in the
analysis

vars_rG <- c("Height_AV", "BMI_AV", "Health_AV", "ea_
pgs3_edu3_ssgac18", "Age_AV", "raedyrs", "raevbrn",
"rabyear", "ragender", "pc1_5a", "pc1_5b","pc1_5c",
"pc1_5d", "pc1_5e", "pc6_10a", "pc6_10b", "pc6_10c",
"pc6_10d", "pc6_10e")

#### Create subset of data
data_rG <- hrs_GePhen_uni[vars_rG]
```

Third, since the PGS for years of education is standardized, we also standardize the other variables. *Standardization* is the process of putting different variables on a comparable scale, so in the regression models, we can compare genetic and phenotypic associations.

```
#### Standardize variables

data _ rG$Height _ AV _ scaled = scale(data _ rG$Height _ AV)
data _ rG$BMI _ AV _ scaled = scale(data _ rG$BMI _ AV)
data _ rG$raedyrs _ scaled = scale(data _ rG$raedyrs)
data _ rG$Health _ AV _ scaled = scale(data _ rG$Health _ AV)
data _ rG$raevbrn _ scaled = scale(data _ rG$raevbrn)
```

Next, we regress the four standardized phenotypes on the 2018 years of education PGS (models: M_rG*) and educational attainment (M_r*) in separate models, including the same covariates as before.

```
###### Regression models

M _ rGHeight <- lm(Height _ AV _ scaled~ ea _ pgs3 _ edu3 _
ssgac18+rabyear+ragender +pc1 _ 5a+pc1 _ 5b+pc1 _ 5c+pc1 _ 5d
+pc1 _ 5e+pc6 _ 10a+pc6 _ 10b+pc6 _ 10c+pc6 _ 10d+pc6 _ 10e,data
=data _ rG)

M _ rHeight <- lm(Height _ AV _ scaled~ raedyrs _ scaled
+rabyear+ragender +pc1 _ 5a+pc1 _ 5b+pc1 _ 5c+pc1 _ 5d+pc1 _ 5
e+pc6 _ 10a+pc6 _ 10b+pc6 _ 10c+pc6 _ 10d+pc6 _ 10e,data=d
ata _ rG)

M _ rGBMI <- lm(BMI _ AV _ scaled~ ea _ pgs3 _ edu3 _
ssgac18+rabyear+ragender +pc1 _ 5a+pc1 _ 5b+pc1 _ 5c+pc1 _ 5d
+pc1 _ 5e+pc6 _ 10a+pc6 _ 10b+pc6 _ 10c+pc6 _ 10d+pc6 _ 10e,data
=data _ rG)

M _ rBMI<- lm(BMI _ AV _ scaled~ raedyrs _
scaled+rabyear+ragender +pc1 _ 5a+pc1 _ 5b+pc1 _ 5c+pc1 _ 5d+
pc1 _ 5e+pc6 _ 10a+pc6 _ 10b+pc6 _ 10c+pc6 _ 10d+pc6 _ 10e,data=
data _ rG)

M _ rGFert <- lm(raevbrn _ scaled~ ea _ pgs3 _ edu3 _
ssgac18+rabyear+ragender +pc1 _ 5a+pc1 _ 5b+pc1 _ 5c+pc1 _ 5d
```

```
+pc1 _ 5e+pc6 _ 10a+pc6 _ 10b+pc6 _ 10c+pc6 _ 10d+pc6 _ 10e,data
=data _ rG)

M _ rFert<- lm(raevbrn _ scaled~ raedyrs _
scaled+rabyear+ragender +pc1 _ 5a+pc1 _ 5b+pc1 _ 5c+pc1 _ 5d+
pc1 _ 5e+pc6 _ 10a+pc6 _ 10b+pc6 _ 10c+pc6 _ 10d+pc6 _ 10e,data=
data _ rG)

M _ rGHealth <- lm(Health _ AV _ scaled~ Age _ AV+ea _ pgs3 _
edu3 _ ssgac18+rabyear+ragender +pc1 _ 5a+pc1 _ 5b+pc1 _ 5c+p
c1 _ 5d+pc1 _ 5e+pc6 _ 10a+pc6 _ 10b+pc6 _ 10c+pc6 _ 10d+pc6 _ 10
e,data=data _ rG)

M _ rHealth<- lm(Health _ AV _ scaled~ raedyrs _
scaled+rabyear+ragender +pc1 _ 5a+pc1 _ 5b+pc1 _ 5c+pc1 _ 5d+
pc1 _ 5e+pc6 _ 10a+pc6 _ 10b+pc6 _ 10c+pc6 _ 10d+pc6 _ 10e,data=
data _ rG)
```

Finally, we want to examine the results. If you run the script, you will get a visualization of the regression coefficients, sorted by phenotypes and showing the nongenetic versus genetic associations between education and the phenotypes.

```
### First install the R packages needed for visualization

install.packages("tidyverse")
library(tidyverse)
install.packages("broom")
library(broom)
install.packages("stargazer")
library(stargazer)

stargazer(M _ rGHeight, M _ rHeight, M _ rGBMI, M _ rBMI,
M _ rGFert,M _ rFert, M _ rGHealth, M _ rHealth, type =
"text", single.row=T)

m1 <- tidy(M _ rGHeight) %>% mutate(var = "height", var2 =
"genetic") %>% filter(term == "ea _ pgs3 _ edu3 _ ssgac18")
%>% mutate(up=estimate+std.error*1.96, low = estimate
- std.error*1.96)

m2 <- tidy(M _ rHeight) %>% mutate(var = "height", var2 =
'social') %>% filter(term == 'raedyrs _ scaled') %>%
```

```
mutate(up=estimate+std.error*1.96, low = estimate - std.
error*1.96)

m3 <- tidy(M _ rGBMI) %>% mutate(var = "bmi", var2 = 'gene-
tic') %>% filter(term == 'ea _ pgs3 _ edu3 _ ssgac18') %>%
mutate(up=estimate+std.error*1.96, low = estimate - std.
error*1.96)

m4 <- tidy(M _ rBMI) %>% mutate(var = 'bmi', var2 =
'social') %>% filter(term == 'raedyrs _ scaled') %>%
mutate(up=estimate+std.error*1.96, low = estimate - std.
error*1.96)

m5 <- tidy(M _ rGFert) %>% mutate(var = 'fertility', var2 =
'genetic') %>% filter(term == 'ea _ pgs3 _ edu3 _ ssgac18')
%>% mutate(up=estimate+std.error*1.96, low = estimate
- std.error*1.96)

m6 <- tidy(M _ rFert) %>% mutate(var = 'fertility', var2 =
'social') %>% filter(term == 'raedyrs _ scaled') %>%
mutate(up=estimate+std.error*1.96, low = estimate - std.
error*1.96)

m7 <- tidy(M _ rGHealth) %>% mutate(var = 'health', var2 =
'genetic') %>% filter(term == 'ea _ pgs3 _ edu3 _ ssgac18')
%>% mutate(up=estimate+std.error*1.96, low = estimate
- std.error*1.96)

m8 <- tidy(M _ rHealth) %>% mutate(var = 'health', var2 =
'social') %>% filter(term == 'raedyrs _ scaled') %>%
mutate(up=estimate+std.error*1.96, low = estimate - std.
error*1.96)

df <- rbind(m1,m2,m3,m4,m5,m6,m7,m8)
df

df %>% ggplot(aes(x = var2, y = estimate)) +
geom _ bar(stat = 'identity') + facet _ grid(~var) + theme _
bw() +
geom _ errorbar(aes(ymin=low, ymax=up), width=.1,
position=position _ dodge(0.1))
```

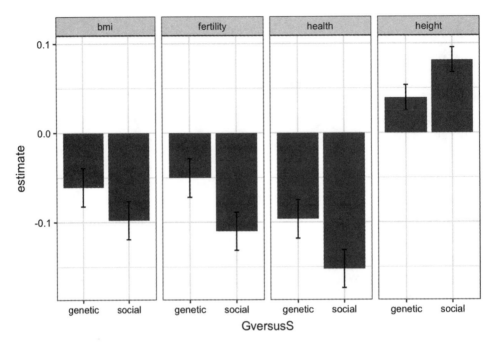

Figure 11.2
Genetic versus nongenetic (social) variables.

Examining figure 11.2, we see that all phenotypes are associated with educational attainment. Individuals with higher education have a lower BMI, lower number of children, better health (fewer reported problematic health conditions), and are taller. This analysis also reveals that this is also true at the genetic level. Genetic variants associated with educational attainment significantly predict lower BMI, lower number of children, better health, and taller stature. Since all variables are standardized, we can compare the regression coefficients and see that the genetic associations are weaker than the phenotypic ones.

Genetic correlations across traits are ubiquitous and well-studied (see also chapter 12 for methods using summary results). An important question, however, is to understand the nature of the association [22]. There are indeed different possibilities. We previously discussed direct and indirect pleiotropy in chapter 5 (section 5.4.3). As we explored in chapter 6, gene-environment correlation can also be responsible for this observation. Different underlying reasons for the observed genetic associations can have very different implications. If we observed direct pleiotropy, for example, we would expect that the phenotypic association is confounded by shared genetic effects across traits, which represent a common cause for the two phenotypes.

In the next section, we study an example from the literature where genetic variants confound an association between phenotypes across generations. Before doing so, however, we suggest that you clean up your workspace.

```
### Workspace hygiene

rm(data _ rG,df,M _ rBMI,M _ rGBMI,M _ rFert,M _ rGFert,M _
rGHealth,
M _ rHealth,M _ rGHeight, M _ rHeight, m1,m2, m3, m4,m5,
m6,m7,m8,health,bmi,
BMI _ AV _ scaled, BMI _ AV,Height _ AV _ scaled, Height _ AV,
height, vars _ rG)
```

11.2.3   Genetic confounding

Genetic confounding might bias estimates of social, health, or lifestyle factors. In previous chapters (e.g., chapter 5, section 5.5.1), we discussed (genetic) confounding in more in
detail, and particularly its role in intergenerational research. One well-studied example is
the intergenerational transmission of educational attainment [23, 24]. Education is a main
predictor of economic success and health across the life course, and therefore the existing
parent–offspring link signals inheritance of social and health inequalities [25]. While for
a long time mainly studied by sociologists and economists [26], the nongenetic inheritance
of education and other traits received (quite) recently major attention in genetic research—
particularly because of gene-environment correlation [25]. With the aid of PGS, we can
model genes and the environment jointly and therefore potentially disentangle their effect.
Let us first take a simple example where we estimate the intergenerational resemblance in
education in the HRS data, which has been done previously [27, 28]. We first extract the
usual control variables from our data and the years of education PGS and years of education of the mother (rameduc) and the father (rafeduc) of the respondent.

```
#### Define the variables for subset

vars _ confounding <- c("raedyrs", "ameduc", "rafeduc",
"ea _ pgs3 _ edu2 _ ssgac16", "ea _ pgs3 _ edu3 _ ssgac18",
"rabyear", "ragender", "pc1 _ 5a","pc1 _ 5b","pc1 _ 5c","pc1 _
5d","pc1 _ 5e", "pc6 _ 10a","pc6 _ 10b","pc6 _ 10c","pc6 _ 10d","p
c6 _ 10e")

#### Create a subset of data

data _ confounding <- hrs _ GePhen _ uni[vars _ confounding]
```

We then estimate a simple model of the intergenerational transmission of educational
attainment, where we jointly fit maternal and paternal effects. The basic model can be
written as:

$$EduYears_{respondent} = \mu + \beta_{edu\_father}Edu\_father + \beta_{edu\_mother}Edu\_mother + \gamma Z + \varepsilon$$

The code to examine this is below:

```
### Intergenerational transmission educational attainment
### Jointly fitting maternal and paternal effects

M _ intergen <- lm(raedyrs~ rafeduc+rameduc+
rabyear+ragender +pc1 _ 5a+pc1 _ 5b+pc1 _ 5c+pc1 _ 5d+pc1 _ 5e
+pc6 _ 10a+pc6 _ 10b+pc6 _ 10c+pc6 _ 10d+pc6 _ 10e,data=d
ata _ confounding)

summary(M _ intergen)
```

```
Call:
lm(formula = raedyrs ~ rafeduc + rameduc + rabyear +
ragender +
    pc1 _ 5a + pc1 _ 5b + pc1 _ 5c + pc1 _ 5d + pc1 _ 5e +
    pc6 _ 10a + pc6 _ 10b + pc6 _ 10c + pc6 _ 10d +
    pc6 _ 10e, data = data _ confounding)

Residuals:
Min           1Q          Median       3Q          Max
-13.6403      -1.4620      -0.1743      1.6051      6.1814

Coefficients:
              Estimate    Std. Error   t value    Pr(>|t|)
(Intercept)   5.176943    4.577607     1.131      0.258
rafeduc       0.163079    0.009717     16.783     < 2e-16 ***
rameduc       0.199456    0.011671     17.089     < 2e-16 ***
                    [control variables omitted]
---
Signif. codes:  0 '***' 0.001 '**' 0.01 '*' 0.05 '.' 0.1 ' ' 1

Residual standard error: 2.195 on 7305 degrees of freedom
    (1131 observations deleted due to missingness)
Multiple R-squared: 0.2158,   Adjusted R-squared: 0.2143
F-statistic: 143.6 on 14 and 7305 DF, p-value: < 2.2e-16
```

We see that both maternal (rameduc) and paternal education (rafeduc), independently, have a positive effect on the respondent's education. For each additional year in school of the father, the respondent has around two months more education $(1.95 = 12 \times 0.163)$, which is slightly higher for mothers $(2.4 = 12 \times 0.199)$.

The next pertinent question is whether this intergenerational transmission of status is genetic or environmental—or perhaps a combination of both. To investigate this, we include the 2018 years of education PGS in the model. The model is written as:

$$EduYears_{respondent} = \mu + \beta_{edu\_father}Edu\_father + \beta_{edu\_mother}Edu\_mother + \beta_{PGS\_edu}PGS_{Edu}$$
$$+ \gamma Z + \varepsilon$$

```
### Model of whether intergenerational transmission of
education is
### Genetic, environmental, or a combination of both

M _ intergen _ PGS2 <- lm(raedyrs~ rafeduc+rameduc+ea _ pgs3 _
edu2 _ ssgac16+ rabyear+ragender+pc1 _ 5a+pc1 _ 5b+pc1 _ 5c+pc1
_ 5d+pc1 _ 5e+pc6 _ 10a+pc6 _ 10b+pc6 _ 10c+pc6 _ 10d+pc6 _ 10e,data
=data _ confounding)

summary(M _ intergen _ PGS2)
```

```
Call:
lm(formula = raedyrs ~ rafeduc + rameduc + ea _ pgs3 _
    edu2 _ ssgac16 + rabyear + ragender + pc1 _ 5a + pc1 _ 5b +
    pc1 _ 5c + pc1 _ 5d + pc1 _ 5e + pc6 _ 10a + pc6 _ 10b +
    pc6 _ 10c + pc6 _ 10d + pc6 _ 10e, data = data _ confounding)

Residuals:
Min         1Q        Median     3Q        Max
-13.7316    -1.4019    -0.1178    1.5241    6.4910

Coefficients:
                         Estimate  Std. Error  t value  Pr(>|t|)
(Intercept)              -2.324933  4.519081    -0.514   0.60694
rafeduc                   0.149716  0.009578    15.631   < 2e-16 ***
rameduc                   0.188384  0.011483    16.406   < 2e-16 ***
ea _ pgs3 _ edu2 _ ssgac16  0.436280  0.026600  16.402   < 2e-16 ***
            [control variables omitted]
---
Signif. codes:  0 '***' 0.001 '**' 0.01 '*' 0.05 '.' 0.1 ' ' 1

Residual standard error: 2.156 on 7304 degrees of freedom
  (1131 observations deleted due to missingness)
Multiple R-squared:  0.2436,  Adjusted R-squared:  0.2421
F-statistic: 156.8 on 15 and 7304 DF,  p-value: < 2.2e-16
```

The respondent's PGS for years of education predicts education independent of parental education. A 1 standard deviation increase in the PGS is associated with over 5 months more schooling ($= 12 \times 0.436$).

Another central research question is how the coefficients for our phenotype change or are reduced due to genetic confounding. In our example, the question is whether the parental transmission of educational attainment changes or is reduced because of genetic confounding. We find that the maternal transmission of the education coefficient reduces by around 5% (from 0.199 to 0.188) and the paternal coefficient by around 8% (from 0.163 to 0.150). In a next step, we now repeat the same estimate, but for this time using the 2018 educational attainment PGS.

```
### Estimate of genetic confounding with the 2018 educational score

M _ intergen _ PGS3 <- lm(raedyrs~ rafeduc+rameduc+ea _ pgs3 _ edu3 _
ssgac18+ rabyear+ragender +pc1 _ 5a+pc1 _ 5b+pc1 _ 5c+pc1 _ 5d+pc1 _ 5e+
pc6 _ 10a+pc6 _ 10b+pc6 _ 10c+pc6 _ 10d+pc6 _ 10e,data=data _ confounding)

summary(M _ intergen _ PGS3)
```

```
Call:
lm(formula = raedyrs ~ rafeduc + rameduc + ea _ pgs3 _
    edu3 _ ssgac18 + rabyear + ragender + pc1 _ 5a + pc1 _ 5b +
    pc1 _ 5c + pc1 _ 5d + pc1 _ 5e + pc6 _ 10a + pc6 _ 10b + pc6 _ 10c
    + pc6 _ 10d + pc6 _ 10e, data = data _ confounding)

Residuals:
Min             1Q          Median      3Q          Max
-13.6763        -1.3805     -0.0945     1.4975      6.1370

Coefficients:
                        Estimate    Std. Error  t value     Pr(>|t|)
(Intercept)             -5.406952   4.496587    -1.202      0.229226
rafeduc                 0.145356    0.009518    15.271      < 2e-16 ***
rameduc                 0.183529    0.011410    16.085      < 2e-16 ***
ea _ pgs3 _ edu3 _ ssgac18   0.519509    0.026687    19.466      < 2e-16 ***
                        [control variables omitted]
---
Signif. codes:  0 '***' 0.001 '**' 0.01 '*' 0.05 '.' 0.1 ' ' 1

Residual standard error: 2.14 on 7304 degrees of freedom
  (1131 observations deleted due to missingness)
Multiple R-squared:  0.2544,    Adjusted R-squared:  0.2529
F-statistic: 166.2 on 15 and 7304 DF,  p-value: < 2.2e-16
```

As one might expect, for the 2018 PGS for years of education, the reduction is slightly stronger, namely around 8% (from 0.199 to 0.184) for maternal transmission and around 11% for paternal transmission (from 0.163 to 0.145).

These results suggest that between 5% and 11% of the parental transmission of education is due to genetic inheritance, and between 89% and 95% is attributed to social or environmental inheritance. However, as discussed previously in chapters 1 and 5, PGSs are imperfect measures of the heritability of a phenotype and therefore these estimates represent the lower genetic boundaries. One solution to this problem is to rescale results under the strong assumption that yet undiscovered (or hidden) genetic effects work exactly the same ways as the ones that we already capture. If we do so, assuming that SNP heritability is 0.20 [10], we have a rescale factor for the 2016 PGS of 1.88 (= 0.2 / 0.052964 / 2) and for the 2018 PGS of 1.39 (= 0.2 / 0.0720726 / 2). This would suggest that maternal transmission is confounded by genes by around 9.5% based on the 2016 PGS estimate and around 11% based on the 2018 PGS. For paternal effects, it would be around 15% for 2016 PGS and 15.3% for the 2018 PGS.

Note that the rescaled results should theoretically be the same when we look at paternal effects by mother or father. In fact, the similarity for paternal effects is quite striking and both PGSs suggest that the inheritance of father's education is 15% due to genetic inheritance. This suggests that the rescaling might be a reasonable approach to tackle the missing heritability issue. However, for mothers the rescaling looks noisier and therefore we recommend caution when interpreting such an approach. Recall also that this is a cohort of individuals aged 50 and older, so the educational attainment of their mothers followed very different patterns. Especially for cross-trait predictions (e.g., [15, 29]), pleiotropic effects might be different between discovered and yet undiscovered SNPs. Thinking of the missing heritability puzzle, one might also want to go one step further and rescale results of PGSs according to SNP heritability, but twin heritability—which would be twice as large, would be accompanied by more strong assumptions.

11.3 Gene-environment interaction

We now build on our theoretical discussion of gene-environment interplay in chapter 6. In this section we provide some examples of applications of GxE interaction, since it has the potential to influence virtually all potential applications of PGSs and beyond. This includes all of the examples shown until now but also Mendelian Randomization (MR) (see chapter 13) as well as genetic discovery and missing heritability [11].. As briefly discussed in chapter 10, there are also some inherent complications concerning the construction of the PGS for GE but also in the discovery of GxE at the SNP level.

Once again we first provide a simple example based on the well-established GxE phenomenon of BMI and birth cohort from the literature [30]. This illustrates how GxE research is currently conducted using PGS and also provides some insights into the potential challenges of such an approach.

11.3.1 Application: BMI × birth cohort

BMI is a central health determinant and is moderately heritable [31]. At the same, it is well-established that genetic effects on BMI are sensitive to environmental factors, for example demographics such as sex, age, and birth cohort and also social factors such as socioeconomic status [30, 32, 33]. One example is that the effect of genetics on BMI increased throughout the twentieth century in the United States, in tandem with the emergence of the so-called obesity epidemic. The theory behind this hypothesis is that more recent birth cohorts became increasingly exposed to high caloric intake and environments at earlier ages. We first examine the HRS data and define and extract the relevant variables.

```
### Define and extract variables for analysis

bmi <-c("r1bmi","r2bmi","r3bmi","r4bmi","r5bmi","r6bmi","r7bmi",
"r8bmi","r9bmi",
"r10bmi","r11bmi","r12bmi")
hrs _ GePhen _ uni$BMI _ AV = rowMeans(hrs _ GePhen _
uni[,bmi], na.rm = TRUE)
summary(hrs _ GePhen _ uni$BMI _ AV)

age <-c("r1agey _ b","r2agey _ b","r3agey _ b","r4agey _ b","r5a
gey _ b","r6agey _ b","r7agey _ b","r8agey _ b","r9agey _ b","r10
agey _ b","r11agey _ b","r12agey _ b")
hrs _ GePhen _ uni$Age _ AV = rowMeans(hrs _ GePhen _
uni[,age], na.rm = TRUE)
summary(hrs _ GePhen _ uni$Age _ AV)

vars _ GxE
<- c("BMI _ AV","Age _ AV","ea _ pgs3 _ bmi _
giant15","rabyear","ragender",
"pc1 _ 5a","pc1 _ 5b","pc1 _ 5c","pc1 _ 5d","pc1 _ 5e",
"pc6 _ 10a","pc6 _ 10b","pc6 _ 10c","pc6 _ 10d","pc6 _ 10e")

data _ GxE <- hrs _ GePhen _ uni[vars _ GxE]
```

Additive model fitting birth cohort and BMI-PGS   We then estimate a model in which we fit both birth cohort and the BMI-PGS [34] additively and look at the results. The model is written as follows:

$$BMI = \mu + \beta_{PGS\_BMI} PGS_{BMI} + \beta_{birth\_cohort} birth\_cohort + \gamma Z + \varepsilon$$

```
### Estimate additive model of birth cohort and BMI PGS

M _ G _ bc <- lm(BMI _ AV~ ea _ pgs3 _ bmi _ giant15 +rabyear + Age _ AV
+ ragender +pc1 _ 5a+pc1 _ 5b+pc1 _ 5c+pc1 _ 5d+pc1 _ 5e+pc6 _ 10a+pc6
_ 10b+pc6 _ 10c+pc6 _ 10d+pc6 _ 10e,data=data _ GxE)
summary(M _ G _ bc)
```

```
Call:
lm(formula = BMI _ AV ~ ea _ pgs3 _ bmi _ giant15 + rabyear +
    Age _ AV + ragender + pc1 _ 5a + pc1 _ 5b + pc1 _ 5c + pc1 _ 5d +
    pc1 _ 5e + pc6 _ 10a + pc6 _ 10b + pc6 _ 10c + pc6 _ 10d +
    pc6 _ 10e, data = data _ GxE)

Residuals:
Min              1Q          Median       3Q          Max
-13.5765         -3.3792      -0.6808      2.6121       30.8307

Coefficients:
                            Estimate   Std. Error   t value   Pr(>|t|)
(Intercept)                 -86.65704  35.89010     -2.415    0.015777 *
ea _ pgs3 _ bmi _ giant15    1.50357   0.05815       25.855   < 2e-16 ***
rabyear                      0.06086   0.01777        3.424   0.000620 ***
                            [Control variables omitted]
---
Signif. codes:  0 '***' 0.001 '**' 0.01 '*' 0.05 '.' 0.1 ' ' 1

Residual standard error: 5.033 on 8433 degrees of freedom
  (3 observations deleted due to missingness)
Multiple R-squared:  0.1251,    Adjusted R-squared:  0.1237
F-statistic: 86.16 on 14 and 8433 DF,  p-value: < 2.2e-16
```

As the output from the results above show, we see that both birth cohorts and the BMI-PGS have independent effects on BMI. A 1-standard-deviation increase in the BMI-PGS is associated with a BMI increase of 1.5. For each year later that an individual is born, their BMI increases on average by 0.06.

Interaction model of birth year by BMI-PGS An additional question is whether the effect of BMI depends on the year an individual is born. To test this, we enter a multiplicative interaction term (birth year $\times$ BMI-PGS) to the equation:

$$BMI = \mu + \beta_{PGS\_BMI}PGS_{BMI} + \beta_{birth\_cohort}birth\_cohort + \beta_{PGS\_BMI\_birth\_cohort}PGS_{BMI} \times birth\_cohort + \gamma Z + \varepsilon$$

```
### Estimate model of multiplicative interaction term of
### Birth year x BMI-PGS

M_Gxbc <- lm(BMI_AV~ ea_pgs3_bmi_giant15 +rabyear + ea_
pgs3_bmi_giant15*rabyear +ragender+Age_AV++pc1_5a+pc1_5b+pc
1_5c+pc1_5d+pc1_5e+pc6_10a+pc6_10b+pc6_10c+pc6_10d+pc6_10e
,data=data_GxE)

summary(M_Gxbc)
```

```
Call:
lm(formula = BMI_AV ~ ea_pgs3_bmi_giant15 + rabyear +
    ea_pgs3_bmi_giant15 * rabyear + Age_AV + ragender +
    pc1_5a + pc1_5b + pc1_5c + pc1_5d + pc1_5e + pc6_10a +
    pc6_10b + pc6_10c + pc6_10d + pc6_10e, data = data_GxE)

Residuals:
Min            1Q       Median     3Q         Max
-13.9415      -3.3807   -0.6661    2.6011     30.9945

Coefficients:
                         Estimate    Std. Error   t value   Pr(>|t|)
(Intercept)              -82.334643  35.833357    -2.298    0.021603  *
ea_pgs3_bmi_giant15      -47.653719  8.741973     -5.451    5.15e-08  ***
rabyear                  0.058750    0.017745     3.311     0.000934  ***
ea_pgs3_bmi_             0.025339    0.004506     5.623     1.93e-08  ***
  giant15:rabyear
                [Control variables omitted]
---
Signif. codes:  0 '***' 0.001 '**' 0.01'*' 0.05 '.' 0.1 ' ' 1

Residual standard error: 5.024 on 8432 degrees of freedom
  (3 observations deleted due to missingness)
Multiple R-squared:  0.1284,     Adjusted R-squared:  0.1269
F-statistic: 82.81 on 15 and 8432 DF,  p-value: < 2.2e-16
```

Examining the output we find that, indeed, there is a significant interaction effect between birth year×BMI-PGS (2.506e-02). However, we also observe something slightly odd. While both of the main effects are still statistically significant, and for birth year it is almost identical across model specifications, the effect size of BMI-PGS on BMI turned strongly negative—which seems highly implausible.

Interaction dependence on scaling metric   Recall from table 6.2 in chapter 6 that one of the central challenges of G×E analysis is that the interaction is dependent on the scaling metric. The problem is due to the fact that the coefficients for the main effects that we observe in the G×E model depend on the scale of the interactive variables. An estimate is presented that assumes the other variable to be zero. Since the BMI-PGS is standardized around the value 0, the estimate for birth year does not change a great deal compared to the model without interaction term. However, birth year is measured in years ranging from 1905 to 1980. The interaction term of 0.025339 indicates that for every year an individual is born later, the effect of BMI-PGS on BMI increases by around 0.025. If we take the effect estimate from the GE model (−47.653719) and project an estimate around the middle of our HRS birth cohorts (1940), we nearly perfectly restore the estimate of the additive model: $−47.653719 + 1940 \times 0.025339 = 1.503941$.

To correct for this—and also for predictions—we can center the variables around 0. Below we center birth year, age and gender and then re-estimate the model.

```
### Center interaction terms and re-estimate model
# Again, we install a helper package first
install.packages('jtools')
library(jtools)

data _ GxE$rabyear _ center <- center(data _ GxE$rabyear)
data _ GxE$Age _ AV _ center <- center(data _ GxE$Age _ AV)
data _ GxE$Gender _ center <-  center(data _ GxE$ragender)

M _ Gxbc _ center <- lm(BMI _ AV~ ea _ pgs3 _ bmi _ giant15 +rabyear _
center + ea _ pgs3 _ bmi _ giant15*rabyear _ center + Age _ AV _ cen-
ter + Gender _ center +pc1 _ 5a+pc1 _ 5b+pc1 _ 5c+pc1 _ 5d+pc1 _ 5e+pc
6 _ 10a+pc6 _ 10b+pc6 _ 10c+pc6 _ 10d+pc6 _ 10e,data=data _ GxE)
summary(M _ Gxbc _ center)
```

```
Call:
lm(formula = BMI _ AV ~ ea _ pgs3 _ bmi _ giant15 + rabyear _ center +
    ea _ pgs3 _ bmi _ giant15 * rabyear _ center + Age _ AV _ center +
    Gender _ center + pc1 _ 5a + pc1 _ 5b + pc1 _ 5c + pc1 _ 5d +
    pc1 _ 5e + pc6 _ 10a + pc6 _ 10b + pc6 _ 10c + pc6 _ 10d +
    pc6 _ 10e, data = data _ GxE)

Residuals:
Min         1Q        Median     3Q        Max
-13.9415    -3.3807    -0.6661    2.6011    30.9945
```

```
Coefficients:
                        Estimate Std. Error t value Pr(>|t|)
(Intercept)            27.746759 0.054692    507.328 < 2e-16 ***
ea_pgs3_bmi_giant15     1.500320 0.058053     25.844 < 2e-16 ***
rabyear_center          0.058750 0.017745      3.311 0.000934 ***
           [Control variables omitted]
---
Signif. codes:  0 '***' 0.001 '**' 0.01 '*' 0.05 '.' 0.1 ' ' 1
Residual standard error: 5.024 on 8432 degrees of freedom
 (3 observations deleted due to missingness)
Multiple R-squared:  0.1284,    Adjusted R-squared:  0.1269
F-statistic: 82.81 on 15 and 8432 DF,  p-value: < 2.2e-16
```

Two things change in this model specification. First, as you can see, the estimate for the BMI-PGS is nearly perfectly restored. Second, the intercept (which gives the average BMI for individuals in case all variables would be zero) changed from –82.334643 to 27.746759 and is now basically identical with the true sample mean of 27.75.

To visualize how the effect of the BMI-PGS changes depending on birth cohort, we can examine the change in the coefficient using the R-package interplot.

```
### Install R packages for visualization
install.packages('arm')
library(arm)
install.packages('lme4')
library(lme4)
install.packages('interplot')
library(interplot)

### Produce plot of GxE interaction

interplot(m = M_Gxbc_center, var1 = "ea_pgs3_bmi_
giant15", var2 = "rabyear_center")+
  xlab("Year born") +
  ylab("Estimated Coefficient for BMI genetic score") +
  theme_bw() +
  ggtitle("GxE interaction between PGS for BMI and birth
year") +
  theme(plot.title = element_text(face="bold"))+
  geom_hline(yintercept = 0, linetype = "dashed")
```

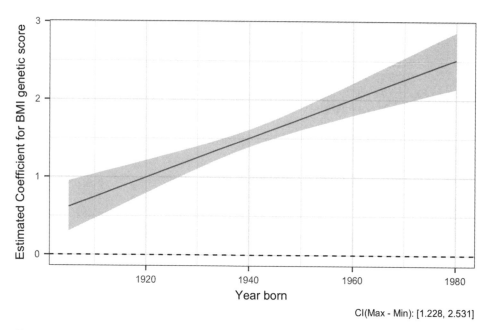

CI(Max - Min): [1.228, 2.531]

Figure 11.3
G×E interaction between the BMI-PGS and birth year.

In figure 11.3, you can nicely visualize how the BMI-PGS is a significant predictor of BMI across all birth years and more easily identify the estimates for the birth year and cohort-specific effects. Note that this plot would look identical in the model where variables are not centered, emphasizing that the rescaling mainly makes interpretation of the table easier.

However, note that implications of the scaling may largely depend on the shape of the interaction and the effect sizes. While in this example all estimates for BMI are above zero, it is perfectly possible that for other examples they may even switch signs based on the environmental treatment (see also the differential susceptibility model in chapter 6). In that case, for some values of E, the PGS would have no significant effect on the outcome. Depending on the scaling of E, a regression table might suggest no association between PGS and the phenotype, while this is only true for the specific location on the E scale, for which the estimate is shown. Drawing a full picture is thus very important for the interpretation of interaction effects.

Showing interaction effects by different subgroups   Another common way to present interaction effects is to show the predicted values of the outcome variable based on different subgroups of the interaction term. Below we use the example where we predict BMI based on different BMI-PGS values, which are then differentiated by birth cohort.

```
### Install R package to produce graph
install.packages("jtools")
library(jtools)
install.packages("interactions")
library(interactions)

### Create interaction plot of predicted BMI by BMI-PGS x
Birth Year

interact _ plot(M _ Gxbc, pred = "ea _ pgs3 _ bmi _ giant15",
modx = "rabyear", interval = TRUE,int.width = 0.95,
x.label="BMI-PGS", y.label="Predicted value BMI", main.title
= "Predicted BMI by birth year", legend.main="Birth year")
```

In figure 11.4, we see the predicted value of BMI for individuals based on changes in the BMI-PGS—holding everything else constant at its mean. If we, for example, assume that the BMI-PGS is 2 standard deviations below the mean, we would expect a BMI of around 25. If we observe an individual with a BMI-PGS that is 2 standard deviations above the mean, we would expect them to have a BMI between around 29 and 33; more importantly, this graph shows that this is dependent on the individuals' birth year.

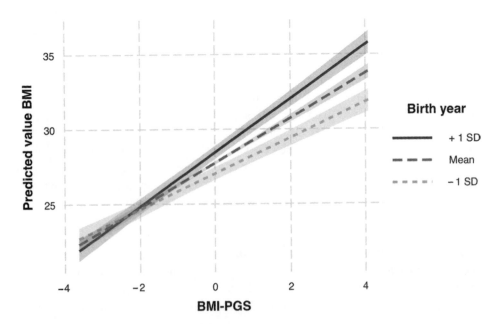

Figure 11.4
Predicted BMI value by BMI-PGS × birth year.

If we recall the theory on G×E from chapter 6, this analysis and the figure above now supports the diathesis-stress or contextual triggering model, namely that the exposure to a high caloric environment of more recent birth cohorts has triggered the predisposition (diathesis) for high BMI risk. Finally, note that an interaction always travels both ways. Another way of looking at this is to ask whether the effects of birth cohorts—the obesity epidemic—had a differential effect on individuals based on their genotypes. To test this, we simply need to exchange the predictor and the moderating variables in our plot.

```
### Create interaction plot of predicted BMI by Birth
Year x BMI-PGS

interact _ plot(M _ Gxbc, pred = "rabyear", modx = "ea _
pgs3 _ bmi _ giant15", interval = TRUE, int.width = 0.95,
x.label="Birth year", y.label="Predicted value BMI", main.
title = "Predicted birth year by BMI-PGS", legend.
main="BMI-PGS")
```

In figure 11.5, see that it also works the other way around. People born in more recent birth years have a higher predicted BMI. However, this trend affected individuals at elevated genetic risk for obesity (i.e., higher BMI-PGS) worse than those with average or low risk for high BMI.

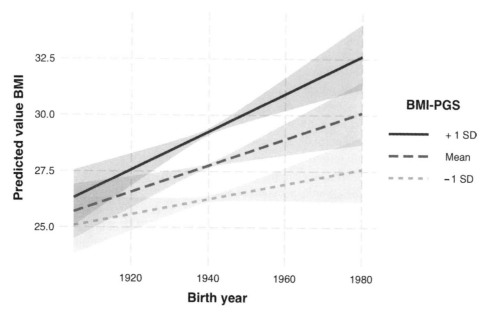

Figure 11.5
Predicted BMI value by birth year × BMI-PGS.

11.4   Challenges in gene-environment interaction research

In table 6.2 we presented a list of challenges in GxE research, partly focusing on the discovery studies of GxE effects (e.g., lack of power), but most of the concerns extend to the application of PGSxE models and some are PGS specific. In the following, we discuss selected challenges in detail and readers should refer to chapter 6 for a more general discussion.

Measuring the environment

The complexity of measuring environmental exposure is also viewed as one of the central challenges of GxE research. First, as discussed previously in chapter 6, environmental exposures are highly interrelated and complex. Second, exposures are also highly heterogeneous and may be continuous or discrete measures. Third, they are often multilevel—meaning that measurements may be at the level of the country, community, school, or household, which includes very different considerations regarding the level of measurement error. Fourth, since environment is multi-temporal, environmental exposure and their biological impact can vary across the life course and historical period. Researchers, therefore, need to be aware of these differences when interpreting results. Environmental measures such as tax policies on cigarettes or air pollution levels vary considerably across time, space, and geographical location. Finally, as we elaborated upon in chapter 6, environmental conditions and exposure are often highly interrelated and correlated with one another. This makes it often very difficult to definitely identify the influence of one particular and independent environmental factor.

Statistically controlling for the environment and confounders   Especially the latter point concerning correlations of environmental factors—and also gene-environment correlation—need our attention for the statistical models. The causal models presented in chapter 2 are ideal types, and we are mainly interested in, for example, how the environment modifies genetic effects. In reality, however, the environment might also have a direct effect on the outcome of interest, which might be correlated with genetic effects. In such a scenario (which is normally the case), it is imperative to include both main effects as well as the interaction term in the model, which is of course a standard.

Gene-environment interaction can furthermore be driven by endogeneity, and specifically omitted variable bias, so the gene-environment interaction might be confounded by third variables—just as direct associations as we discuss in detail in chapter 13. Readers interested in this issue should refer to the excellent article by Keller [35], which addresses this problem using examples from the literature. The simple solution to this problem is that you need to enter the covariate-by-environment and the covariate-by-gene interaction in the same model that tests the GxE interaction and not only control for main effects of the confounders for the interaction. Ideally, you can think of it as an exogenous environmental treatment, which is clearly identified and therefore interpretation is straightforward.

Sample size requirements   G×E studies demand considerably larger sample sizes than those that examine only direct effects of genetics or environmental effects [36]. Often a central goal of G×E studies is to improve the power to detect associations. Some have discussed different approaches in the area of case-control studies as alternatives to traditional G×E tests [37]. This includes approaches such as case-only studies, Bayesian model averaging, joint tests, and two-step approaches, which go beyond the auspices of this book but should be consulted especially if you engage in G×E discovery.

Environmental robustness of GWAS meta results

A concern in PGS-based studies and those utilizing results from large-scale GWASs in general is that those results are often based on dozens of datasets from various countries and birth cohorts—in short, they are averaged across various environments. This has important implications for the use of such results for follow-up G×E research. Likely, this explains why there are still many null-findings in G×E research using polygenic scores [38]. In fact, one of our studies [39] has shown that heterogeneity of genetic effects across GWAS discovery sample partly explains missing (more specific hidden) heritability for traits such as fertility, educational attainment, and BMI. If we focus on average genetic effects across heterogeneous environments, the prediction based on genetic effects will therefore be smaller than within environments. Furthermore, the detected common genetic effects across environments are potentially relatively robust to G×E research. Again, one solution might be to find a truly exogenous environmental interaction term (such as a country-specific school reform [4]) or potentially to use large homogeneous data sources such as the UK Biobank for the discovery and study of G×E in more heterogeneous prediction environments, whereas the interpretation in such studies will be potentially different to the general framework we present here.

Lack of diversity of genetic samples

As described in table 6.2 and also previously in more detail in chapter 4 and elsewhere [7], the majority of results from GWASs as of the end of 2018 have been dominated by European-ancestry participants, nonrepresentative healthy individuals, often from high socioeconomic environments. This is particularly disconcerting considering the fact that there are high socioeconomic gradients in health behavior and many of the traits under study [40]. Furthermore, as outlined previously, key G×E theories also hypothesize and demonstrate a differential impact of genetic predisposition in relation to the environment.

Although the use of representative populations is commonplace in social science and demographic research, the focus on studying particular case-control disease groups or drive to obtain the largest sample possible has resulted in highly selective populations forming the basis of GWAS research. As Boardman, Daw, and Freese [41] and McAllister et al. [42] noted, this selectivity is particularly problematic for G×E research (see box 11.1). In fact, as of 2018, one of the largest publicly available datasets for genetic research

Box 11.1
Why would sample selectivity and lack of diversity in genetic data be a problem for G×E research?

Disease or the development of detrimental health behavior such as smoking, sedentary life-style, or high levels of alcohol consumption are often a key area of research in genetic and G×E studies. If we consider the key theoretical models and research questions of G×E, one theory, for example, is that there is differential susceptibility to genetic loci that put people at risk for detrimental health behavior such as smoking or drinking. If we would like to study smoking or detrimental health behavior and outcomes in a study such as the UK Bio-bank or others where few people smoke cigarettes and are largely healthy, we may falsely conclude that a particular allele might be protective or beneficial for health.

Conversely, if we used data from a sample where smoking was widespread and individu-als were generally less healthy, those with a particular allele might in fact be the ones that either smoke the most or have the poorest health outcomes. If we used a broad and represen-tative sample, we may not be able to observe any clear association because they would can-cel each other out. As Boardman and colleagues [41] conclude, "Without a complete representation of the individuals across the full range of environments, researchers can only tell one part of the story" (p. S68). Since there is an increasing move to use PGS for policy applications and interventions and the results of GWA studies for pharmaceutical develop-ment, this lack of representative samples and environments requires serious attention.

is the UK Biobank, which consists of around 500,000 individuals and has a 5.5% response rate, including largely healthy, older, and high socioeconomic status participants [43].

11.5 Conclusion and future directions

G×E analysis is perhaps one of the most exciting frontiers in this area of research. Since many complex traits and behaviors have a strong social and environmental component, we anticipate that this area of research may hold the key in unlocking further understand-ing. Just as genetic research has rapidly evolved from classic family designs to whole-genome data with increasing precision, so too have environmental measures. We urge interdisciplinary researchers not to "reinvent the wheel" when devising environmental measures and refer to the rich body of nongenetic research within the social sciences that has developed precision measures of the social environment. This includes basic socio-economic measures such as income, educational level, and labor market participation but also environmental measures such as neighborhood deprivation.

As noted throughout this book, the future of this area of research and particularly G×E research relies on moving toward more representative and diverse populations and in partic-ular not only non-European ancestry groups but also beyond the "healthy volunteer" bias and concentration of research in high socioeconomic samples and environments in wealthy industrialized countries. Next to general statistical and econometric concerns that need to be addressed in G×E studies, certain challenges specific to these kinds of interaction models remain, such as low power and potential environmental robustness of GWAS meta results.

Exercises

Explore the following questions in the HRS data:

1. What is the incremental R^2 of the BMI_PGS predicting the BMI phenotype (GWAS-heritability)? Refer to the out-of-sample prediction exercise in section 11.2.1.

2. Does the BMI-PGS predict other phenotypes in the HRS data? Refer to the cross-trait prediction discussion in section 11.2.2.

3. Is the phenotypic association between education and health conditions confounded by genes for education? Refer to the confounding exercise in section 11.2.3.

4. Finally, translate the gene-environment interaction exercise from 11.3.1 to the following question: Does birth cohort modify the effect of genes on educational attainment?

Further reading

The main examples we use here relate to research conducted in references 5, 15, 30, and 32.

References

1. A. Sonnega et al., Cohort profile: The Health and Retirement Study (HRS). *Int. J. Epidemiol.* **43** (2014), doi:10.1093/ije/dyu067.

2. J. Euesden, C. M. Lewis, and P. F. O'Reilly, PRSice: Polygenic risk score software. *Bioinformatics* **31**, 1466–1468 (2014).

3. N. R. Wray et al., Pitfalls of predicting complex traits from SNPs. *Nat. Rev. Genet.* **14**, 507–515 (2013).

4. A. Okbay et al., Genome-wide association study identifies 74 loci associated with educational attainment. *Nature* **533**, 539–542 (2016).

5. J. J. Lee, R. Wedow, and A. Okbay, Gene discovery and polygenic prediction from a genome-wide association study of educational attainment in 1.1 million individuals. *Nat. Genet.* **50**, 1112–1121 (2018).

6. D. Bloome, S. Dyer, and X. Zhou, Educational inequality, educational expansion, and intergenerational income persistence in the United States. *Am. Sociol. Rev.* **83**, 1215–1253 (2018).

7. M. C. Mills and C. Rahal, A scientometric review of genome-wide association studies. *Commun. Biol.* **2** (2019), doi:10.1038/s42003-018-0261-x.

8. T. Polderman et al., Meta-analysis of the heritability of human traits based on fifty years of twin studies. *Nat. Genet.* **47**, 702 (2015).

9. A. R. Branigan, K. J. McCallum, and J. Freese, Variation in the heritability of educational attainment: An international meta-analysis. *Soc. Forces* **92**, 109–140 (2013).

10. C. A. Rietveld et al., GWAS of 126,559 individuals identifies genetic variants associated with educational attainment. *Science* **340**, 1467–1471 (2013).

11. F. C. Tropf et al., Hidden heritability due to heterogeneity across seven populations. *Nat. Hum. Behav.* **1**, 757–765 (2017).

12. S. M. Purcell et al., Common polygenic variation contributes to risk of schizophrenia and bipolar disor-
     der. *Nature* **460**, 748–752 (2009).

13. S. H. Lee et al., Genetic relationship between five psychiatric disorders estimated from genome-wide
     SNPs. *Nat. Genet.* **45**, 984–994 (2013).

14. F. C. Tropf et al., Human fertility, molecular genetics, and natural selection in modern societies. *PLoS
     One* **10**, e0126821 (2015).

15. J. P. Beauchamp, Genetic evidence for natural selection in humans in the contemporary United States.
     Proc. Natl. Acad. Sci. USA **113**, 7774–7779 (2016).

16. A. Kong et al., Selection against variants in the genome associated with educational attainment. *Proc.
     Natl. Acad. Sci. USA* **114**, E727–E732 (2017).

17. B. Bulik-Sullivan et al., An atlas of genetic correlations across human diseases and traits. *Nat. Genet.* **47**,
     1236–1241 (2015).

18. W. Johnson, K. O. Kyvik, A. Skythe, I. J. Deary, and T. I. A. Sørensen, Education modifies genetic and
     environmental influences on BMI. *PLoS One* **6**, e16290 (2011).

19. P. K. E. Magnusson, F. Rasmussen, and U. B. Gyllensten, Height at age 18 years is a strong predictor of
     attained education later in life: Cohort study of over 950,000 Swedish men. *Int. J. Epidemiol.* **35**, 658–
     663 (2006).

20. M. C. Mills, R. R. Rindfuss, P. McDonald, and E. te Velde, Why do people postpone parenthood? Rea-
     sons and social policy incentives. *Hum. Reprod. Update* **17**, 848–860 (2011).

21. G. Conti, J. Heckman, and S. Urzua, The education-health gradient. *Am. Econ. Rev.* **100**, 234–238 (2010).

22. J. K. Pickrell et al., Detection and interpretation of shared genetic influences on 42 human traits. *Nat.
     Genet.* **48**, 709–717 (2016).

23. S. E. Black, P. J. Devereux, and K. G. Salvanes, Why the apple doesn't fall far: Understanding intergen-
     erational transmission of human capital. *Am. Econ. Rev.* **95**, 437–449 (2005).

24. M. van Doorn, I. Pop, and M. H. J. Wolbers, Intergenerational transmission of education across European
     countries and cohorts. *Eur. Soc.* **13**, 93–117 (2011).

25. A. Kong et al., The nature of nurture: Effects of parental genotypes. *Science* **359**, 424–428 (2018).

26. R. Breen and J. Jonsson, Inequality of opportunity in comparative perspective: Recent research on edu-
     cational attainment and social mobility. *Annu. Rev. Sociol.* **31**, 223–243 (2005).

27. D. Conley et al., Is the effect of parental education on offspring biased or moderated by genotype? *Sociol.
     Sci.* **2**, 82–105 (2015).

28. H. Liu, Social and genetic pathways in multigenerational transmission of educational attainment. *Am.
     Sociol. Rev.* **83**, 278–304 (2018).

29. A. Kong et al., Selection against variants in the genome associated with educational attainment. *Proc.
     Natl. Acad. Sci. USA* **114**, E727–E732 (2017).

30. S. Walter, I. Mejia-Guevara, K. Estrada, S. Y. Liu, and M. M. Glymour, Association of a genetic risk
     score with body mass index across different birth cohorts. *JAMA* **316**, 63–69 (2016), doi:10.1001/
     jama.2016.8729.

31. C. M. A. Haworth et al., Increasing heritability of BMI and stronger associations with the FTO gene over
     childhood. *Obesity* **16**, 2663–2668 (2008).

32. H. Liu and G. Guo, Lifetime socioeconomic status, historical context, and genetic inheritance in shaping
     body mass in middle and late adulthood. *Am. Sociol. Rev.* **80**, 705–737 (2015).

33. T. W. Winkler et al., The influence of age and sex on genetic associations with adult body size and shape:
     A large-scale genome-wide interaction Study. *PLOS Genet.* **11**, e1005378 (2015).

34. A. E. Locke et al., Genetic studies of body mass index yield new insights for obesity biology. *Nature* **518**,
     197–206 (2015).

35. M. C. Keller, Gene × environment interaction studies have not properly controlled for potential confound-
     ers: The problem and the (simple) solution. *Biol. Psychiatry* **75**, 18–24 (2014).

36. H. Aschard, A perspective on interaction effects in genetic association studies. *Genet. Epidemiol.* **40**, 678–688 (2016).

37. W. J. Gauderman et al., Update on the state of the science for analytical methods for gene-environment interactions. *Am. J. Epidemiol.* **186**, 762–770 (2017).

38. D. Conley, The challenges of G×E: Commentary on "Genetic endowments, parental resources and adult health: Evidence from the Young Finns Study." *Soc. Sci. Med.* **188**, 201–203 (2017).

39. F. C. Tropf et al., Hidden heritability due to heterogeneity across seven populations. *Nat. Hum. Behav.* **1**, 757–765 (2017).

40. F. B. Pampel, R. F. Krueger, and J. T. Denney, Socioeconomic disparities in health behaviors. *Annu. Rev. Sociol.* **36**, 349–370 (2010).

41. J. D. Boardman, J. Daw, and J. Freese, Defining the environment in gene-environment research: Lessons from social epidemiology. *Am. J. Public Heal.* **103**, S64–S72 (2013).

42. K. McAllister et al., Current challenges and new opportunities for gene-environment interaction studies of complex diseases. *Am. J. Epidemiol.* **186**, 753–761 (2017).

43. A. Fry et al., Comparison of sociodemographic and health-related characteristics of UK Biobank participants with those of the general population. *Am. J. Epidemiol.* **186**, 1026–1034 (2017).

12

Applying Genome-Wide Association Results

Objectives

- Know where to find *association summary statistics* for multiple phenotypes
- *Create Manhattan plots* to illustrate genome-wide association results
- Use LocusZoom to *examine regions of interests* in association results
- Estimate *heritability and genetic correlation* from association results using linkage disequilibrium score regression
- Understand how to *examine multitrait associations from genome-wide association results* using Multi-Trait Analysis of Genome-wide association summary statistics (MTAG)

12.1 Introduction

As genome-wide association studies (GWASs) became the standard for genetic discovery, there has been an explosion of studies on a huge variety of diseases and traits, ranging from anthropometric, diseases, socioeconomic, and behavioral traits. The GWAS wave was met with a data-sharing revolution, making it one of the most collaborative efforts across the sciences. The scientific community recognized the importance of sharing the summary results from GWA studies to the wider community to ensure replication and further extensions of studies. In chapter 7, section 7.3.3, we describe where researchers can obtain summary statistics. This includes the GWAS Catalog (https://www.ebi.ac.uk /gwas/summary-statistics), Atlas of GWAS summary statistics (http://atlas.ctglab.nl), and the Ben Neale Lab of GWASs on virtually all traits in the UK Biobank (http://www .nealelab.is/uk-biobank/).

The aim of this chapter is to provide you with the tools to actively work with genome-wide association results. We first illustrate various techniques to produce Manhattan plots, which are the standard GWAS figures that plot genome-wide associations. This is then followed by instructions on how to plot and interpret regional association plots. We then produce and describe Quantile-Quantile (Q-Q) plots and the associated λ (lambda) statistic. The next section demonstrates how you can estimate heritability from these GWAS summary statistics, followed by how to estimate genetic correlations and heatmaps. We conclude with a description of how to run and interpret a multitrait association GWAS (MTAG) analysis.

Before you start you will need to ensure that you have following data installed in the appropriate directory that we will use in this chapter:

- EA2_results.txt.gz
- Giant_Height2018.txt.gz
- eur_w_ld_chr.tar.bz2
- w_hm3.snplist.bz2
- GWAS_EA_example.txt
- GWAS_CP_example.txt
- LD-Hub_genetic_correlation_example.txt

You will also need to ensure that you have an Internet connection, because we will use R to download additional files. We also recommend having Git installed to download the up-to-date version of the LD score regression and MTAG software. Please refer to appendix 1 with details on how to install Git.

12.2   Plotting association results

12.2.1   Manhattan plots

Association results from GWA studies are in the form of lists of summary statistics for millions of genetic variants. Results are "clustered" in blocks corresponding to genetic loci with high LD. For this reason, it is impossible to open the file and browse the results by looking at the summary statistics. Several graphical tools have been developed to visualize the entire list of results. The most widely used, recognizable visual tool to explore genome-wide association statistics is the Manhattan plot, which we discussed already in chapter 4 on GWASs.

To remind readers, the Manhattan plot is a type of scatterplot where genomic coordinates are displayed along the X-axis, and the negative logarithm of the association p-value for each genetic variant along the Y-axis. The first results from a GWAS were reported in 2005 and 2006 [1]. Since variants are in LD, we often observe groups of points with similar *p*-values, resembling high towers. Well-powered GWASs of polygenic traits show many loci with highly significant *p*-values, (i.e., very small *p*-values), and thus high skyscapers. The resemblance to the Manhattan's skyline gave the name to this graph. By looking at a Manhattan plot, it is easy to recognize immediately if any genetic loci are highly significant. It is standard to place a line on top of a Manhattan plot to indicate hits that have reached the genome-wide significance level ($p \leq 5 \times 10^{-8}$). For an explanation of why this significance level is chosen, refer to our previous discussion on correcting for multiple testing in GWASs (chapter 4, section 4.3.3).

To create a Manhattan plot, we only need the genomic position (chromosome number and base-pair position) and the association *p*-value. If you use the following code in R, you can plot the association results of a GWAS of educational attainment from Okbay et al. 2016 [2], using the following command:

```
# Load the library
install.packages("qqman")
library(qqman)
```

When the package is installed, we can then download the association results and import them in R.

```
# Use R to download the summary statistics
download.file ("http://ssgac.org/documents/EduYears _ Main
.txt.gz", dest= "EA2 _ results.txt.gz")

# Import the summary statistics in R
gwasResults<-read.table("EA2 _ results.txt.gz", header=T)
```

Lastly, we use the function `manhattan` to produce the plot. This can take some time since we are plotting millions of points. We recommend that you save the figure as an external .png file. The following command saves the Manhattan plot into a new file called `manhattan _ without _ highlights.png`. The function reads the association file and uses information on *p*-values, chromosome numbers, and position to draw the Manhattan plot shown in figure 12.1.

```
# Save the figure into an external png file format
png(file= "manhattan _ without _ highlights.png", width =
1200, height = 600)

manhattan(gwasResults, chr= "CHR", bp= "POS", snp= "Marker-
Name", p= "Pval",suggestiveline=F)

dev.off()
```

If you would like to emphasize the top hits from a GWAS, it is possible to highlight the SNPs that exceed the genome-wide significance threshold together with the surrounding

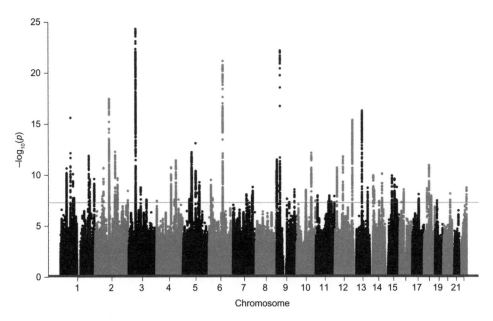

Figure 12.1
Manhattan plot of the 2016 GWAS of educational attainment (EA2) *without* highlights.

SNPs (we chose a +/– 500 Kb window around the top hits) using the code below. Please note that these are not independent loci, but rather it is a convenient and visual way to highlight the top results.

The following code is used to select all of the SNPs in the specified vicinity of genome-wide association results. We use a loop where we scan all and pick up all of the SNPs within 500 Kb of the GWAS-significant hits.

```
gwasResults<-subset(gwasResults, Pval<0.05)
hits<-gwasResults[gwasResults$Pval<5e-08,]

# Create new variable with SNPs that we want to highlight
gwasResults$highlight.snps<-0

for ( i in 1:  dim(hits)[1]){
    chr<-hits[i,2]
    loc _ min<- hits[i, 3]-5000
    loc _ max<- hits[i, 3]+5000

neighbors.snps<-gwasResults$MarkerName[gwasResults$CHR==chr &
                          gwasResults$POS>loc _ min &
                          gwasResults$POS<loc _ max]
```

```
gwasResults$highlight.snps[gwasResults$MarkerName %in%
                              neighbors.snps]   <- 1
}
```

Once we have created a new variable with the SNPs that we want to highlight, we can pass this information to the manhattan function using the option highlight.

```
png(file= "manhattan _ with _ highlights.png", width = 1200,
height = 600)

# Add highlight command to the Manhattan plot
manhattan(gwasResults, chr= "CHR", bp= "POS", snp= "Marker-
Name", p= "Pval", highlight=gwasResults$MarkerName[gwas
Results$highlight.snps==1],suggestiveline=F)
dev.off()
```

The result for the highlighted plot is shown in figure 12.2 (see the light gray shading), with the full color version of the results available on our companion website (http://www.intro -statistical-genetics.com).

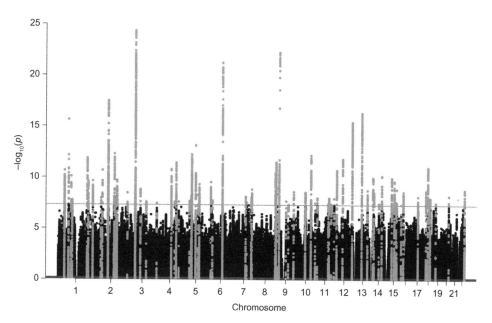

Figure 12.2
Manhattan plot of 2016 GWAS of educational attainment (EA2) *with* highlights.

12.1.2 Regional association plots

Manhattan plots show the statistical associations of all genetic variants but conceal a considerable amount of important information. Regional associations are often used to visualize LD patterns between SNPs that are significant and SNPs in nearby regions that have been previously reported in other GWASs. To visualize the association results of a specific area in the genome, we can use the online tool LocusZoom (http://locuszoom .org/) and produce a regional association plot. This type of plot essentially zooms in or magnifies part of a Manhattan plot in a specific area. Using the online tool (figure 12.3), we investigated the results of rs11130222, the top hit on chromosome 3 from the 2016 educational attainment GWAS. Figure 12.4 provides more details around this SNP. The x-axis and the y-axis are the same as in a Manhattan plot (genomic position and nega- tive logarithm of association p-value). Above that a regional association plot shows which genes are in that region of the genome in the bottom panel. SNPs are plotted in different colors, based on the correlation (r^2) with rs11130222, the reference SNP in this example. A color version of this plot can be found at the LocusZoom webpage or our companion website to this book. The dots in red (upper right) indicate SNPs that are in high LD with the reference one, while blue SNPs (generally lower in the plot) are inde- pendent variants.

12.1.3 Quantile-Quantile plots and the λ statistic

Another typical figure associated with a GWAS is the Quantile-Quantile (Q-Q) plot, which is examined together with the λ (lambda) statistic. We described the Q-Q plot earlier in our statistical introduction in chapter 2 (section 2.2.1). Q-Q plots compare the distribution of observed *p*-values (logarithm scale) with the expected *p*-value distribu- tion under the null hypothesis (i.e., no association between genotypes and phenotype). It is a tool that is used to visualize the appropriate control of population substructure and the presence of an association. Figure 12.5 shows the Q-Q plot for the same results of educational attainment from the 2016 educational attainment (EA2) GWAS. Under the null hypothesis, points would lie on the horizontal line ($y=x$). Data that generally fall on the line suggest that there is no deviation from the null hypothesis, meaning absence of association. Any deviation on the left (as clearly this figure shows) or in other words in the upper right tail from the $y=x$ line, indicates a substantial deviation from the null hypothesis, indicating compelling evidence for genomic association. The R command to produce Q-Q plots from genome-wide association results is contained in the qqman package.

```
# Produce Q-Q plots
qq(gwasResults$Pval)
```

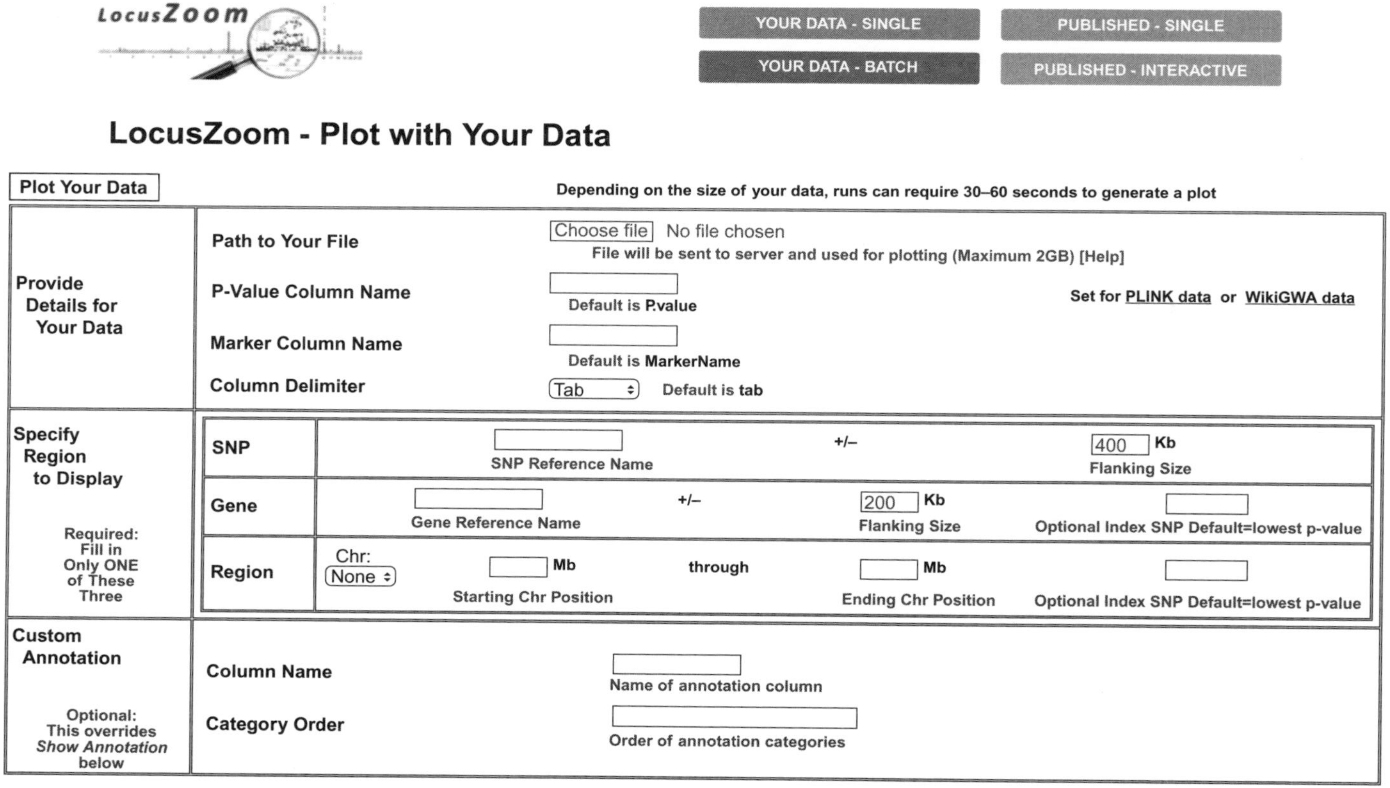

LocusZoom - Plot with Your Data

YOUR DATA - SINGLE

YOUR DATA - BATCH

PUBLISHED - SINGLE

PUBLISHED - INTERACTIVE

Plot Your Data

Depending on the size of your data, runs can require 30-60 seconds to generate a plot

Provide Details for Your Data	Path to Your File	Choose file No file chosen
		File will be sent to server and used for plotting (Maximum 2GB) [Help] Set for PLINK data or WikiGWA data
	P-Value Column Name	Default is P-value
	Marker Column Name	Default is MarkerName
	Column Delimiter	Tab ⁒ Default is tab

Specify Region to Display	SNP		SNP Reference Name	+/–	
Required: Fill in Only ONE of These Three	Gene		Gene Reference Name	+/–	200 Kb Flanking Size
	Region	Chr: None ⁒	Mb Starting Chr Position	through	Mb Ending Chr Position 400 Kb Flanking Size
					Optional Index SNP Default=lowest p-value
					Optional Index SNP Default=lowest p-value

Custom Annotation	Column Name	Name of annotation column
Optional: This overrides Show Annotation below	Category Order	Order of annotation categories

Figure 12.3
Screenshot of the LocusZoom online tool.

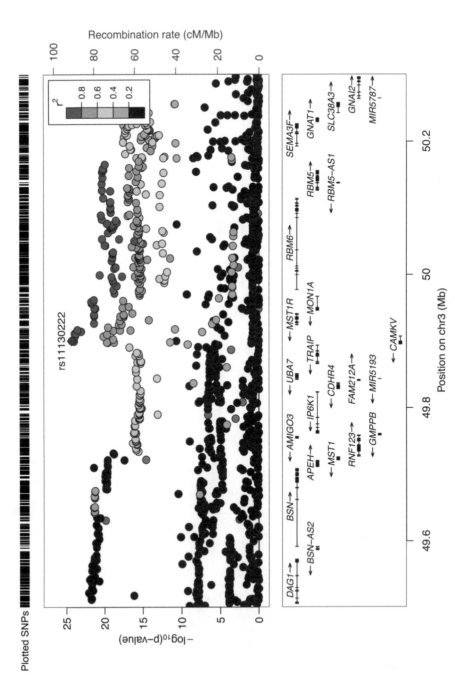

Figure 12.4
LocusZoom plot of SNP rs11130222 using summary statistics for 2016 educational attainment (EA2) GWAS.

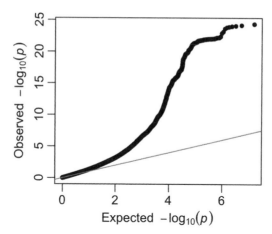

Figure 12.5
Q-Q plot of educational attainment GWAS.

The degree of deviation from the red line is formally measured by the λ-statistic, also called genomic control. A λ value can be calculated from Z-scores, chi-square statistics, or p-values, depending on the output you have from the association analysis.

The example below shows how to calculate the λ-statistic using R.

```
lambda <- round(median((gwasResults$Beta/
gwasResults$SE)^2) / 0.454, 3)
```

```
lambda
[1] 1.239
```

As artificial differences in allele frequencies due to population stratification, cryptic relatedness, and genotyping errors will affect all SNPs, the test statistic will be inflated across the entire genome. For this reason, λ is also used to detect possible inflation due to population stratification. A value close to 1 suggests that data have been properly adjusted for the population substructure. If $\lambda > 1.2$, this suggests the presence of stratification. Sometimes it is possible to correct for population stratification by dividing all of the test statistics by the value of lambda. Many GWASs do it at the meta-analyses stage to reduce the risk of false positive results. In this example, a λ of 1.239 suggests the possible presence of population stratification. However, several recent studies show that highly polygenic traits (such as educational attainment or height) exhibit a λ statistic over 1.2 due to the fact that the genetic basis of the trait is spread across the entire genome. We therefore recommend caution in the interpretation of λ as a test for population stratification.

12.2   Estimating heritability from summary statistics

Heritability of a polygenic trait can be estimated with a certain degree of accuracy from summary statistics. In polygenic traits, SNPs that are highly correlated with many other SNPs (i.e., have a high LD score) are more likely to "tag" a causal SNP (i.e., it is correlated). In a GWAS, these SNPs with a high LD score are expected to have higher association statistics (Chi-squared test statistics) than SNPs with a lower LD score. By exploiting this relationship, it has been shown that it is possible to estimate heritability in highly polygenic traits [3]. In particular, we can estimate SNP-heritability by running a linear regression of all the association test statistics from a GWAS on the LD scores of each SNP. This method for estimating SNP-heritability is called LD score regression (LDSC) and was discussed already in detail in chapter 5. In contrast with the GREML method, it does not require individual genetic data. Compared to GREML, LDSC considerably reduces the computational speed, but with the cost of a higher standard error (SE) of estimates. For LDSC, the standard errors of the variance component estimates are typically larger (usually by 50% or more) than those of a GREML analysis for the same sample size. However, LDSC is usually applied to large datasets from GWAS consortia that generate smaller SEs.

To estimate SNP-heritability from summary statistics, we need to install the Python command LDSC (see appendix 1). Once LDSC is installed, we require three sources of data: (1) summary statistics from a GWAS, (2) a list of SNPs, and (3) LD scores for each SNP. The location of where to obtain these summary statistics was discussed previously and in chapter 7 (section 7.3.3). LD scores can be calculated using LDSC, but if using GWAS results from European or Asian populations, it is possible to directly download the LD scores for HapMap 3 SNPs from the LDSC website (https://github.com/bulik /ldsc).

There are several aspects that need to be taken into account before using LDSC. First, because we are using summary statistics, often from different datasets (i.e., cohorts or data sources), all of the quantities we are estimating (e.g., heritability) have substantial measurement errors. It is also important to reiterate that LD structure is population specific, and that LD scores are not stable across populations. Results from LDSC are therefore population specific because the method is entirely based on LD scores. A second aspect to keep in mind is that the LDSC method assumes that heritability is spread across the entire genome. The consequence is that LDSC is a method that works well for highly polygenic traits.

The following code provides a guide on how to estimate the SNP-heritability for adult height on individuals with European ancestry. The first step is to download the data from the summary statistics of the latest GWAS on adult height [4] and a list of LD scores for Europeans based on 1000 Genomes. You can download the three files from the companion website to this book (http://www.intro-statistical-genetics.com). Before we start the analysis, we need to extract the zipped files as follows:

```
bunzip2 eur _ w _ ld _ chr.tar.bz2
tar -xvf eur _ w _ ld _ chr.tar
bunzip2 w _ hm3.snplist.bz2
```

Once you have obtained the files, it is possible to start to prepare the data. LDSC is software written in Python 2. You need to ensure that you have Python installed on your system (see appendix 1).

To install all the packages needed to run LDSC, it is possible to create a new Python environment using the following command:

```
conda env create—file ldsc/environment.yml
```

In (older versions of) Windows or those not using Ubuntu (see appendix 1):

```
activate ldsc
```

In Linux and macOS:

```
source activate ldsc
```

If LDSC is correctly installed on your system, when you type the following command shown below, it will display various options of how to use the commands and list all of the possible arguments.

```
python ldsc/ldsc.py -h
```

Once LDSC is correctly installed on your computer, we can start working with the data. The first necessary step is to ensure that the data are in the correct format. There is a command in LDSC that is designed to prepare the summary statistics, which is called munge _ sumstats.py.

The necessary data for munge _ sumstats are: an input file containing summary statistics (--sumstats), an external file with the list of SNPs and their alleles that will be used in the analysis (this is one of the files that we downloaded earlier: --merge-alleles), and the name of the output file (--out). Additional arguments can be used to specify the information contained in each column. In the example below we have: `--snp`, indicating the column with the genetic markers; `--N--col`, indicating the columns with sample sizes; `--a1` and `--a2`, with the two alleles of each marker; `--p`, indicating the column with *p*-values; and `--frq`, indicating allele frequencies.

```
python ldsc/munge_sumstats.py \
--sumstats Giant_Height2018.txt.gz \
--snp SNP \
--N-col N \
--a1 Tested_Allele \
--a2 Other_Allele \
--p P \
--frq Freq_Tested_Allele_in_HRS \
--out height \
--merge-alleles w_hm3.snplist
```

The output will tell you how many SNPs have been read from LDSC and how many have been filtered out due to low allele frequencies or incorrect information in other columns. Finally, LDSC provides some descriptive statistics including how many genome-wide significant SNPs are saved to the output file called `height.sum-stats.gz`.

```
Read 2334001 SNPs from --sumstats file.
Removed 1315341 SNPs not in --merge-alleles.
Removed 0 SNPs with missing values.
Removed 0 SNPs with INFO <= 0.9.
Removed 44 SNPs with MAF <= 0.01.
Removed 48 SNPs with out-of-bounds p-values.
Removed 0 variants that were not SNPs or were
strand-ambiguous.
1018568 SNPs remain.
Removed 0 SNPs with duplicated rs numbers (1018568 SNPs
remain).
Removed 0 SNPs with N < 472784.666667 (1018568 SNPs
remain).
Median value of BETA was 0.0, which seems sensible.
Removed 0 SNPs whose alleles did not match
--merge-alleles (1018568 SNPs remain).
Writing summary statistics for 1217311 SNPs (1018568 with
nonmissing beta) to height.sumstats.gz.
```

```
Metadata:
Mean chi^2 = 8.986
Lambda GC = 3.61
Max chi^2 = 1424.989
61727 Genome-wide significant SNPs (some may have been
removed by filtering).

Conversion finished at Wed Feb 6 16:10:09 2019
Total time elapsed: 33.34s
```

The second step involves the calculation of LD score regression and the estimation of SNP-heritability. This can be done using the command ldsc.py.

```
python ldsc/ldsc.py \
--h2 height.sumstats.gz \
--ref-ld-chr eur _ w _ ld _ chr/ \
--w-ld-chr eur _ w _ ld _ chr/ \
--out height _ h2
```

The results are stored in the newly created file height _ h2.log. The log file reports how many SNPs were provided by the summary statistics and how they merged with the LD score file.

```
***********************************************************
*******
* LD Score Regression (LDSC)
* Version 1.0.0
* (C) 2014-2015 Brendan Bulik-Sullivan and Hilary Finucane
* Broad Institute of MIT and Harvard / MIT Department of
Mathematics
* GNU General Public License v3
***********************************************************
*******
Call:
./ldsc.py \
--h2 height.sumstats.gz \
--ref-ld-chr eur _ w _ ld _ chr/ \
--out height _ h2 \
```

```
--w-ld-chr eur _ w _ ld _ chr/

Beginning analysis at Sat Mar 2 23:31:21 2019
Reading summary statistics from height.sumstats.gz …
Read summary statistics for 1018568 SNPs.
Reading reference panel LD Score from eur _ w _ ld _ chr/
[1-22] …
Read reference panel LD Scores for 1290028 SNPs.
Removing partitioned LD Scores with zero variance.
Reading regression weight LD Score from eur _ w _ ld _ chr/
[1-22] …
Read regression weight LD Scores for 1290028 SNPs.
After merging with reference panel LD, 1013762 SNPs
remain.
After merging with regression SNP LD, 1013762 SNPs
remain.
Using two-step estimator with cutoff at 30.
```

Finally, we have an estimate of heritability (h^2) for height. The estimate is $h^2 = 0.46$ with a standard error of 0.019. The estimate is substantially lower than the estimate of heritability from twin studies that is around 0.80 [5], as this is an estimate based on SNP-heritability. Refer to chapter 1 (section 1.6.3) where we discuss the reasons for these disparities in heritability estimates.

```
Total Observed scale h2: 0.4552 (0.0193)
Lambda GC: 3.6129
Mean Chi^2: 8.8736
Intercept: 1.8968 (0.0452)
Ratio: 0.1139 (0.0057)
Analysis finished at Sat Mar 2 23:31:32 2019
Total time elapsed: 11.33s
```

12.3   Estimating genetic correlations from summary statistics

LD score regression was introduced in chapter 5 in our introduction to polygenic scores. This method can also be used to estimate the degree of genetic correlation of multiple traits [6]. Recall from this earlier chapter that genetic correlation is an estimate of the additive genetic effect that is shared between a pair of traits. For example, height and educational attainment are both heritable traits that are usually positively correlated. However, it could be interesting to examine if there is also genetic correlation, meaning that they are likely to share the same genes. LDSC can work with multiple traits, and by exploiting the LD structure of the data it can be used to estimate the degree of genetic correlation.

We start by preparing the data for educational attainment. For this example we once again use the data from the 2016 GWAS study of educational attainment (EA2). As before, we use the command munge _ sumstats.py to prepare the data. The command is identical to the one used previously, with the only difference being that we adapt the name of the columns to the new summary statistics file. The output file is Educ.sumstats.gz.

```
python ldsc/munge _ sumstats.py \
--sumstats EA2 _ results.txt.gz \
--snp SNP\
--N 300000 \
--a1 A1 \
--a2 A2 \
--p Pval \
--frq EAF \
--out Educ \
--merge-alleles w _ hm3.snplist
```

Genetic correlations can be calculated very easily using ldsc.py. Contrary to the calculation of heritability, we use the option --rg (genetic correlation) followed by the two (or more) traits for which we estimate genetic correlation.

```
python ldsc/ldsc.py \
--rg height.sumstats.gz,Educ.sumstats.gz \
--ref-ld-chr eur _ w _ ld _ chr/ \
--w-ld-chr eur _ w _ ld _ chr/ \
--out height _ educ
```

The output file is height _ educ.log. The first part of the output is dedicated to log how many SNPs are merged with LD scores and how many are removed from the analysis. This is important information because it can warn you if too many SNPs are removed from the analysis. The second part of the output reports the genetic heritability of both traits.

In this example, the heritability of the first trait is 0.51 (height) and 0.12 for the second trait (educational attainment). Note that the estimate for h^2 of height slightly differs from the previous estimate. This is because, by merging SNPs from two traits, the set of genetic variants used in the analysis may differ.

```
Heritability of phenotype 1
---------------------------
Total Observed scale h2: 0.5052 (0.0236)
Lambda GC: 3.6129
Mean Chi^2: 8.8826
Intercept: 1.5332 (0.1104)
Ratio: 0.0676 (0.014)

Heritability of phenotype 2/2
---------------------------
Total Observed scale h2: 0.1223 (0.0045)
Lambda GC: 1.4853
Mean Chi^2: 1.6516
Intercept: 0.9322 (0.0107)
Ratio < 0 (usually indicates GC correction).
```

The last portion of the output reports the estimates of genetic covariance and correlation.

The estimate of genetic correlation between height and educational attainment is 0.14, meaning that there is evidence of a shared genetic basis between the two traits. At least part of the correlation between the two traits may be explained by the fact that the same genes are associated with the traits. We discussed this covariance in traits in detail in chapter 5, particularly in relation to pleiotropy.

```
Genetic Covariance
---------------------------
Total Observed scale gencov: 0.035 (0.0035)
Mean z1*z2: 0.3942
Intercept: 0.0712 (0.0115)

Genetic Correlation
---------------------------
Genetic Correlation: 0.1408 (0.0128)
Z-score: 10.9712
P: 5.2582e-28
```

Finally, a summary of all of the genetic correlations is presented in the output. LDSC also reports the *p*-value of a statistical test where the null hypothesis is the absence of correlation. The *p*-value is 5.2×10^{28}, hence a very small *p*-value indicating statistical departure from the null hypothesis.

```
Summary of Genetic Correlation Results
p1        p2      rg       se       z          p h2 _ obs
height    educ    0.1408   0.0128   10.9712    5.2582e-28
```

As we discussed in chapter 5, it is possible to estimate genetic correlations for multiple traits, to examine the genetic overlap of across traits. Recall figure 5.2, where we illustrated the genetic overlap between two reproductive behavior traits with 27 other phenotypes.

Genetic correlations can also be calculated without downloading results but by using a recent online tool called LDHub (http://ldsc.broadinstitute.org/) [7]. LD Hub is a centralized database of summary-level GWAS results for hundreds of diseases/traits from different publicly available resources/consortia and a web interface that automates the LD score regression analysis pipeline. It is possible to explore genetic correlation for the traits of interests by either exploring existing correlations or by uploading the summary statistics of the trait of interest.

A nice way to visualize multiple genetic correlations is to draw a heatmap. Heatmaps are graphs that can be used to visualize a data matrix (often also used to show LD blocks in genotype data). This type of graph uses colors to visualize the intensity of genetic correlations between traits. In the example below, we use data from genetic correlations of 49 traits from Bulik-Sullivan et al. (2015) [3], downloaded from LD Hub (figure 12.6). Darker shading on the heatmap represent, stronger positive or negative genetic correlations. The color version of this graph is shown on the companion website to this book, where positive genetic correlations are in bright red and negative genetic correlations are pictured in blue. By looking at multiple traits, it is possible to have a first indication of shared etiology among traits.

```
# Import data on genetic correlation
data _ rg<-read.table("LD-Hub _ genetic _ correlation _
example.txt",fill =T, sep="\t", header=T, quote="")
# load ggplot2 library
library(ggplot2)

# Draw heatmap
ggplot(data = data _ rg, aes(Trait1, Trait2, fill = rg))+
```

```
      geom _ tile(color = "white")+
      scale _ fill _ gradient2(low = "blue", high = "red", mid =
                           "white",  midpoint = 0, limit =
                           c(-1.1,1.1), space = "Lab",
                           name="Genetic\nCorrelation") +
      theme _ minimal()+
      theme(axis.text.x = element _ text(angle = 45, vjust = 1,
           size = 8, hjust = 1))+
   coord _ fixed()
```

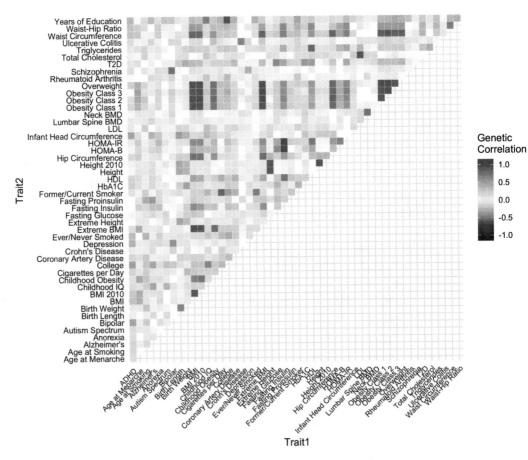

Figure 12.6
Heatmap of genetic correlation for 49 traits.

As we noted in chapter 5, pleiotropy—which is when a single gene affects multiple traits—is ubiquitous. The genetic correlations that we are studying are indicative of shared genetic basis, but they do not definitively imply pleiotropy. Two causal genetic loci for different traits may be very close to each other on the genome and therefore be in linkage disequilibrium. If that is the case, you would see a genetic correlation even in absence of pleiotropy. Refer to our detailed discussion of pleiotropy in chapter 5 (section 5.4.3) that delves into this topic in more detail. Since this is a very advanced topic, we do not explore additional practical examples in this introductory textbook.

12.4   MTAG: Multi-Trait Analysis of Genome-wide association summary statistics

As we have noted on multiple occasions throughout this book, one limitation of the GWAS approach is that you need a very large sample size to have sufficient statistical power to detect signals that may have very small effects (see box 4.1). For this reason, GWASs have privileged phenotypes that are measured consistently in many cohorts, often at the expense of precision in the measurement of the phenotype of interest. For example, we might be interested in measuring cognitive impairment, but since it may be very costly to design a study that collects precise measurements of the trait that we are interested in, we rely on a *proxy* phenotype that is measured in a larger sample, such as educational attainment.

As described in detail in chapter 5 and as demonstrated in the previous sections, phenotypes often share a common genetic basis. For this reason we now apply the multi-trait analysis of GWASs developed by Turley and colleagues in 2018 called MTAG (Multi-Trait Analysis of Genome-wide association summary statistics), which allows analysis of GWASs of multiple correlated traits using summary statistics [8]. As we noted previously in chapter 5, this method enables the joint analysis of multiple traits, boosting statistical power to detect genetic associations for each trait that is analyzed. In this way, we can leverage the sample size of a large study sample with a GWAS on a different but related trait to increase the detection of genetic loci in both traits. MTAG results can also be used in further analysis to construct polygenic scores.

Here we present a simple example of how to use MTAG. Further documentation is available at https://github.com/omeed-maghzian/mtag/wiki. MTAG is a software written in Python 2 and may be downloaded by cloning this GitHub repository (see appendix 1 for further instructions on how to install Git and MTAG):

```
git clone https://github.com/omeed-maghzian/mtag.git
```

As for LDSC, the software developers recommend the installation of the Python Anaconda distribution (see appendix 1). In particular, mtag needs the following Python packages installed on your system:

- numpy (≥1.13.1)
- scipy
- pandas (≥0.18.1)
- argparse
- bitarray (for ldsc)
- joblib

To install all of the packages needed to run MTAG, it is possible to create a new Python environment using the following command:

```
conda env create --file mtag/environment.yml
```

In (older versions of) Windows in a non-Ubuntu environment (see appendix 1):

```
activate mtag
```

In Linux and macOS:

```
source activate mtag
```

To test that the tool has been successfully installed, type:

```
python mtag/mtag.py -h
```

In the example below, we use the data from the more recent 2018 educational attainment GWAS (EA3) by Lee et al. [9]. To combine results from a GWAS on educational attainment (sample size N = 1,131,811) with a smaller GWAS on cognitive performance (sample size N = 257,828). We use the following files containing summary statistics of 10,000 SNPs: GWAS _ EA _ example.txt and GWAS _ CP _ example.txt.

The summary statistics are passed to mtag using the option --sumstats. As is customary, we specify the suffix of the output files using the option --out. The rest of the options specify the column names of: names of genetic variants (--snp _ name), allele names (--a1 _ name) and (--a2 _ name), allele frequencies (--eaf _ names), name of effect sizes (--beta _ name), standard errors (--se _ names), and sample size (--n _ name). As we are using betas and standard errors and not Z-scores, we need to include the option --use _ beta _ se.

```
python mtag/mtag.py  \
    --sumstats GWAS _ EA _ example.txt,GWAS _ CP _ example.txt \
    --out ./EA _ CP _ mtag \
    --snp _ name MarkerName \
    --a1 _ name a1 \
    --a2 _ name a2 \
    --eaf _ name EAF \
    --use _ beta _ se \
    --beta _ name Beta \
    --se _ name SE \
    --n _ name n
```

MTAG then generates five output files in your current directory:

- EA _ CP _ mtag.log provides a description of the different steps taken by mtag.py.

- EA _ CP _ mtag _ sigma _ hat.txt reports the estimated residual covariance matrix.

- EA _ CP _ mtag _ omega _ hat.txt reports the estimated genetic covariance matrix.

- EA _ CP _ mtag _ trait _ 1.txt and EA _ CP _ mtag _ trait _ 2.txt are tab-delimited results files corresponding to the MTAG-adjusted effect sizes and standard errors for the Educational Attainment and Cognitive Performance summary statistics, respectively.

The end of the log file (EA _ CP _ mtag.log) also reports some additional measures that can be used to assess the gain in statistical power using MTAG. In particular, MTAG reports the increases in power from MTAG by approximating the sample size needed to achieve the same mean chi-square value in a standard GWAS. This information is reported in the column GWAS equiv. (max) N.

The equivalent GWAS sample size in this example is over 2 million individuals for both the traits. For GWAS conducted in small samples, using MTAG can tremendously increase the statistical power of a study.

```
Summary of MTAG results:
---------------------------

                                      GWAS         MTAG        GWAS equiv.
Trait                  N (max)  N (mean) ... mean chi^2 mean chi^2 (max) N
1 GWAS_EA_example.txt  1131811  1131811  ... 1.377      1.694      2082199
2 GWAS_CP_example.txt  2578282  2578282  ... 2.020      2.089      2752964
```

MTAG is a very flexible tool that can be used with GWAS results for overlapping cohorts. The key assumption of the method is that all SNPs share the same variance-covariance matrix of effect sizes across traits. This assumption may be violated, in particular, when some SNPs influence only a subset of the traits. The researchers who developed this method show that, however, even if this assumption is not satisfied, MTAG is a consistent estimator and always improves the precision of the estimates compared to single-trait GWASs.

12.5 Conclusion

Recent years have seen a proliferation of GWASs and the increased reporting and availability of summary statistics. This has opened up the possibility to conduct meaningful genetic analyses without having direct access to individual-level genotyped data, but rather the myriad of summary statistics from GWASs. Researchers are now able to download a plethora of genetic association results on virtually any trait and use the many and growing number of online resources. Given the great effort, interdisciplinary expertise, and resources that are required to lead and produce a large GWAS, it is not adviseable for those new to the field to attempt such a large task without the considerable expertise or guidance from experts within this area of research. Our hope is that this chapter opens up the use of these results, however, to allow researchers to feel confident to explore and work with existing summary results. The aim is that you are now able to produce various plots that explore these results, such as the Manhattan plot and regional association plots, but also evaluate results using Q-Q plots and related statistics. As we elaborated upon in detail in chapter 5, it is likewise useful to be able to estimate heritability and genetic correlations, and engage in multitrait analysis.

Exercises

1. Download the summary statistics of the most recent GWAS on Age at First Birth from the Atlas of GWAS summary statistics.
2. Use the data from exercise 1 and draw a Manhattan plot using RStudio.
3. Use LocusZoom to investigate the region around snp rs2777888.
4. Calculate the SNP heritability of age at first birth and the genetic correlation with educational attainment using LD score regression.

Further reading and resources

GWAS summary statistics

The Atlas of GWAS summary statistics includes a database of publically available GWAS summary statistics: https://atlas.ctglab.nl/.

The NHGRI-EBI Catalog also contains summary statistics: https://www.ebi.ac.uk/gwas/summary-statistics.

UK Biobank results by the Ben Neale Lab can be found at: http://www.nealelab.is/uk-biobank.

Web references

LD Hub, a centralized database of summary-level GWAS results and web interface for LD Score regression: http://ldsc.broadinstitute.org/.

LD Score Regression (LDSC) on GitHub: https://github.com/bulik/ldsc.

LocusZoom, used to create visualizations of GWAS results: http://locuszoom.org/.

MTAG Python command line tool on GitHub: https://github.com/omeed-maghzian/mtag.

References

1. M. C. Mills and C. Rahal, A scientometric review of genome-wide association studies. *Commun. Biol.* **2** (2019), doi:10.1038/s42003-018-0261-x.

2. A. Okbay et al., Genome-wide association study identifies 74 loci associated with educational attainment. *Nature* **533**, 1467–1471 (2016).

3. B. K. Bulik-Sullivan et al., LD score regression distinguishes confounding from polygenicity in genome-wide association studies. *Nat. Genet.* **47**, 291–295 (2015).

4. L. Yengo et al., Meta-analysis of genome-wide association studies for height and body mass index in ~700,000 individuals of European ancestry. *Hum. Mol. Gen.* **27**, 3641–3649 (2018).

5. A. Jelenkovic et al., Genetic and environmental influences on height from infancy to early adulthood: An individual-based pooled analysis of 45 twin cohorts. *Sci. Rep.* **6**, 28496 (2016).

6. B. K. Bulik-Sullivan et al., An atlas of genetic correlations across human diseases and traits. *Nat. Genet.* **47**, 1236–1241 (2015).

7. J. Zheng et al., LD Hub: A centralized database and web interface to perform LD score regression that maximizes the potential of summary level GWAS data for SNP heritability and genetic correlation analysis. *Bioinformatics* **33**, 272–279 (2017).

8. P. Turley et al., Multi-trait analysis of genome-wide association summary statistics using MTAG. *Nat. Genet.* **50**, 229–237 (2018).

9. J. J. Lee, R. Wedow, and A. Okbay, Gene discovery and polygenic prediction from a genome-wide association study of educational attainment in 1.1 million individuals. *Nat. Genet.* **50**, 1112–1121 (2018).

13

Mendelian Randomization and Instrumental Variables

Objectives

- Understand the challenge of *endogeneity* in drawing *causal inference* from *observational studies*
- Understand the basic principal of using *randomized control trials* to study causality
- Be able to define the technique of *Mendelian Randomization (MR)*
- Understand the fundamentals of the *Instrumental Variable (IV) approach in an MR framework*
- Grasp the *main assumptions* of the IV approach in an MR framework
- Be aware of *extensions of MR and advances* in the field
- Be introduced to several *substantial applications of MR* in different areas of research

13.1 Introduction

Until now the majority of our applications have focused on examining correlational associations between genotypes and phenotypes. But association does not necessarily imply causation. Studying causality is perhaps one of the most vexing problems in research. While description, classification, and associations are of upmost importance in scientific research, the central goal is often to uncover cause and effect. The problem we regularly face is that it is rarely possible to pinpoint the variable that is the actual cause or the effect or determine whether it is another third unobserved factor that affects both variables and they, in turn, impact one another. (See also causal models in chapter 2.) Yet understanding causation is central for developing evidence-based clinical or policy interventions.

Various techniques have been introduced in the literature to deal with the direction or relationship of causation, confounding variables, and endogeneity through twin designs [1], or using natural experiments such as educational reforms [2]. The main problems that we grapple with are related to *endogeneity*, namely *reverse causation*, *omitted variable bias due to confounding*, *measurement error*, and *bidirectional causality* (see box 13.1 for

Box 13.1
Case study: The four causal models linking age at first birth and educational attainment

To understand why causality is a problem, let us use the simple example of understanding the causal relationship between fertility (specifically the timing of first birth) and educational attainment of an individual. Since the 1970s, there has been a massive postponement in the age at which individuals in many countries have their first child. This phenomenon typically coincides with educational expansion, particularly of women [3].

Four causal models can be been used to describe this relationship. The first and most frequently studied causal mechanism is that *educational attainment has a causal impact on the age of first birth*. Achieving higher education (particularly for women) operates to postpone the timing of first birth [4]. The mechanisms of later fertility for the highly educated are multiple and outlined elsewhere [3, 4]. These include the higher educated having better knowledge of contraceptive use, being more likely to participate in the labor market and thus higher earnings, and opportunity costs to leave the labor market. But they also have more secularized views, higher consumption aspirations, and due to increased costs of children, delaying until they feel they can "afford" children. The second hypothesis is *reverse causation, where age at first birth has a causal impact on educational attainment*. In other words, an early or teenage pregnancy impedes higher educational attainment [5]. Multiple studies have demonstrated that early and particularly teenage childbirth curtails educational attainment and later life course outcomes (e.g., [6]). There is also an average of a motherhood wage penalty of around 7% per child, with each year delay in motherhood increasing women's earnings by 9% [7]. A third model is that it is in fact *bidirectional* and that both age at first birth and educational attainment both affect one another, which we describe shortly in an MR framework. Finally, it may be that both the timing of first birth and educational attainment are *both influenced by common omitted variables* such as personality, fertility preferences, parental environment, attitudes, or genes [8–10].

an applied example). *Endogeneity* refers to the case where the explanatory variable is correlated with the error term in the regression model. The most important source of endogeneity is *omitted variable bias*, which means that we neglect a control variable in our model that is a common cause for X (predictor) and Y (outcome) (see also chapter 2) and therefore confounds the observed association between X and Y. Given the large amount of tutorial materials already available to apply these methods (see the "Further reading" section), we only introduce this topic and the main challenges and do not present applied examples.

In this chapter we first introduce Randomized Control Trails (RCTs) as the gold standard in causal analysis but clarify that they are often not feasible. We then introduce a possible solution in the form of Mendelian Randomization (MR) and the Instrumental Variable (IV) approach, and describe MR in the context of the IV framework. This includes outlining the main premises of the model and key statistical assumptions. This is followed by a collection of substantive applications across various topics to allow the reader to understand the broader applicability of these techniques. We conclude with a discussion of future directions.

13.2 Randomized control trials and causality

The ideal strategy to identify a causal effect is to conduct an experiment in the form of a randomized control trial (RCT). In RCTs, individuals are assigned nonsystematically—namely randomly—into a treatment and control group. Imagine, for example, that you wanted to examine the impact of a selected variable (e.g., smoking) on health and that you divided a hundred people into two groups by tossing a coin. Since the results are random, you would not expect significant differences in, for example, the average health status between both groups. Within both groups, there will still, however, be variation with some healthier and unhealthier individuals, but there would be no systematic reason to expect an average difference in health status between the groups except by chance, which we would be able to quantify. Then we could introduce a treatment to this experiment. Instead of only assigning individuals to random groups 1 and 2, we would ask individuals with heads (group 1) to smoke cigarettes and those with tails (group 2) to refrain from smoking. If we would then track the health status of these groups, the only reason that we would expect differences in health outcomes would be due to the treatment of the experiment, which is smoking behavior.

In real-life research, it is difficult and unethical to introduce different types of treatment effects in the form of an RCT. If we did opt to introduce a treatment that held considerable risks, such as asking people to smoke in the example above, but offered it to individuals only on an opt-in or opt-out basis, we would run into additional selection issues. The group that opted to participate in a higher-risk behavior such as smoking could be more strongly incentivized by the potential reward for participation, which in turn would introduce *selection bias* into our results. This would make it difficult to generalize results beyond this sample group. For this reason, MR was introduced as a possible solution to study causality.

13.3 Mendelian Randomization

Mendelian Randomization (MR) relates directly to RCTs (see also figure 13.1 [11]). MR is a statistical technique that uses the measured variation in genes to examine the causal effect of one phenotype on another. The design was coined as Mendelian Randomization in the early 1990s by Gray and Wheatley [12]. They were interested in the effectiveness of bone marrow transplants in leukemia patients. The odds, however, of getting such a transplant were not determined by chance but were rather correlated with the prognosis for the patient. The authors focused on the compatibility of siblings to be donors, a process that is

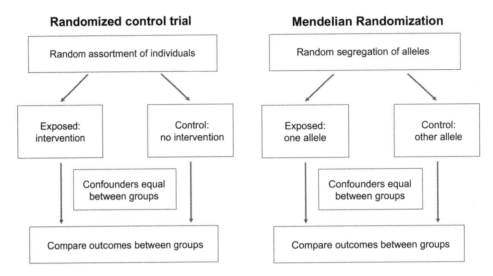

Figure 13.1
Parallel research designs: Classic randomized control trial (*left*) versus Mendelian Randomization (*right*).
Source: Figure is adapted from Smith (2005) [13].

determined by genetic inheritance. They then used this experiment to study the patient's preference for this therapy within leukemia cases in children. This allowed them to obtain unbiased causal estimates in the absence of conducting an RCT. Since this early experiment, there have been multiple extensions.

MR capitalizes on the random assignment of an individual's genotype that occurs at conception [14]. Recall our discussion in chapter 1 (section 1.2) about Mendel's laws, sexual reproduction, and genetic recombination. There we discussed in particular the random "genetic reshuffling" that takes place during meiosis when recombination occurs. In figure 1.2, we illustrated in a stylized picture how this recombination occurs. Here we see that during meiosis a random shuffle of genetic recombination takes place that is akin to a RCT. Figure 13.2 provides a simple explanation of how genotypes are randomly assigned from parents. Here the father has a genotype with two alleles of CT and the mother has CC. During conception there is an equal chance of passing one of the two alleles to offspring.

This is also referred to as Mendel's second law [15]. Given the true reshuffling of genetic information when a child is conceived, Fisher already pointed out in 1951 that "*Genetics is indeed in a peculiarly favoured condition in that providence has shielded the geneticist from many of the difficulties of a reliably controlled comparison. The different genotypes possible from the same mating have been beautifully randomised by the meiotic process. A more perfect control of conditions is scarcely possible, than that of different genotypes appearing in the same litter*" [16]. Genetic recombination is thus a natural experiment that allows researchers to engage in causal modeling.

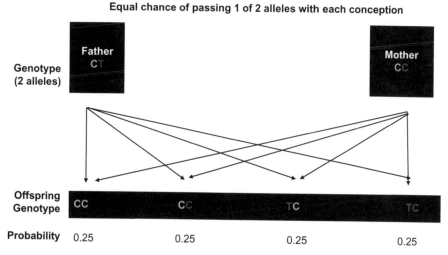

Figure 13.2
Genotypes are assigned randomly from parents.

The probabilities in figure 13.2 are stylized in a fictional scenario. The actual genetic relatedness between siblings is indeed only on *average* 50%. As shown in figure 13.3, Visscher and colleagues [17] demonstrated that the actual variation in siblings is around 0.374 to 0.617 (mean 0.498, SD 0.036) in the shape of a thin normal distribution.

By the turn of the millennium, several studies began to utilize this natural experiment in order to assess causality in observational studies, however, in a different way [14, 18, 19] (see the right panel in figure 13.1). MR as we know it today was described in 2003 by George Davey Smith, who remains one of the driving forces of its development and applications [14]. The underlying idea of the approach is that if we find that a genetic variant is deemed causal for an exposure variable and we can plausibly argue that this variant does not have a direct causal effect on the outcome of interest, an association between the genetic variant and the outcome variable can only be observed via the causal effect of exposure on outcome variable.

13.4 Instrumental variables and Mendelian Randomization

13.4.1 The IV model in an MR framework
Within a statistical framework, MR relates directly to the IV approach utilized in econometrics to assess causality [20]. A general formulation of this approach would be that an IV is correlated with X but not with omitted variables (U) and not with Y—except through X. Typical IVs used in econometrics research are lotteries, political reforms such as variation in tobacco prices [21], natural disasters such as Hurricane Katrina that force the relocation of individuals [22], or birth month, which can influence education [23].

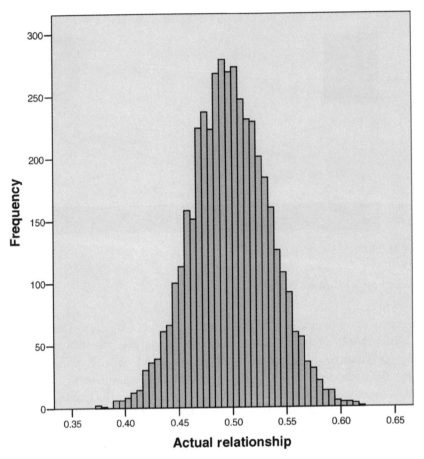

Figure 13.3
Empirical distribution of genome-wide additive genetic relationships of full sibling pairs estimated from genetic markers.
Source: Visscher et al. (2006) [17]. Open Access.

Due to MR, genes can now be used as IVs, under the condition that the gene is causal for *X* but not correlated with omitted variables. Just as in an experiment, we can focus only on the fraction of variance in the exposure variable, which is generated by the IV and analyze whether this variance causes differences in the outcome of interest. If so, we observe a causal relationship, just as we would if we found systematic differences between groups defined in randomized control trial experiments, where the variance we analyze would be generated by the researcher (e.g., by a coin toss).

Figure 13.4 illustrates the idea and assumptions behind the IV approach. If we have a valid *IV* for a variable *X*, we only observe a covariation between *IV* and *Y* if *X* has a causal effect on *Y*. In the depicted path model, effects multiply along causal paths (expressed in a

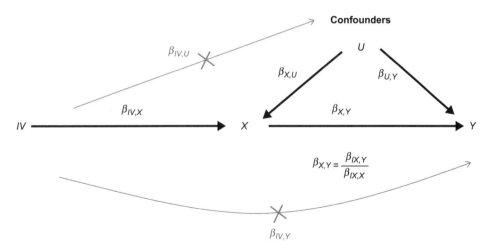

Figure 13.4
Illustration of the Instrumental Variable approach in order to identify a causal effect of X on Y.
Note: This Instrumental Variable (IV) is expected to causally influence X and allows us to identify the causal effect of X on Y independent of any unknown confounders (U). Crossed paths depict the necessary IV-model assumptions that there is no causal effect of $\beta_{IV,U}$ nor of $\beta_{IV,Y}$.

one-headed arrow). This means that the association between IV on $Y(\beta_{IV,Y})$ is equal to the effect of IV on X times the effect of X and Y $((\beta_{IV,Y}) = (\beta_{IV,X}) \times (\beta_{X,Y}))$. In the IV approach, we do not assume that we observe a true causal relationship between $\beta_{X,Y}$ in the data due to endogeneity. We assume, however, that we can obtain an unbiased estimate for $\beta_{IV,Y}$ and $\beta_{IV,X}$. We can thus simply calculate $\beta_{X,Y}$ by dividing the observed association between the IV and Y by the estimate of IV and X $((\beta_{X,Y}) = (\beta_{IV,Y}) / (\beta_{IV,X}))$.

Typically, a two-stage least squares regression is estimated in this approach. In the first stage, we regress X on the IV.

$$X = \mu_1 + \gamma IV + \varepsilon_1$$

In the second stage, we regress Y on the predicted value of $X (\hat{X})$ based on the first stage.

$$Y = \mu_2 + \beta_{2SLS} \hat{X} + \varepsilon_2$$

The reduced form regresses Y directly on the IV:

$$Y = \mu_3 + \rho IV + \varepsilon_3$$

The coefficient of interest (β_{2SLS}) is the ratio of the regression coefficient of Y on Z (the *reduced form*) to the regression coefficient of X on Z (*the first stage*), that is equivalent to

$$\frac{Cov_{y,z}}{Cov_{x,z}}.$$

Two-stage least squares estimation methods are included in several standard software packages, which also accommodate and clarify potential issues in the model as well as estimate correct standard errors and produce two-stage estimates in one step.

As we noted earlier, sources of endogeneity are *simultaneity bias* (reversed or bidirectional causality) and *measurement error* (imprecise measures). Using the example of the relationship between educational attainment and age at first birth (see box 13.1), *omitted variables* that are associated with both educational attainment (X) and age at first birth (Y) include personality, cognitive ability, fertility preferences, socioeconomic status, or parental environment. We provide an example of how an IV might be estimated in box 13.2 using the same example in box 13.1 of the relationship between educational attainment and age at first birth.

Box 13.2
Example of MR to disentangle relationships of age at first birth and educational attainment

We previously discussed the possible causal mechanisms linking age at first birth to educational attainment (box 13.1). To model this question in an MR framework we could opt for the following approach, which is graphed below. As described in the body of the text, it simply means that we introduce a third variable (i.e., *IV*) of a polygenic score (PGS) for educational attainment. This PGS instrument is then assumed to (partly) determine the level of the "treatment," which is observed educational attainment but does not have a direct or indirect effect on our outcome of age at first birth other than through its effect on the observed educational level. This instrument is thus exploited to enable us to make causal inferences about the effect of the level of education on fertility.

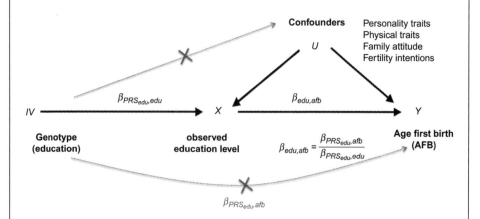

Here the bottom line is that any difference in Y associated with the IV indicates a causal effect of X on Y. To reiterate, the figure above shows that: (1) the IV is associated with the exposure of interest X, (2) the IV is independent of the confounding factors U (i.e., that confound the X-Y association), and (3) genotype is related to the outcome only via its association with the modifiable exposure.

13.4.2 Violation of statistical assumptions of the IV approach

There are several statistical assumptions underlying the MR approach that need to be considered in order to evaluate whether genetic variants or PGSs are good instruments [24]. We summarize these in table 13.1 and discuss them briefly here.

Weak instrument A central criterion for an appropriate IV is that it is not a *weak instrument*. In other words, it needs to be correlated with the treatment and generate sufficient variation in the independent or exposure variable. A violation is often a general problem in IV studies [25]. As elaborated upon throughout this book, individual SNPs in the form of candidate genes or a selection of single markers have very little predictive value and thus are often weak instruments (with some exceptions such as in relation to the FTO gene, discussed elsewhere). If we used these we would run the risk that as weak instruments, single variants would not only generate little variation in the model but also potentially cause substantial bias. Recall that a simple IV approach has the prediction of the instrument ($\beta_{IV,x}$) in the denominator. Very small and imprecise estimates may become problematic and can potentially lead to immense bias in the instance that there is a direct effect of *IV* on *Y* ($\beta_{IV,y}$) that inflates the numerator (in case the exclusion restriction assumption is violated, which we discuss shortly).

Table 13.1
Summary of the main assumptions in MR research.

Assumption	Description	Solutions	Limitations
Weak instrument	IV needs to be strongly correlated with the treatment (X) variable and if not, can lead to noise and biased estimates	Use PGSs rather than single genetic varaints. Test strength of association of PGSs in independent samples	Lose knowledge about the instruments and potentially introduce pleiotropy
Exclusion restriction	Association between the *IV* and outcome (Y) variables is entirely due to a causal effect of the exposure on the outcome	Modeling pleiotropic effects, detecting and excluding pleiotropic SNPs, controlling for direct genetic effects on Y independent of X	Quantitative solutions always inherit the shortcomings of observational studies (e.g., measurement error)
Independence assumption	*IV* and Y do not have a common cause	Control for population stratification	Control might be imperfect; see also chapter 3
No genetic assortative mating	Partners are not mating based on the same genotypes for X or across X and Y	Controlling for parental genotype	Measurement error
Monotonicity assumption	Genes influence the exposure in the same direction for all individuals	Exclude SNPs with non-monotonic effects	Difficult to identify SNPs with non-monotonic effects due to low power

This returns to our earlier discussions in chapter 5 of the construction and predictive power or R^2 of PGSs on observed outcomes. Here we noted that when we sum up effects of multiple genes into a single PGS, it has higher predictive power than individual genetic variants. For example, the PGS for educational attainment in the latest 2018 study [26] explains over 10% of the variation in education in the National Longitudinal Study of Adolescent to Adult Health. The most recent PGS for height has an R^2 of 0.30 [27]. MR can only be employed with PGSs or genetic variants that have been shown to affect the outcome.

In the analysis, the first step is therefore to demonstrate that the genetic markers predict the outcome. Using the example in box 13.2, for instance, you would test whether the PGS for educational attainment would predict years of education in the data that you are using. Other strategies include using multiple genes as independent IVs, and meta-analysing results, which also improve the precision of the estimation. In such approaches—in parallel to the use of PGSs—we face the mechanism-prediction trade-off in a sense that we might have stronger instruments but are at a higher risk to violate the exclusion restriction assumption (see the next point). While we might have sufficient predictive power for a strong genetic instrument, the mechanisms causing the association are less clear, with variants included that are potentially unrelated to the outcome of interest.

Exclusion restriction assumption   The second important assumption in this type of analysis is the exclusion restriction assumption. This assumes that the IV does not influence the outcome of interest directly or that it is exogenous to the outcome of interest (see also figure 13.4, crossed lines). This scenario directly relates to the *mechanisms-prediction trade-off* that we discussed in relation to the ubiquity of pleiotropy (chapter 5). When using IVs in a genetic framework, the exclusion restriction may be violated by four central reasons. First, biological processes could bias causal mechanisms due to *canalization*. Using the example in box 13.2, genetic variants associated with educational attainment might have a biological function in regulating fertility. Note that in a previous study the genetic correlation (LD score regression) was 0.70, suggesting that this might be plausible [28].

Second, genetic variants associated with the PGS *IV* (e.g., educational attainment) might be in *linkage disequilibrium (LD)* with genetic variants that are associated with the outcome *Y* (e.g., age at first birth). Although Mendel's second law states that the inheritance of one trait is independent of the inheritance of another, some variants may be in fact co-inherited (i.e., LD). LD could therefore potentially bias our estimates if our genetic marker instrument (IV) is in LD with another SNP that directly affects our outcome (*Y*).

A third and related problem is *direct pleiotropy*, which refers to the situation where one genetic variant has multiple functions (see chapter 5). Using our example in box 13.2 again, it would be problematic for our model if SNPs for educational attainment are also affecting the timing of first birth directly.

A final concern related to the violation of the exclusion restriction is that certain behaviors or conditions related to the parents may depend on their (inherited) genotype and at the same time influence their children [29]. If a parental phenotype based on the genetic scores in the IV influences the Y of their children, the exclusion restriction assumption would also be violated.

Assumption of independence   Another statistical condition required for a valid causal interpretation of the IV estimate is the *assumption of independence*, which is related to population stratification. In general, it is assumed that the IV and the outcomes of interest do not share a common cause. In reference to population stratification, this would refer to the situation where the relationship between a genetic variant for educational attainment and the age at first birth is observed differently over different subpopulations. We discussed the importance of controlling for principal components in many places throughout this book (see sections 3.3.1, 5.3.3, and 9.4) and therefore do not repeat it here.

No genetic assortative mating   Assortative (or nonrandom) mating refers to the situation when partners are more alike in specific characteristics than random pairs of individuals in the population. For instance, we know that that there is strong homogamy in that people often marry or partner with someone who is of a similar educational level. There is evidence that assortative mating on educational attainment, for instance, leads to genetic spousal resemblance in their genetic scores [30]. There does, however, remain discussion about whether these types of effects are related to population stratification. Hartwig et al. [31] have shown that both single-trait genetic assortative mating (for example, on educational level) and cross-trait genetic assortative mating (for example, higher educated women mate with taller men) can bias MR estimates.

Monotonicity assumption   Another important assumption is that the instrument influences X in the same direction for all individuals—it does not, for example, increase educational attainment for some individuals and decrease it for others. This assumption is challenged mainly by an interaction effect between genes (epistasis) and genes and the environment (see also chapter 11 for discussion of $G \times E$). However, empirical evidence for epistasis is scarce, and given that associations used in the MR framework are typically based on meta-analyses across multiple environments, it seems reasonable to assume the result to be relatively robust against $G \times E$ [32, 33].

13.5   Extensions of standard MR

Several strategies have been proposed to improve standard MR analysis, in particular with respect to raising the predictive power of the instrument and at the same time taking

Table 13.2
Selected Mendelian Randomization (MR) approaches based on genome-wide association study (GWAS) summary statistics only or in combination with individual-level data.

Based on summary results

Method	Short description	Software	Requirements	Main advantage	Main disadvantages	Paper
MR-Egger	• Uses multiple individual genetic variants as instruments in separate models • Meta-analyzing results • Adds an intercept indicating bias due to pleiotropy	https://raw.github.com/remlapmot/mrrobust/master/	GWAS summary statistics on X and Y for independent individuals	Models pleiotropy—relaxes exclusion restriction assumption	Assuming no correlation between genetic effects on X and pleiotropic effects (InSIDE) assumption)	Bowden et al. 2015 [36]
MR-Presso	• Uses multiple individual genetic variants as instruments • Meta-analyzing results • Detects and removes pleiotropic SNPs	https://github.com/rondolab/MR-PRESSO	Top SNPs from GWAS summary statistics on X and Y (ideally for independent individuals)	Outlier detection—improvement to MR-Egger	Requires InSIDE assumption, < 50% of the SNPs are assumed to be pleiotropic	Verbanck et al. 2018 [37]
GSMR/ Heidi	• Uses multiple individual genetic variants as instruments • Removes pleiotropic SNPs • Meta-analysing results • Detects and removes pleiotropic SNPs • Takes LD into account	http://cnsgenomics.com/software/gsmr/	Top SNPs from GWAS summary statistics on X and Y (ideally for independent individuals) AND LD reference panel based on individual data	Outlier detection—more powerful than MR-Presso (taking LD into account)	Requires at least 10 SNPs with p-value $< 5 \times 10^{-8}$	Zhu et al. 2018 [38]

Based on individual-level data

Method	Short description	Software	Requirements	Main advantage	Main disadvantages	Paper
GIV	• Uses PGSs for X as IV • Controls for PGS for Y conditional on X • Removes measurement error for PGS Y	For score construction (e.g., PRSice, LDpred; see chapter 10). For analysis, any data analysis software (e.g., R; see appendix 1)	Individual-level genotype and phenotype data and GWAS summary statistics for X and Y excluding the prediction sample	Can control for environmental confounders or gene-environment correlation in regression framework	Might be biased due to environmental confounders unrelated to genetic effects	DiPrete et al. 2018 [39]

direct pleiotropy into account. We summarize some of these in table 13.2, but for a more exhaustive review of techniques, refer to other extensive reviews [34, 35]. Specifically, we focus on three extensions that use, namely, multiple genetic variants as independent instruments, multiple polygenic scores for a regression-based IV, and a bidirectional MR. We further differentiate approaches that use multiple genetic markers from those that try to model pleiotropy and those that try to exclude pleiotropic SNPs.

13.5.1   Using multiple markers as independent instruments

Bowden and colleagues [36] developed a method that uses multiple genetic variants for independent MR models and meta-analyzed results across these studies taking pleiotropy into account. In general, MR approaches using multiple variants as instruments regress the variants' effects on an outcome on the variant's effect on the independent variable or exposure in a linear model [37, 40]. However, the slope of this regression model, which estimates the causal effect of independent variable on outcome, might be biased when the exclusion restriction assumption is violated, such as in the case of pleiotropy.

Bowden et al. adapted an approach from a previously developed meta-analysis model, which was originally designed to detect potential publication bias in meta-analyses, called Egger-Regression [41]. In the context of MR analysis, they proposed an approach, which not only models the slope of increased effects of stronger instruments on the outcome of interest Y, but also an intercept quantifying potential pleiotropic effects. Using this, the slope indicating the causal effect of X on Y is supposedly corrected for potential bias due to pleiotropy. While this approach claims to relax the exclusion restriction assumption by modeling pleiotropy, it is limited by a new assumption stating that pleiotropic effects are not associated with increasing the strength of the instrument.

Another series of methods described in table 13.2 aim to identify and remove pleiotropic outliers from the meta-analysis such as, for example, by calculating the difference between the causal effect of the SNP and the predicted causal effect of the SNP from a model, which excludes the respective SNP. Larger distances between predicted and actual effects indicate pleiotropic effects [37]. More complex and powerful approaches of the same basic logic also try taking LD into account [38].

13.5.2   Using polygenic scores as IVs

Another approach is an extension of a standard IV regression based on individual-level data and utilizing PGSs [32, 39]. If we sum up the information across multiple genetic variants in a polygenic score for our independent variable X, again, we increase predictive power of the instrument but struggle with the exclusion restriction assumption due to potential pleiotropy.

The proposed genetic IV regression approach by DiPrete and colleagues [39] introduces a series of corrections to overcome this issue. First, they propose to control for potential pleiotropy by controlling for a polygenic score for Y. However, one issue is that the genetic discovery for Y includes effects, which go via X, since it does not discriminate between direct pleiotropy, which would violate the exclusion restriction assumption, and indirect pleiotropy, which occurs because X is causal for Y. Only the former represents a violation of the model. However, controlling for the GWAS result on Y includes both types of pleiotropy and therefore also the effect of the instrument on Y via the causal relationship between X and Y (i.e., what we are interested in). The authors therefore propose to condition the genetic discovery for Y on X or condition the PGS for Y on X and therefore to create a score that only captures direct pleiotropic effects.

Although this approach comes closer to solving the pleiotropy problem, it remains challenging since our PGS measures remain noisy. Missing heritability and pleiotropic effects may only be entirely eliminated if the PGS for Y conditional on X would be the true one. One idea to solve this would be to conduct two independent GWASs for Y conditional on X and generate two quasi-independent scores from the results of these. Assuming that the measurement error between both of these scores is independent, it is possible to use one as an instrument for the other. This is a classic approach to remove measurement error and, in this case, restores heritability estimates. This so-called genetic IV regression (GIV-regression) framework is a promising extension to the classic MR approach but also introduces new challenges where there are strong nongenetic confounders between X and Y and furthermore tends to underestimate the causal effect of X on Y [32].

13.5.3 Bidirectional MR analyses

Another approach that has emerged is a bidirectional MR that allows researchers to look at reverse causation and unobserved residual confounding (i.e., endogeneity). One example is a recent study that was able to demonstrate that it was obesity that determined vitamin D deficiency and not the other way around [42]. In principle, the MR approach is first performed in the direction of X on Y and then of Y on X to test for reverse causality. This approach assumes single causal directions of effects. However, in reality, causal feedback loops are possible and structural equation models may assist in modeling in these circumstances [34].

13.6 Applications of MR

In this section, we provide some illustrative examples of MR applications from the literature. Specifically, we discuss examples from effect of cause studies on the consequences of alcohol consumption on blood pressure, body mass index (BMI) on mortality, and cause of effect studies on the risk of dementia and Alzheimer's disease.

13.6.1 Consequences of alcohol consumption

Alcohol intake is associated with hypertension—high blood pressure—and potentially a causal risk factor for these health outcomes. In turn, hypertension is a major precursor for stroke and coronary heart disease, responsible for one in eight deaths per year worldwide [43]. It is therefore important for us to establish whether individuals who consume alcohol die earlier than if they would not, particularly since alcohol intake is a modifiable behavior. Alcohol intake is a prime example of an exposure for which a randomized control trial would not be feasible.

MR can be useful since a common polymorphism in aldehyde dehydrogenase 2 (ALDH2) encodes a major enzyme that is implicated in the metabolism of alcohol. Individuals with two copies of the null-variant experience more "adverse symptoms of drinking" (i.e., hangovers) compared to individuals with alternative genotypes. The null-variant genotype is therefore also associated with less drinking. Assuming that the same polymorphism does not directly influence blood pressure, we can interpret an association between the ALDH2 polymorphism and blood pressure as evidence of a causal effect of alcohol metabolism on blood pressure since this association can only emerge due to a causal effect of the metabolism of alcohol on blood pressure.

Chen and colleagues [44] meta-analyzed eight studies on the effect of ALDH2 on blood pressure and hypertension and found significant evidence that alcohol intake has detrimental effects on both outcomes. Making their case for a good instrument, among others, the authors show that this effect is seen for males and not females, who drink less in general. If direct pleiotropy would cause the association—and the exclusion restriction assumption would therefore be violated—we should, however, observe the same effect for both sexes.

13.6.2 Body mass index and mortality

Another study examined the well-established relationship between BMI and mortality [45]. Not only can severe obesity (BMI ≥ 35 kg (77 lbs)/m^2, for example > 113 kg/250 lbs for a body length of 1.80 m/5′9″) increase the risk of death, already pre-obesity (BMI ≥ 25 kg (55 lbs)/m^2, for example > 80 kg/176 lbs for a body length of 1.80 m/5′9″) is associated with higher mortality due to elevated risk of vascular mortality, diabetes, or cancer. However, some studies also suggest [46] that higher BMI might be protective against all-cause mortality—death by any cause—resulting in an obesity paradox [45]. Potential explanations are that being underweight might also have detrimental health consequences, however, it is also possible that the observational studies report partly spurious findings due to confounding.

Wade and colleagues [45] conducted an MR study on around 335,000 individuals from the UK Biobank, of which around 10,000 had died due to various causes. The authors constructed a weighted PGS based on 77 SNPs. They found suggestive evidence that 1 kg/m^2 higher BMI increases the risk of death by around 3% across ages as well as evidence for various specific death causes due to, for example, cardiovascular diseases. Importantly, they provided a set of robustness analyses including MR-Egger, adjust for various covariates and exclude specific genetic variants in the score construction to check for pleiotropy.

13.6.3   Causes of dementia and Alzheimer's disease

Dementia is a health threat affecting around 35 million older adults worldwide in 2010, with recent projections forecasting that it will more than triple by 2050 [47]. While the development of adequate treatment is a topic of ongoing medical research, an alternative strategy to *inter*vention is *pre*vention, namely facing this challenge by reducing the exposure to known determinants of the disease [48], such as BMI, physical inactivity, and smoking. However, again, it is imperative to ensure that those factors that we aim to change are indeed the actual causes of the disease and not merely correlates, otherwise the prevention will be ineffective.

One of the best-established correlates of dementia is educational attainment. Higher educational attainment is associated with better cognitive performance in old age and substantially reduced risk of developing dementia. It is intuitive that cognitive performance in younger ages and cognitive health in older ages might be influenced by common causes. But it would be important to understand if the relationship between education and dementia is causal.

Several studies apply IV and MR to test for this causality. A paper by Nguyen and colleagues [49] combines classic IV approaches and MR in the study of education and dementia using data from the Health and Retirement Study in the United States. They first use a PGS for educational attainment. Second, in a more classic manner, they use variation in compulsory school laws as an IV. School laws generate variation in the educational attainment of the individuals but should not directly affect their risk for dementia. Their study finds evidence that education protects from dementia in a causal manner in both research designs. However, sensitivity analyses suggest that compulsory school laws operate as a more valid instrument than genes, due to potential violations of IV assumptions.

In relation to Alzheimer's disease, Zhu and colleagues [38] applied a more advanced MR approach, which includes multiple SNPs as independent instruments, striving to remove pleiotropic outliers. Their study supports the finding of a protective effect of education on Alzheimer's disease and on general health measured in terms of disease counts. In a broader cause of effect perspective, a meta-analysis by Kuźma et al. [48] aimed to isolate the main determinants of Alzheimer's disease based on several existing MR studies. The meta-analysis includes 18 MR studies covering causes such as lifestyle factors, cardiovascular factors and related biomarkers, diabetes-related and other endocrine factors, and telomere length. The strongest evidence for an increased risk of Alzheimer's disease was based on shorter telomeres, with further evidence pointing toward smoking quantity, vitamin D, homocysteine, systolic blood pressure, fasting glucose, insulin sensitivity, and high-density lipoprotein cholesterol as causal determinants of Alzheimer's disease.

Importantly, studies applied various MR approaches deriving causal estimates by various methodological approaches using both single variants or combinations of multiple variants. While the number of included determinants and the reported variety of approaches continues to expand, the main concerns of these studies remain in lack of statistical power, overrepresentation of European ancestry groups, and violations of the weak instrument problem, and the exclusion restriction assumption.

13.7 Conclusion

MR has immense potential to establish causal inference. As we outlined in tables 13.1 and 13.2, there are both specific challenges but also a growing number of solutions. One recurrent problem that is not isolated to MR but also implicit in standard regression models is that large sample sizes are required. Furthermore, although the use of PGSs increases the predictive power of IVs and counters the weak instrument challenge, it introduces new problems related to the exclusion restriction, independence, and other biases.

A crucial factor for all approaches is that they will inherit potential problems from the original GWAS summary statistics in which they are based. Those include winner's curse and bias due to gene-environment correlation (see also chapter 6). The aforementioned GIV approach might be able to correct for some gene-environment correlation given that PGSs are incorporated in flexible modeling approaches, confounders might be taken into control and within-family analyses are feasible.

As we also noted in chapter 5 in our discussion of PGSs, the challenge that remains is coping with pleiotropy and the related exclusion restriction assumption in the IV framework. In the current chapter we also discussed the main methodological advances, such as modeling the potential bias in MR-Egger or attempting to remove SNPs with pleiotropic effects. It is likewise essential to note the prediction-mechanism trade-offs especially when it comes to MR and the selection of included SNPs in the construction of a PGS (see chapter 10). Further challenges also include gene-environment interaction, which we address elsewhere (see also chapters 6 and 11).

Methodological developments also continue within this area of research. One rather advanced approach, which goes beyond MR models, introduces a latent variable. This variable mediates causal effects across traits, estimating a genetic causality proportion based on mixed marginal effect sizes for each outcome [50]. This method requires even fewer assumptions, and it is potentially important since GWAS samples may overlap and it enables the research to draw from all of genetic information obtained from a GWAS. Finally, we note that although quantitative solutions will take us very far, understanding the biological mechanisms and including more specific SNPs and effects will be an important frontier.

Exercises

1. Devise a phenotypic causal relationship which is important in your own research.

2. Look up GWAS summary statistics for both the exposure and the outcome variable (see chapter 7 on how to obtain these).

3. Find the main hit from the respective GWASs from the exposure variable, gather information about this variant (see also chapter 12), and think about whether you expect the exclusion restriction assumption is reasonable.

4. Make a case as to which of the following approaches would be most suitable for your research question: using a single variant as instrument, using summary results, or using polygenic scores.

5. Think about a potential nongenetic IV for your research question.

Further reading

A lot of MR development and training happens at the University of Bristol in the MRC Integrative Epidemiology Unit at the University of Bristol: http://www.bristol.ac.uk/integrative-epidemiology/.

An online tool and R packages for MR can be found at http://www.mrbase.org/.

"A Two Minute Primer on Medelian Randomisation," by George Davey Smith: https://www.youtube.com/watch?v=LoTgfGotaQ4.

Updates on Twitter @Mendelian_lit and @mendel_random.

References

1. J. L. Rodgers et al., Education and cognitive ability as direct, mediating, or spurious influences on female age at first birth: Behavior genetic models fit to Danish twin data. *AJS* **114**, S202 (2008).

2. S. H. Barcellos, L. S. Carvalho, and P. Turley, Education can reduce health differences related to genetic risk of obesity. *Proc. Natl. Acad. Sci. USA* **115**, E9765–E9772 (2018).

3. M. C. Mills, R. R. Rindfuss, P. McDonald, and E. te Velde, Why do people postpone parenthood? Reasons and social policy incentives. *Hum. Reprod. Update* **17**, 848–860 (2011).

4. N. Balbo, F. C. Billari, and M. C. Mills, Fertility in advanced societies: A review of research. *Eur. J. Popul.* **29**, 1–38 (2013).

5. D. M. Upchurch, L. A. Lillard, and C. Panis, Nonmarital childbearing: Influences of education, marriage, and fertility. *Demography* **39**, 311–329 (2002).

6. D. M. Fergusson and L. J. Woodward, Teenage pregnancy and female educational underachievement: A prospective study of a New Zealand birth cohort. *J. Marriage Fam.* **62**, 147–161 (2000).

7. M. J. Budig and P. England, The wage penalty for motherhood. *Am. Sociol. Rev.* **66**, 204 (2001).

8. A. Alvergne, M. Jokela, and V. Lummaa, Personality and reproductive success in a high-fertility human population. *Proc. Natl. Acad. Sci.* **107**, 11745–11750 (2010).

9. D. A. Briley, F. C. Tropf, and M. C. Mills, What explains the heritability of completed fertility? Evidence from two large twin studies. *Behav. Genet.* **47**, 36–51 (2017).

10. F. C. Tropf and J. J. Mandemakers, Is the association between education and fertility postponement causal? The role of family background factors. *Demography* **54**, 71–91 (2017).

11. D. A. Lawlor, R. M. Harbord, J. A. C. Sterne, N. Timpson, and G. D. Smith, Mendelian randomization: Using genes as instruments for making causal inferences in epidemiology. *Stat. Med.* **27**, 1133–1163 (2008).

12. R. Gray and K. Wheatley, How to avoid bias when comparing bone marrow transplantation with chemotherapy. *Bone Marrow Transpl.* **7**, 9–12 (1991).

13. G. Davey Smith, What can mendelian randomisation tell us about modifiable behavioural and environmental exposures? *BMJ* **330**, 1076–1079 (2005).

14. G. D. Smith and S. Ebrahim, "Mendelian randomization": Can genetic epidemiology contribute to understanding environmental determinants of disease? *Int. J. Epidemiol.* **32**, 1–22 (2003).

15. G. Mendel and W. Bateson, Experiments in plant hybridization (EN transl.). *J. R. Hortic. Soc.* (1865).

16. R. Fisher, Statistical methods in genetics. *Int. J. Epidemiol.* **39**, 329–335 (2010).

17. P. M. Visscher et al., Assumption-free estimation of heritability from genome-wide identity-by-descent sharing between full siblings. *PLoS Genet.* **2**, e41 (2006).

18. B. Keavney et al., Fibrinogen and coronary heart disease: Test of causality by "Mendelian randomization." *Int. J. Epidemiol.* **35**, 935–943 (2006).

19. D. Clayton and P. M. McKeigue, Epidemiological methods for studying genes and environmental factors in complex diseases. *Lancet* **358**, 1356–1360 (2001).

20. W. W. H. Greene, *Econometric analysis*, 7th ed. (Boston: Prentice-Hall, 2012).

21. J. P. Leigh and M. Schembri, Instrumental variables technique: Cigarette price provided better estimate of effects of smoking on SF-12. *J. Clin. Epidemiol.* **57**, 284–293 (2004).

22. D. S. Kirk, A natural experiment of the consequences of concentrating former prisoners in the same neighborhoods. *Proc. Natl. Acad. Sci. USA* **112**, 6943–6948 (2015).

23. C. A. Rietveld and D. Webbink, On the genetic bias of the quarter of birth instrument. *Econ. Hum. Biol.* **21**, 137–146 (2016).

24. V. Didelez and N. Sheehan, Mendelian randomization as an instrumental variable approach to causal inference. *Stat. Methods Med. Res.* **16**, 309–330 (2007).

25. J. D. Angrist and A. B. Keueger, Does compulsory school attendance affect schooling and earnings? *Q. J. Econ.* **106**, 979–1014 (1991).

26. J. J. Lee et al., Gene discovery and polygenic prediction from a genome-wide association study of educational attainment in 1.1 million individuals. *Nat. Genet.* **50**, 1112–1121 (2018).

27. L. Lello et al., Accurate genomic prediction of human height. *Genetics* **210**, 477–497 (2018).

28. N. Barban et al., Genome-wide analysis identifies 12 loci influencing human reproductive behavior. *Nat. Genet.* **48**, 1–7 (2016).

29. A. Kong et al., The nature of nurture: Effects of parental genotypes. *Science* **359**, 424–428 (2018).

30. D. Hugh-Jones, K. J. H. Verweij, B. St. Pourcain, and A. Abdellaoui, Assortative mating on educational attainment leads to genetic spousal resemblance for polygenic scores. *Intelligence* **59**, 103–108 (2016).

31. F. P. Hartwig, N. M. Davies, and G. Davey Smith, Bias in Mendelian randomization due to assortative mating. *Genet. Epidemiol.* **42**, 608–620 (2018).

32. D. Conley and S. Zhang, The promise of genes for understanding cause and effect. *Proc. Natl. Acad. Sci. USA* **115**, 5626–5628 (2018).

33. F. C. Tropf et al., Hidden heritability due to heterogeneity across seven populations. *Nat. Hum. Behav.* **1**, 757–765 (2017).

34. J. Zheng et al., Recent developments in Mendelian randomization studies. *Curr. Epidemiol. Reports* **4**, 330–345 (2017).

35. J. B. Pingault et al., Using genetic data to strengthen causal inference in observational research. *Nat. Rev. Genet.* **19**, 566 (2018).

36. J. Bowden, G. D. Smith, and S. Burgess, Mendelian randomization with invalid instruments: Effect estimation and bias detection through Egger regression. *Int. J. Epidemiol.* **44**, 512–524 (2015).

37. M. Verbanck, C. Y. Chen, B. Neale, and R. Do, Detection of widespread horizontal pleiotropy in causal relationships inferred from Mendelian randomization between complex traits and diseases. *Nat. Genet.* **50**, 693 (2018).

38. Z. Zhu et al., Causal associations between risk factors and common diseases inferred from GWAS summary data. *Nat. Commun.* **9**, 224 (2018).

39. T. A. DiPrete, C. A. P. Burik, and P. D. Koellinger, Genetic instrumental variable regression: Explaining socioeconomic and health outcomes in nonexperimental data. *Proc. Natl. Acad. Sci.* **115**, E4970–E4978 (2018).

40. S. Burgess, R. A. Scott, N. J. Timpson, G. D. Smith, and S. G. Thompson, Using published data in Mendelian randomization: A blueprint for efficient identification of causal risk factors. *Eur. J. Epidemiol.* **30**, 543–552 (2015).

41. M. Egger et al., Bias in meta-analysis detected by a simple, graphical test. *BMJ* **315**, 629–634 (1997).

42. K. S. Vimaleswaran et al., Causal relationship between obesity and vitamin D status: Bi-directional Mendelian randomization analysis of multiple cohorts. *PLoS Med.* **10**, e1001383 (2013).

43. H. Arima, F. Barzi, and J. Chalmers, Mortality patterns in hypertension. *J. Hypertens.* **29**, S3–S7 (2011).

44. L. Chen, G. D. Smith, R. M. Harbord, and S. J. Lewis, Alcohol intake and blood pressure: A systematic review implementing a mendelian randomization approach. *PLoS Med.* **5**, e52 (2008).

45. K. H. Wade, D. Carslake, N. Sattar, G. Davey Smith, and N. J. Timpson, BMI and mortality in UK Biobank: Revised estimates using Mendelian randomization. *Obesity* **26**, 1796–1806 (2018).

46. K. M. Flegal, B. K. Kit, H. Orpana, and B. I. Graubard, Association of all-cause mortality with overweight and obesity using standard body mass index categories a systematic review and meta-analysis. *JAMA* **309**, 71–82 (2013).

47. M. Prince et al., The global prevalence of dementia: A systematic review and metaanalysis. *Alzheimer's Dement.* **9**, 63–75 (2013).

48. E. Kuźma et al., Which risk factors causally influence dementia? A systematic review of Mendelian randomization studies. *J. Alzheimer's Dis.* **64**, 181–193 (2018).

49. T. T. Nguyen et al., Instrumental variable approaches to identifying the causal effect of educational attainment on dementia risk. *Ann. Epidemiol.* **26**, 71–76 (2016).

50. L. J. O'Connor and A. L. Price, Distinguishing genetic correlation from causation across 52 diseases and complex traits. *Nat. Genet.* **50**, 1728–1734 (2018).

14

Ethical Issues in Genomics Research

Objectives

- Raise awareness of the *multiple ethical issues* in genomics research
- Inform researchers to *guard against genetic determinism*, highlighting the use of PGSs in monogenic versus polygenic traits and issues of missing heritability
- Provide an overview of the *clinical use of polygenic scores* including family history, screening, intervention and life planning, drug development, and public understanding
- Reiterate the *lack of diversity* in current genomic research and implications for the wider applicability of results
- Explore *privacy, consent, legal issues, and insurance* in relation to genomic data

14.1   Introduction

In 2018, genetic data was used to solve the cold case of the Golden State Killer, a man who had evaded police for decades after raping more than 50 women and killing around a dozen people. In that same year, Erlich and colleagues [1] showed that around 60% of Americans of European ancestry could be identified using genetic data via one of their relatives. These examples demonstrate some of the potentials of genetic data use, but they also expose the *risks* and *privacy issues*.

Meanwhile, prominent medical scientists are increasingly endorsing PGSs to be used for personalized medicine [2]. Others, such as well-known behavioral geneticist Robert Plomin, are advocating for the use of the intelligence and educational attainment polygenic scores (PGSs) to develop targeted educational policy [3, 4]. Plomin has argued, for instance, that the nature–nurture war is over and that "nature wins, hands down," rendering the influence of families or the school environment irrelevant [5]. This harkens back to the basic and repeated discussions surrounding *genetic determinism and nature versus nurture* (see box 14.1).

Box 14.1
The challenges of using individual polygenic scores to predict complex outcomes: The case of intelligence

There has been considerable discussion both in the scientific literature [2] and public debate [6] surrounding the use of individual PGS as effective predictors of disease and behavior and how lay individuals interpret these scores. This debate has been sparked further by recent moves by prominent behavioral genetic researchers such as Robert Plomin and colleagues [3, 4] to argue that the genetic scores derived for intelligence and educational attainment could be applied for personalized educational policy, akin to personalized medicine.

A vivid example of why measuring genetic intelligence at the individual level remains currently very difficult—and actually useless—is described by *New York Times* writer Carl Zimmer after he compared his personal genetic results on the open data platform of DNA. land [7]. At the time of writing, if you upload your genetic data to DNA.land, it compares your own genome with the 52 genes isolated in the 2017 educational attainment study [8] of 78,000 people (which had a predictive power of almost 5%) against the reference group of everyone in the DNA.land dataset. (It may very well be that the site will be updated in the future using more recent results, such as the 2018 updated publication of educational attainment [9].)

The effect that each variant has on the PGS is added up to determine the total impact, which as we know is normally distributed (see chapter 5, box 5.1). Not only did the *New York Times* writer end up on the left side of the curve, but so too did the brilliant developer of the code and the well-known clever computational bioinformatics and genetics researcher Yaniv Erlich. In this discussion, Zimmer raises the vital point that the use of the words *precision* and *prediction* provides the illusion of more certainty than is actually the case. In the case of individual PGSs, particularly for behavioral traits, we need to remember that these are not accurate individual forecasts. For this reason, the use of individual PGSs alone for prediction without considering GxE or other influences may be premature, particularly for behavioral phenotypes. Or, as Erlich warns in his interview with Zimmer in relation to the focus on genetics over environmental factors in the area of intelligence: "I'm afraid that policy makers won't focus on the real things that bother me about inequality and education."

On other fronts, in early 2019, in an unprecedented move, the Editorial Board of the *New York Times* [6] wrote a stern warning about the potential risks of direct-to-consumer health tests from the direct-to-consumer genetics company 23andMe. The warning focused on the content of what the information in the PGSs contained, but also the *interpretation of genetic risk by individuals*. Concerns and debates continue to rage surrounding the *legal issues* of genetic data and their availability and *use by insurance companies, employers, or the government*.

As we demonstrated in a publication in early 2019 [10], the results of genetic discoveries—and thus the PGSs derived from them—continue to be dominated by the study of only European-ancestry groups. This ranges from around 95% of these groups in 2007 to 88% in 2017, with selection also occurring along other lines such as socioeconomic status, country of origin, and key demographic variables. This raises further ethical questions surrounding the *equity of genomics* and whether using these highly selective results actually reinforces existing health and social inequalities.

These expansive, serious, and constantly evolving questions cannot fully be tackled in detail in this introductory textbook. Yet ethics is particularly a key part of this research and will emerge across all levels and stages of your work, and it is pivotal to communicate the correct interpretation of results. In this chapter we first discuss genetic determinism, which is a key ethical concern that emerges from this type of research. Within this we look at variation in monogenic and polygenic traits and the ability to use PGSs as predictors and also address related issues of missing heritability. In the next section we focus on the growing assertions that PGSs can be used for clinical applications. Here we focus on family history, how these scores can be used for screening, intervention and life planning, pharmacogenics, and the public understanding of their own genetic information. As highlighted previously in our chapter 4 on genome-wide association studies (GWASs), we reiterate the current lack of diversity in the data and implications for research. We then focus on the commonly raised concerns regarding privacy, consent, legal issues, and insurance followed by a brief reflection on future directions.

14.2   Genetics is not destiny: Genetic determinism

14.2.1   Variation in traits and ability to use individual PGSs as predictors
As discussed throughout this book, results from genetic analyses should always be considered with great care and technical scrutiny. Some may be tempted to overstate results whereas others might adopt a deterministic view that genetics seals your fate and is destiny [4]. As we illustrated in figure 1.3 in chapter 1, there is a wide spectrum of the genetic contribution to phenotypes. While it is true that for a handful of rare genetic diseases genetics is clearly and highly predictive, for the majority of complex common diseases and behavioral traits that we study, genetics only explains a fraction of the variance of the outcome. There are exceptions such as monogenetic diseases or the APOE genes for Alzheimer's, for instance, which explains around 13% [11] of the heritability in the phenotype. At this time, such cases, however, represent more the exception than the rule. Carriers of the risk mutations for the BRCA1 and BRACA2 genes, for example, have an increase of risk to develop breast cancer in their lives of 65% and 45%, respectively [12], while the genetic variants account only for 5% of the breast cancer cases [13].

While genetics may be very relevant for some rare genetic diseases or mutations, when examining complex behavioral phenotypes or diseases that have a highly polygenic or behavioral component, they are often only partially explained by genetics. The controversial phenotype of intelligence, for instance, currently has a polygenic score that predicts around 4% of the variance of intelligence [3]. As we note in more detail in our later section on the potential clinical applications of PGSs, additional factors are also key, including the socioeconomic environment and other behaviors such as lifestyle and exercise. If a simplistic and purely fatalistic genetic deterministic view was promulgated as a single solution for interventions, this could undermine policy solutions that may play a more important role. Although there is considerable disagreement in this area of research, at this time due to the issues of polygenicity and pleiotropy discussed at length in this book, it seems difficult to conclude that PGSs alone could be effectively used as the only predictors for interventions on complex behavioral traits such as social or educational policies (see box 14.1 in relation to intelligence). Later in this chapter we address the broader applicability to disease-related traits that also have a behavioral component, such as obesity.

14.2.2 Heritability and missing heritability

As we described in a previous chapter, we often interpret the relationship of genetics with a particular phenotype in terms of the heritability of a disease or behavioral outcome. Recall that this is the proportion of variance of the phenotype that can be explained by genetic variance within a population. Moreover, we can also measure how much of that heritability can *currently* be explained by known genetic variants, which as we show in chapter 4 on GWASs continues to increase over time. As we also outlined previously, many outcomes have a heritability that can range anywhere from 5% to 80%.

In this context, it is also important to understand the distinct features of PGSs in contrast to heritability studies. While PGSs might explain only a fraction of the heritability, they provide a type of risk stratification to identify individuals who are at elevated risk for a disease or outcome. Low variance explanations imply that many more factors influence the phenotype. While a PGS does not tell us the full story, it is still potentially able to identify individuals with an elevated risk for a disease or a predisposition for a phenotype in general. Heritability estimates, in contrast, cannot identify individuals with high genetic risk and do not provide absolute measures of genetic distinction between cases and controls or across the spectrum of a continuous variable.

As we discussed previously in the section on misnomers of heritability, there is often the incorrect notion that low heritability or genetic prediction means that genetics are not important. The chance of winning a lottery, for instance, would have no genetic component. (Some may argue that buying the lottery ticket in the first place may play a role, but we leave this selection issue aside for now.) Other complex outcomes such as developing lung cancer have a low heritability of around 8%, and thus a relatively small genetic component [14]. Rather, there are other behavioral aspects related to smoking often grounded in deeper socioeconomic environmental patterns, interrelated with correlated risks for other traits such as obesity, alcohol use, or type 2 diabetes. Environmental factors and attempting to separate genetic and confounding environmental influences remain pivotal, which we also explored in chapters 6 and 11.

As noted throughout this book, a majority of the statistical tests show an "association" between certain genetic alleles or loci with a phenotypic outcome, which is not causation.

Although we have often isolated thousands of genetic variants and their associations, only the downstream biological research or other methods such as Mendelian Randomization (chapter 13) can actually show "causality," or in other words in relation to biology, describe what the genetic loci actually does.

We also encounter the problem of missing or hidden heritability. In a trait such as height, although we know that it is 55% to 81% heritable [15], as of late 2018 genome-wide significant SNPs explained around 25% of the heritability of human height [16]. Or in other words, at this time the known genetic loci discovered in existing GWASs that are commonly associated with height cannot explain the heritability of height.

Another issue is the interplay between common and rare variants, concepts introduced in chapter 1. Mendelian phenotypes, where a single genetic mutation is often pinpointed as the reason for a disease or outcome, are often rare in populations but common in families since they are passed down through the generations. For the majority of genetic outcomes we study, however, they are complex with many variants of relatively small individual effects contributing to the phenotype (recall box 1.2, discussing NBA star Shawn Bradley). Some researchers now suspect that the genomic basis of many complex phenotypes is a mix of common variants with small effects and rare variants with large effects [17]. A large technical challenge that will undoubtedly be the focus of many upcoming studies is understanding and mapping the specific roles of common and rare variation (see chapter 15).

14.3   Clinical use of PGSs

Typically, clinical risk predictions have relied on family history and demographic characteristics such as age and sex. But it has also drawn on lifestyle information, such as smoking status, drinking, exercise, and on general health information and biomarkers, such as body mass index and blood pressure, as well as on the health history of family members [2]. The health history of family members is especially important in the context of genetic diseases.

14.3.1   Genetics and family history

Primary genetic information can be of great value both in combination and independent of family history. Family history is especially important for family-based diseases that might be rare and based on strong effects of single genes. In this context, however, due to early mortality often related to the disease in question, family information might be incomplete. Also, family history may justify genetic screening. If breast cancer, for example, runs in the family, this might motivate genetic screening for the risk mutations on the BRCA1 and BRCA2 genes [18]. However, if we are interested in common diseases, which are influenced by common variants related to behavior or complex behavioral traits such as obesity or smoking, they may not necessarily be strongly linked to family health history but represent population-wide phenomena for which polygenic information is especially important. It has been shown that PGSs provide information independent of family history, for example, for coronary artery diseases [19].

14.3.2 Genetic scores for screening, intervention, and life planning

According to Torkamani and colleagues [2], we can differentiate between three main intervention stages, where PGSs can and already have made important contributions, especially in combination with classic clinical measures. Those stages are: screening, intervention, and life planning. In general, early detection is a strong motivation for the consideration of PGSs and a precursor of intervention strategies and lifestyle management. Genetics are often thought of as something static throughout the life course—which is also true. However, that does not mean that they do not control or influence processes that have dynamic effects on changes in phenotypes or the age of the onset of diseases, demographic events, or other conditions. Genetics may influence hormones or the heritability in the age at menopause, menarche, or the age at first birth, but also on the age at onset of Alzheimer's disease [20]. Timely detection is also not only important for mid- or late-life conditions but given that genes may provide the earliest evidence for a condition, it might also provide information, for example, about a disposition toward dyslexia, which necessarily is always diagnosed too late, namely when a child has already started learning. Early detection may enable early and therefore often more effective intervention.

As mentioned previously, a family history of breast cancer can motivate genetic screening for the BRCA genes. Many readers may be familiar with the case of actress Angelina Jolie, who underwent genetic testing and a double mastectomy to reduce her risk of developing breast cancer. He mother died at age 56, after years of battling cancer. Jolie had the BRCA1 variant, which has very high risks of developing breast and ovarian cancer. Given genetic screening, it has been shown that the number of women necessary to screen to detect breast cancer is less than half among those in the top quintile of genetic risk of a polygenic score compared to the bottom quintile of genetic risk. Furthermore, women in the top 5% of the PGS risk range reach the average risk level for breast cancer of 47—when mammogram screening is initiated—already by age 37 [21]. Medical intervention—both invasive or noninvasive—can be costly and risky for the patient. The use of statins for the prevention of primary heart attacks, for example, bears a considerable risk to develop diabetes mellitus. The number of people that could be treated with statins to prevent a heart attack is around a third in the top quintile of a risk PGS compared to the bottom quintile [2].

Khera et al. (2018) [22] recently developed and validated genome-wide PGSs for five common diseases (coronary artery disease, atrial fibrillation, type 2 diabetes, inflammatory bowel disease, and breast cancer). They test different methodologies and different p-value thresholds. Finding a polygenic score that is predictive enough to be considered for clinical recommendation is difficult because many factors are involved. There is no specific rule of the optimal criteria for selecting a PGS that is clinically relevant. For breast cancer, for instance, the authors created a PGS based on 5,218 SNPs based on a p-value threshold of $p < 5 \times 10^{-4}$. In 2019, Khera et al. [23] applied a similar approach to using PGSs for prediction of weight and severe obesity by developing and validating different PGSs, similar to our chapter 10. In middle-aged adults they found a 13-kilogram gradient in weight in the risk of severe obesity across the various PGS deciles.

A key question is whether there are any statistical tests to evaluate the accuracy of these scores. The *area under the curve (AUC)* is a statistical test typically used to evaluate the accuracy of a quantitative diagnostic test. An AUC equal to 0.5 would be the equivalent of tossing a coin, while a prefect diagnostic tool would have an AUC = 1. An AUC of 0.68 would normally be considered a poor test, and certainly not considered for clinical purposes. The aforementioned polygenic score for coronary artery disease [22], calculated using all SNPs, has a rather strong predictive value of AUC of 0.81, although typical values for diagnostic tests have an AUC greater than 0.95. Despite the low accuracy of PGS as clinical diagnostic tools, the idea of using PGS *in conjunction with* other information—for example, family history, lifestyle, and so forth—is gaining popularity in medicine [23]. Considerable research is being developed on how to summarize genetic information in predictive scores that could effectively be used by clinicians. What is not often discussed in these articles is the ethical difficulty in using PGS as a diagnostic tool due to the inability to apply the score beyond European ancestry individuals to different ancestry groups [7]. As we have noted often in this book, PGSs derived from particular ancestry groups cannot be applied to other groups [24].

Finally, a *healthy lifestyle* can dramatically reduce the risk of a disease or reduce the consequences—including behavioral interventions such as targeted education for high-risk dyslexic children. Again, the earlier a precondition is known, the more efficient the lifestyle changes will be. It is important to emphasize that PGSs are one additional piece in the puzzle for the diagnosis and treatment of diseases. They often need to be combined with classic clinical risk factors, where they can modify the eventual decision for intervention. Using genetic scores in isolation as a predictor for a broad range of health and behavioral traits by personalized direct-to-consumer genomics companies, has been a recurrent discussion we return to shortly.

14.3.3   Pharmacogenetics

Pharmacogenetics is a scientific discipline that is interested in heterogeneous reactions to drugs, as it is well-established, for example, in the medication of schizophrenia patients. This field wishes to identify and understand genetic variants that cause so-called unintended or negative side effects of drugs in some individuals, which can then have detrimental consequences and even cause death. Instead of following a "one size fits all" policy, the goal is to establish personalized drugs tailored to individual requirements. Genetic information might be useful to avoid initial shocks from a drug for an identified subgroup of people.

While this field is still in its infancy, preliminary results are promising and a recent GWAS identified 21 SNPs that are partly associated with weight gain given a specific drug [25]. As described in chapter 5 and demonstrated in chapter 10, in this example PGSs effectively sum up genetic loci of individuals more prone to weight gain, which could be one way to better identify people prone to this medication-related problem and intervene earlier. Another study highlights that within schizophrenia cases, variation in the schizophrenia PGSs are associated with the effectiveness of specific drugs [26]. Since around 30% of patients do not respond to typical prescriptions, they need to be medicated and often experience potentially severe side effects—or adverse events—even when a particular treatment might fail. The schizophrenia PGS partly predicts these heterogeneous reactions to drug treatments. This is important information since a delay in treatment of psychosis worsens the prognosis. These examples, albeit conceptually slightly more complex, relate to genetic risk prediction. While these aforementioned studies are early and often weak in statistical power and not ideal in design, they clearly show a similar—or clear potential—of how PGSs may eventually contribute to personalized medicine, particularly in relation to disease or pharmaceutical applications.

14.3.4 Public understanding of genetic information and information risks

There is a growing group of individuals who use personal direct-to-consumer genetics companies to obtain more information on their ancestry or genetic relationship to various outcomes. There are various ethical risks in this respect. Primarily there are *personal information risks* that an individual might learn in relation to their family or individual genetic makeup. First, it may reveal adoptions, infidelity, or nondisclosed family secrets. Some of this information can uncover hidden information, split families, and cause considerable anxiety and discomfort. Second, others are increasingly faced with ancestry or identity conflict, from as the realization that they are not "Irish" or "Italian" to uncovering that the social categorization of being "black" or "African American" that they had understood their entire lives is not aligned with genetic categorizations.

A central confusion with interpreting PGSs in personal genomics lies in the fact that many horoscope-like forecasts for psychological or lifestyle characteristics are paired with more certain predictions on, for example, morphological traits such as height and hair color. There are two problems associated with this. First, genetics are more important for the latter. Second, our genetic knowledge on morphological traits is much more advanced than for other psychological and behavioral traits and most multifactorial health conditions. The psychological reaction to information that a genetics customer is in the highest quintile for depression or a particular potentially fatal disease might be strong. Therefore, personal genomic results need to be interpreted with considerable caution; when it comes to health conditions, they especially need consultation with a genetic counselor or expert.

McBride and colleagues, for instance, examined how information related to genetic discoveries motivated individuals in relation to risk-reducing health behaviors [27]. They found that research that was focused on single-gene variants or those with very low-risk probabilities had very little impact on health-related behavior. In a systematic review examining attitudes toward genetic testing for cancer risk among ethnic minorities, Hann and colleagues [28] found that there was generally low awareness and knowledge of genetic counseling and testing for cancer susceptibility. Although attitudes toward testing were generally positive, particularly in relation to informing close family members, negative attitudes were also present, such as emotional impact of results and concerns about confidentiality and discrimination.

14.4   Lack of diversity in genomics

14.4.1   Lack of diversity in GWASs
In chapter 4 we discussed the lack of diversity currently in GWASs, which we will only briefly reiterate. Here, the central problems are in relation to ancestral diversity, with the majority of GWASs (88–90%) still conducted exclusively in European-ancestry populations. We also noted selectivity in terms of the "healthy volunteer" effect of higher socioeconomic status such as in one of the most prominently used datasets, UK Biobank [29]. Others have noted this bias in relation to mortality selection in datasets such as the Health and Retirement Study [30]. More attention is also required to gender and sex-specific effects as well as how phenotypes and genotype–phenotype interactions change over the life course. This lack of diversity not only has consequences for the inability to apply these PGSs across diverse populations but can serve to exacerbate health inequalities [31].

14.4.2   European ancestry bias related to PGS construction
Construction of the PGS can also lead to bias. As discussed in chapter 10, we undertake LD-based pruning or adjustments when we construct the PGS. There are, however, limited haplotype panels for diverse non-European ancestry populations [32, 33]. Martin and colleagues show how PGSs derived from European-ancestry populations are biased in many and often unpredictable directions when they are applied to non-European cohorts [24]. Another problem is that the commercial genotyping arrays used to sequence data were selected from sequencing data of a small sample of European-ancestry individuals. This ascertainment bias also embedded in the actual tools used to produce PGSs can, in turn, result in the incorrect estimation of disease and other outcomes [34].

14.5   Privacy, consent, legal issues, insurance, and General Data Protection Regulation

14.5.1   Privacy in the age of public genetics: Solving crimes and finding people
A website called GEDmatch is a public genealogy site that allows individuals to upload their own genetic data and link it with names and genealogy to search for relatives. The key difference with GEDmatch that distinguishes it from virtually all other genetic data

is that it is public and linked to actual names. In April 2018, GEDmatch was used to solve the Golden State Killer cold case of a man who raped more than 50 women and killed around a dozen people. Using the website, it was possible to link the killer to a distant relative. As we know from chapter 3, with genetic data it is possible to isolate factors related to geography and ancestry. This case, and the many that have been solved since, raised the ethical question that whenever a relative or distant family member makes their data openly available, they unknowingly share data of their kin as well.

A 2018 study published in *Science* by Erlich and colleagues [1] analyzed 1.28 million anonymous individuals of European ancestry in the direct-to-consumer genetic company MyHeritage. As of 2019, this was one of the largest genetic data companies, after 23andMe and Ancestry.com. They found that around 60% of Americans of European ancestry can be identified via one of their relatives. Contrary to the inequality in medical applications of using PGSs for non-European groups derived from GWASs, the fact that so many European ancestry individuals are overrepresented in GEDmatch and other direct-to-consumer companies means that they are more likely to be located or found in crime-related research. They calculated that once a genetic database covers around 2% of the population, almost any person could be matched to up to at least a third cousin level (i.e., people who share a great-great grandparent). They even showed that they were able to isolate those who shared a great-grandparent (i.e., second cousins) for around 15% of people. They emphasized this point further by showing how they could identify a supposedly anonymous woman living in Utah within one day. The authors urge direct-to-consumer companies to encrypt data to ensure that it is not being gathered and used inappropriately.

Another study published in *Cell* in late 2018 led by Kim et al. [30] concluded that over 30% of people in forensic databases held by law enforcement can be linked to relatives in a consumer database. Linking these databases thus makes it possible to find a suspect when there is a DNA sample, but at this time largely in European ancestry populations. In early 2019, FamilyTreeDNA, which marketed itself as a company that did not sell information to third parties, revealed that it voluntarily shared its database of around 2 million records with the FBI [35]. This was done in order to identify suspects of unsolved murder and rape cases. We anticipate that these ethical issues will continue to be discussed and grow in importance in the years to come.

14.5.2   The changing nature of informed consent in genomic research

Genomic research also challenges many of the established practices and norms of informed consent in research. *Informed consent* is a cornerstone of most research involving human subjects, with the goal to ensure that subjects are aware of the risks and potential benefits in order to make a voluntary decision to participate in research. Advances in technology, the emergence of large-scale public databases, and rise of direct-to-consumer companies challenges many traditional models of informed consent.

Standard informed consent procedures require that participants understand the purpose; the procedures outlining biospecimen collection; additional information collection (e.g., questionnaires); and issues associated with storage, researcher access, recontact, large-scale data sharing, risks and benefits, protecting privacy and confidentiality, commercialization, feedback of research results, and ability to withdraw [36]. Although we cannot cover each aspect, several key challenges can be highlighted.

Most consent involves individuals *understanding the purpose of the research* and increasingly the related aspects of *large-scale data sharing and commercialization* [37]. A central challenge is allowing participants to understand and make an informed decision. Some individuals, for instance, were surprised when they learned that the direct-to-consumer genetics company 23andMe sold their data for $60 million to a pharmaceutical company [38, 39]. This relates as well to the previous point that FamilyTreeDNA voluntarily shared its customer files to the FBI [35]. *Protecting privacy and confidentiality* is another evolving consent issue that this area needs to re-evaluate. Academic, government, and direct-to-consumer data holders have clear policies and procedures to protect identifiable information. This often includes encryption, limited access, nondisclosure agreements, and other protections. Given that it is possible to locate family members from people that have openly provided their genetic data, recent issues suggest that we need to rethink the issue of consent for this topic [1].

As noted by METADAC, a U.K.-based ethics approval board responsible for releasing genetic information, *consent is a process* [40]. In that committee, where we managed data that was sometimes consented in the 1950s, the meaning of consent and clinically relevant information and feedback was unknown. Consent, particularly in long-term longitudinal studies, ensured trust and ensuring participants that their privacy (including confidentiality) will be protected and that uses of the data and samples they contribute will fall within agreed and anticipated parameters. There is also increasingly what has been termed as *broad consent* (within specific stipulations reflecting contemporary social values and scientific practice), which is consent sought for future uses of the data and samples collected.

14.5.3   Insurance and genetics

Fear of discrimination via the use of genetic data by insurance companies or employers is a topic often discussed in this area and is reason for people to be hesitant about providing genetic data. Most countries have very specific laws protecting individuals. As with many regulations surrounding genetics, it is important to note that most regulations are nationally bound and rapidly changing. For this reason, we cannot provide an exhaustive summary. Within the United States, in 2008, the Genetic Information Non-Discrimination Act (GINA) was a U.S. federal law that was passed to prohibit the discrimination in health coverage and employment based on genetic information [41]. It prohibited insurers from using an individual's genetic information in determining eligibility or premiums. It also stopped employers from using genetic information to determine hiring, firing, promotions, and job assignments [42]. GINA, however, does not cover life, disability, or long-term care insurance or prohibit medical underwriting based on current health status [42].

In the United Kingdom, on November 1, 2001, the Moratorium on Genetics and Insurance came into effect, followed by a Concordat on March 14, 2005, and at the time of writing, the policy was in place until November 2019. They represent a voluntary agreement between the U.K. Government and the ABI (Association of British Insurers) [43]. This agreement stops insurance companies from accessing or using genetic test results to make any insurance coverage and rate decisions. This includes genetics tests from consumers and scientific studies. This agreement means that individuals are not required to reveal any genetic results unless it is approved by the Genetics and Insurance Committee (GAIC). At the time of writing this book, only genetic tests for Huntington's disease were approved by life insurance companies.

14.5.4 GDPR and genetics

General Data Protection Regulation (GDPR) is a new European data protection law that replaces the existing European Union (EU) data protection regime under Directive 95/46/EC. GDPR sets out provisions intended to harmonize data protection laws throughout the EU by applying a single data protection law that is binding throughout all member states. GDPR became effective on May 25, 2018. Although it is a EU-based regulation, given the international and collaborative nature of genetic research will influence virtually all researchers in this research area. A review from 2018 showed the different countries and local ethics boards have been interpreting GDPR in various manners, which means that at this time we cannot make one encompassing statement on how it will influence this research area [44]. We can, however, outline the main requirements.

Since its implementation in 2018, GDPR is still actively being interpreted by researchers, institutions, governments, and funders. Particularly key for GDPR in relation to genetic research is clarity for recruitment of participants in addition to informed consent that is produced in the language that is understandable to participants. There is also the thorny question about whether family members, inevitability implicated by genetic results or the public posting of one's own genetic results, are actually able to give informed consent.

Another core GDPR principle is that measures should be taken to avoid risks that vulnerable individuals or groups will not be stigmatized. In relation to potential misuse, researchers are required to engage in a risk assessment and provide details on the measures that they will take to prevent the misuse of their research findings. As a researcher, you should therefore be vigilant and sensitive about reporting results to not only avoid stigmatization or sensationalism but also to present results in simple and accessible language. Many have done this for various studies, such as producing a detailed list of "frequently asked questions" for genetic results related to educational attainment (https://www.thessgac.org/faqs), reproductive behavior (http://www.sociogenome.org/faq-gwa12), or lack of diversity in genetics (https://www.melindacmills.com/faqs).

Another GDPR issue is that researchers must provide details on unexpected findings, which in genetic research might reveal information or law enforcement risks, but also clinically relevant risks. Researchers need to develop a policy for dealing with incidental findings, which, as articulated by the Wellcome Trust and the United Kingdom's Medical Research Council, is often developed on a case-by-case basis (https://wellcome.ac.uk/sites/default/files/wtp056059.pdf).

As is standard with all research, copies of ethical committee approval for the research needs to be obtained in advance of all projects. Here it is important to clarify the protection of personal data and ensure that you are compliant and have contacted your local data protection officer. Each institution generally has a framework in place for data protection by design and default, which includes screening and data protection impact assessment. If you are collecting primary genetic data, a data protection officer must be appointed as a contact point for all data subjects. Other aspects include details on the retention period of data, rights of individuals to withdraw consent, or ability to lodge a complaint if data is perceived as misued.

GDPR also requires that researchers ensure that the rights of the data subjects have been established under the national legislation of the country where the research takes place and in accordance with the national legal framework. Technical and organizational measures must also be put in place to safeguard the rights and freedoms of your research participants.

Another rule is the data minimization principle, which means that as a researcher you only process information relevant and limited to the purposes of the research project. This more narrow approach could raise some tensions with broader machine-learning or deep phenotyping techniques that use and link large amounts of genetic and other types of data and the way that research actually often takes place.

Protection of personal data under GDPR also necessitates that researchers provide a detailed description of the security measures they implement to avoid unauthorized access to the data or equipment used to process it. It is common to develop a sensitive data security plan and security pledge for all students or research staff that have access to the data. Researchers using this type of data will often sign very detailed agreements on how and where they will process the data, including storage or internet related requirements. Sensitive data should not, for example, be stored on external disks that are in an open area or could be misplaced or on nonencrypted or easy-to-access platforms (e.g., Dropbox).

Researchers also need to clarify whether personal data are transferred from a non-EU country to the EU (or another third state) and that such transfers comply with the laws of the country where the data was collected. There are also specific GDPR requirements for non-EU countries and data from these countries when it is imported to or exported from the EU. Considering that this type of research is often collaborative, involving the sharing of summary statistics or data across borders, this remains a very relevant factor that researchers need to consider.

Another requirement is that you provide a description of the anonymization and/or pseudonymized techniques that are implemented when processing the genetic data. A simplified definition of pseudonymization is the act of replacing an identifier with a code for the purpose of avoiding direct identification of the participant, except by persons holding the key linking code and identifier. This reduces the risks to data subjects and helps data processors meet their GDPR compliance. Unless you are collecting primary data, in virtually all cases you will received anonymized data and will not have access to the key code or identifier. Anonymization means that it is unlikely that you would be able to use the data to link it to an identifiable individual.

Profiling is another GDPR concern that needs to be reckoned with in this area of research, particularly due to the increased move toward data automation. Profiling is when automated processing is used to evaluate personal information and in particular to analyze or predict aspects related to performance at work, economic situation, health, personal preferences, interests, reliability, behavior, location, or movements. What is key, however, is how that categorization of groups is used and whether it is used for research or targeted commercialization or other interventions. The underlying GDPR concern is when these automated profiling techniques are used solely to produce automated decisions or interventions with limited human judgment or intervention that may have effects on individuals. Considering the increased impetus to use PGSs for medical interventions using advanced automated, statistical, and machine learning techniques, further considerations in this area are likely warranted.

There are also requirements that data is publicly available, which is not strictly possible for virtually all types of genetic data. In this case it is commonplace to note the data access rules and regulations for each of the data sources that you use. Lawful data processing of previously collected data and evaluation of ethics risks should also be in place, which is generally taken into account during your local ethics review. In particular, researchers who are new to this area should consider all of these aspects and when in doubt also appoint an external ethics advisor.

14.6 Conclusion and future directions

This chapter attempted to show the multifaceted and complex aspects underlying the ethics of using genomic data. We note that the use of genetic data and PGSs could have potentially positive clinical applications for screening, intervention, and life planning, in addition to drug development. But at the same time we also warn against genetic determinism and acknowledgment of the difference between applying monogenic genes versus polygenic genetic scores, where we have little understanding of the underlying biological mechanisms or the role of gene-environment interplay. We likewise once again flag the problem of a lack of diversity currently in genomic data. If we are going to actively use this data, we need to acknowledge that around 90% of the genetic discoveries are applicable to European-ancestry populations only. As we noted elsewhere, although around 76% of the current world population resides in Asia or Africa, 72% of genetic discoveries emanate from participants recruited from only three countries (the United Kingdom, the United States, and Iceland) [7]. Finally, we draw attention to the vexing issue of privacy in the age of very public genetics, the fluid nature of informed consent, and concerns and regulations regarding the usage of genetic data by insurers and employers. The new GDPR regulations introduced in 2018 also provide some guidelines regarding key issues in relation to genetics such as consent, profiling, and data storage and security. We anticipate that even by the time that this book is published, many new ethics and privacy issues will have emerged in this rapidly evolving field.

Futher reading and resources

Ekong, R. et al. Ethical data management: Checklist for gene/disease specific database curators. The Human Variome Project Working Group, 2018 (available at http://www.humanvariomeproject.org/images/documents /HVP_-_Ethical_Data_Management-_Checklist_for_Gene_Disease_Specific_Database_Curators__-_FINAL.pdf).

Shabani, M., and P. Borry. Rules for processing genetic data for research purposes in view of the new EU General Data Protection Regulation. *Eur. J. Hum. Gen.* **26**, 149–156.

Volume 137 of *Human Genetics*, which is a special issue on Genomic Data Sharing published in 2018, contains various country-specific profiles such as these for the United States and the United Kingdom:

Majumder, M. A. United States: Law and policy concerning transfer of genomic data to third countries. *Hum. Gen.* **137**, 647–655 (2018).

Taylor, M. J. et al. United Kingdom: Transfer of genomic data to third countries, *Hum. Gen.* **137**, 637–645 (2018).

References

1. Y. Erlich, T. Shor, I. Pe'er, and S. Carmi, Identity inference of genomic data using long-range familial searches. *Science* **362**, 690–694 (2018).

2. A. Torkamani, N. E. Wineinger, and E. J. Topol, The personal and clinical utility of polygenic risk scores. *Nat. Rev. Genet.* **19**, 581–590 (2018).

3. R. Plomin and S. von Stumm, The new genetics of intelligence. *Nat. Rev. Genet.* **19**, 148–159 (2018).

4. R. Plomin, *Blueprint: How DNA makes us who we are* (Cambridge, MA: MIT Press, 2018).

5. R. Plomin, In the nature–nurture war, nature wins. *Sci. Am.* (2018) (available at https://blogs.scientific american.com/observations/in-the-nature-nurture-war-nature-wins/).

6. New York Times Editorial Board, Why you should be careful about 23andMe's health test. *New York Times*, February 1, 2019 (available at https://www.nytimes.com/interactive/2019/02/01/opinion/23andme -cancer-dna-test-brca.html?mtrref=www.google.com&gwh=486E6E4BA9B5AE0CA119112BEE6A85E 2&gwt=pay).

7. C. Zimmer, Genetic intelligence tests are next to worthless. *Atlantic*, May 29, 2018 (available at https://www.theatlantic.com/science/archive/2018/05/genetic-intelligence-tests-are-next-to-worth less/561392/).

8. S. Sniekers et al., Genome-wide association meta-analysis of 78,308 individuals identifies new loci and genes influencing human intelligence. *Nat. Genet.* **49**, 1107–1112 (2017).

9. W. D. Hill et al., A combined analysis of genetically correlated traits identifies 187 loci and a role for neurogenesis and myelination in intelligence. *Mol. Psychiatry* **24**, 169–181 (2018).

10. M. C. Mills and C. Rahal, A scientometric review of genome-wide association studies. *Commun. Biol.* **2** (2019), doi:10.1038/s42003-018-0261-x.

11. P. M. Adams et al., Assessment of the genetic variance of late-onset Alzheimer's disease. *Neurobiol. Aging* **41**, 200.e13–200.e20 (2016).

12. P. D. P. Pharoah, A. C. Antoniou, D. F. Easton, and B. A. J. Ponder, Polygenes, risk prediction, and targeted prevention of breast cancer. *N. Engl. J. Med.* **358**, 2796–2803 (2008).

13. A. Antoniou et al., Average risks of breast and ovarian cancer associated with BRCA1 or BRCA2 mutations detected in case series unselected for family history: A combined analysis of 22 studies. *Am. J. Hum. Genet.* **72**, 1117–1130 (2003).

14. K. Czene, P. Lichtenstein, and K. Hemminki, Environmental and heritable causes of cancer among 9.6 million individuals in the Swedish family-cancer database. *Int. J. Cancer* **99**, 260–266 (2002).

15. N. Zaitlen et al., Leveraging population admixture to characterize the heritability of complex traits. *Nat. Genet.* **46**, 1356–1362 (2014).

16. L. Yengo et al., Meta-analysis of genome-wide association studies for height and body mass index in ~700,000 individuals of European ancestry. *Hum. Mol. Gen.* **27**, 3641–3649 (2018), doi:10.1101/274654.

17. E. A. Boyle, Y. I. Li, and J. K. Pritchard, An expanded view of complex traits: From polygenic to omnigenic. *Cell* **169**, 1177–1186 (2017).

18. S. Chen and G. Parmigiani, Meta-analysis of BRCA1 and BRCA2 penetrance. *J. Clin. Oncol.* **25**, 1329–1333 (2007).

19. G. Abraham et al., Genomic prediction of coronary heart disease. *Eur. Heart J.* **37**, 3267–3278 (2016).

20. R. S. Desikan et al., Genetic assessment of age-associated Alzheimer disease risk: Development and validation of a polygenic hazard score. *PLoS Med.* **14**, e1002289 (2017).

21. N. Mavaddat et al., Prediction of breast cancer risk based on profiling with common genetic variants. *J. Natl. Cancer Inst.* **107**, cvj036 (2015).

22. A. V. Khera et al., Genome-wide polygenic scores for common diseases identify individuals with risk equivalent to monogenic mutations. *Nat. Genet.* **50**, 1219–1224 (2018).

23. A. V. Khera et al., Polygenic prediction of weight and obesity trajectories from birth to adulthood. *Cell* **177**, 587–596 (2019).

24. A. R. Martin et al., Human demographic history impacts genetic risk prediction across diverse populations. *Am. J. Hum. Genet.* **100**, 635–649 (2017).

25. D. E. Adkins et al., Genomewide pharmacogenomic study of metabolic side effects to antipsychotic drugs. *Mol. Psychiatry* **16**, 321–332. (2011).

26. J. Frank et al., Identification of increased genetic risk scores for schizophrenia in treatment-resistant patients. *Mol. Psychiatry* **20**, 150–151 (2015).

27. C. M. McBride, L. M. Koehly, S. C. Sanderson, and K. A. Kaphingst, The behavioral response to personalized genetic information: Will genetic risk profiles motivate individuals and families to choose more healthful behaviors? *Annu. Rev. Public Health* **31**, 89–103 (2010).

28. K. E. J. Hann et al., Awareness, knowledge, perceptions, and attitudes towards genetic testing for cancer risk among ethnic minority groups: A systematic review. *BMC Public Health* **17**, 503 (2017).

29. A. Fry et al., Comparison of sociodemographic and health-related characteristics of UK Biobank participants with those of the general population. *Am. J. Epidemiol.* **186**, 1026–1034 (2017).

30. J. Kim et al., Statistical detection of relatives typed with disjoint forensic and biomedical loci. *Cell* **175**, 848–858 (2018).

31. A. R. Martin et al., Current clinical use of polygenic scores will risk exacerbating health disparities. *Nat. Gen.* **51**, 584–591 (2019).

32. B. J. Vilhjálmsson et al., Modeling linkage disequilibrium increases accuracy of polygenic risk scores. *Am. J. Hum. Genet.* **97**, 576–592 (2015).

33. J. Euesden, C. M. Lewis, and P. F. O'Reilly, PRSice: Polygenic risk score software. *Bioinformatics* **31**, 1466–1468 (2014).

34. M. S. Kim, K. P. Patel, A. K. Teng, A. J. Berens, and J. Lachance, Genetic disease risks can be misestimated across global populations. *Genome Biol.* **19**, 179 (2018).

35. S. Hernandez, One of the biggest at-home DNA testing companies is working with the FBI. *BuzzFeedNews.com*, January 31, 2019 (available at https://www.buzzfeednews.com/article/salvadorhernandez/family-tree-dna-fbi-investigative-genealogy-privacy).

36. A. L. McGuire and L. M. Beskow, Informed consent in genomics and genetic research. *Annu. Rev. Genomics Hum. Genet.* **11**, 361–381 (2010).

37. NHGRI, Informed consent for genomics research (2019) (available at http://www.genome.gov/27026588).

38. K. Servick, Can 23andMe have it all? *Science* **349**, 1472–1477 (2015).

39. K. Grogan, Roche signs Parkinson's pact with 23andMe. *PharmaTimes*, January 8, 2015 (available at http://www.pharmatimes.com/news/roche_signs_parkinsons_pact_with_23andme_970837).

40. M. J. Murtagh et al., Better governance, better access: Practising responsible data sharing in the META-DAC governance infrastructure. *Hum. Genomics* **12**, 24 (2018).

41. *Genetic Information Nondiscrimination Act of 2008* (2008) (available at https://www.govinfo.gov/content/pkg/PLAW-110publ233/pdf/PLAW-110publ233.pdf).

42. K. L. Hudson, M. K. Holohan, and F. S. Collins, Keeping pace with the times—The Genetic Information Nondiscrimination Act of 2008. *N. Engl. J. Med.* **358**, 2661–2663 (2008).

43. H. G. Abi, *Concordat and Moratorium on Genetics and Insurance* (2014) (available at https://www.abi.org.uk/globalassets/sitecore/files/documents/publications/public/2014/genetics/concordat-and-moratorium-on-genetics-and-insurance.pdf).

44. M. Phillips, International data-sharing norms: From the OECD to the General Data Protection Regulation (GDPR). *Hum. Genet.* **137**, 575–582 (2018).

15

Conclusions and Future Directions

15.1 Summary and reflection

This book engaged in the daunting task of covering a topic that traverses the disparate areas of genetics, biology, statistics, bioinformatics, and social science conceptual thinking. The competences required for researchers to examine the human genome are truly interdisciplinary. Our hope is that we have signaled that a lack of understanding of the human genome and evolution can result in dire errors and misinterpretation. Conversely, those estimating genetic models without consideration of the intricate gene-environment interplay or behavioral aspects can also make grave oversights. We also recognize that leaving our monodisciplinary silos can be taxing. The reader undoubtedly felt stretched when they were required to learn multiple new languages and ways of thinking, from genetics to computing and ways of conceptualizing their phenotype. Due to the introductory nature of this book, we acknowledge that multiple concepts and ideas were introduced fleetingly, inevitably lacking depth at times. Armed with a basic understanding of these topics, however, our intention is that we have made it less daunting and that you now have the basics to enter this field and will know where to practically start to ask questions and seek solutions.

15.2 Future directions

Providing a description of future directions in this field is a formidable and virtually impossible task. Particularly since 2005, genomics has experienced a period of rapid expansion, change, and development. Whereas some research areas inch incrementally along like a steady steam train, genomics is like stepping into a high-speed rail bullet train. Developments are rapid. Even during the process of finalizing this book, new techniques and data emerged on a virtually daily basis. For this reason, here we only sketch some more advanced topics and broad future directions the field might take.

As we noted in chapter 4, the focus for a large part has been gathering ever larger datasets to isolate more genetic variants and increase the explanation of variance. Moving beyond that, we anticipate that *higher-resolution data* will undoubtedly emerge in the next years. By this we mean a move from examining individuals to studying cell types (e.g., from single cells). This is, of course, coupled with the drive to *move from description and explaining causal associations* to *understanding biological mechanisms*. Deeper biology will also likely explore cis regulatory elements and noncoding RNAs, focusing on functional studies. New technologies and approaches will be able to explain the link between genetic variants and gene expression but also how variants influence protein structure, stability, and interactions. Understanding these mechanisms could, in turn, enhance the *clinical and pharmaceutical applications*. As with virtually all scientific disciplines, genomics remains subject to questions of *replicability and transferability* of results.

As data grows in complexity and size, we likewise anticipate the broad emergence of *automated methods*, including *machine learning* and *artificial intelligence*. The statistics and methods-based approach may be shifting to these broader automated tactics to uncover underlying patterns and principles in the large masses of complex data. Although we anticipate change, we likewise predict stability in the *substantive focus on solving diseases and understanding behavior*. Here we also foresee continued and growing interest in *evolutionary questions* that can track which genes have been passed down or are important in complex diseases. Understanding and pinpointing the areas of the genome that have been fundamental to fitness and survival and trade-offs in relation to positive selection and contemporary maladaptation are only a few options. We also expect that there could be a shift beyond single-species genomics (like our book on humans) to *multiple species* comparisons.

Chapter 14 raised the *ethical issues* that will undoubtedly continue to emerge and even grow more important in the next years over data *security, sharing, and ownership*. As the chapter illustrated, genomic data has challenged classic notions about informed consent, data protection, regulation of new technologies, and identification. Blockchain technologies are often mentioned as technologies that may play a role to address at least some of these technical and security factors. Regulations are often built on data protection based on the premise that data are held and managed by universities or the government, yet data are increasingly retained by large commercial companies that often have a profit or pharmaceutical interest. Genomic data is furthermore highly identifying, not only for the individual but also and inadvertently to their family. Here antidiscrimination laws, personal freedom, and data protection needs to be evaluated and openly discussed. This is intertwined with an *individual's understanding* of his or her own genetic information but also basic human rights about self-knowledge and treatment. On another front, precision medicine via genetics is also envisioned as important for employers and government, raising multiple issues and concerns that would need to be communicated carefully to the public. Individual understanding, perceptions, trust, and fear of data usage requires a behavioral understanding—not a technical solution.

Perhaps one of the largest gains (we hope) will be that the lack of *ancestral diversity* will be gradually ameliorated. With this there also needs to be a shift in the reliance on universal genome reference panels, which impairs our understanding of the likely true genetic diversity of the human species. Other promising techniques that have already emerged are the *microbiome, epigenetics*, and *CRISPR/Cas9-based methods* (i.e., gene editing), and *single-cell RNA sequencing*, which are expansive topics we cannot do justice to here.

One topic that we wish that we could have delved into, but that went beyond the scope of this textbook, is epigenetics. *Epigenetics* is a way in which to study the dynamic nature of the genome or, simply put, it examines the biological mechanisms that switch genes on and off. One way to think of it is that DNA is the hardware and epigenetics is the software. Recall from our discussion in chapter 1 about the central dogma of molecular biology, that DNA provides the instructions for functional proteins to be produced inside cells. Epigenetics impacts how genes are read by cells and whether these cells produce relevant proteins. Although certain genes may be present in various types of cells, it may only be expressed in certain types of cells. Epigenetics is also very pervasive because what you eat or how you exercise or sleep could cause chemical modifications around genes that turn genes on or off. Since it is so multifaceted and dynamic it has been difficult to analyze, and at the time of writing this book, still somewhat in its infancy. There are also many analytical challenges. Batch and cell distribution confounds need to be discerned for each analysis, and results are often tissue specific. A fascinating example is this area is Hovarth and Raj's epigenetic clock [1] extended by Hovarth [2]. In this model, DNA methylation levels are used to predict age across a broad array of human tissues and cell types. This is a fascinating area of research that we anticipate will lead to breakthroughs in developmental biology, cancer, aging and longevity research.

The focus in genetics has been in many cases almost exclusively on increasing the quality and precision of genetic data. Relatively less attention has been placed on improving the definition of phenotypes. We are limited by what we can actually measure. A 2012 review found that phenotypes or the traits that are studied in genetics are often "sloppy and imprecise" [3]. A growing area of interest is in the area of deep phenotyping, particularly in the area of precision medicine. *Deep phenotyping* refers to a more comprehensive analysis of the phenotype by providing more specificity, linkage of new types of data and measurements, and new connections between trait subtypes and genetic variations. It is phenotyping in a much more fine-grained and multifaceted way, often using advanced algorithms and data linkage. A recent study, for example, used machine learning models and electronic health records to generate better patient survival and mortality in coronary artery disease [4]. Although the future is hard to predict, what we can say is that is that it will be full of discovery—and many surprises—along the way.

References

1.   S. Hovarth and K. Raj, DNA methylation-based biomarkers and the epigenetic clock theory of ageing. *Nat. Rev. Gen.* **19**, 371–384 (2018).
2.   S. Hovarth, DNA methylation age of human tissues and cell types. *Gen. Bio.* **14**, R115 (2013). See also erratum, *Gen. Bio.* **16**, 96 (2015).
3.   P. N. Robinson, Deep phenotyping for precision medicine. *Hum. Mutat.* **33**, 777–780 (2012).
4.   A. J. Steele et al., Machine learning models in electronic health records can outperform conventional survival models for predicting patient mortality in coronary artery disease. *PLoS One* **13**, e0202344 (2018).

Appendix 1: Software Used in This Book

A1.1 Introduction

We use a variety of different software packages in this book, which are all open source. Although there are a variety of programs to run different models, we have opted for ones such as R and PLINK that contain the most options. This appendix provides a short overview of the software used in this book that you will need to download in order to engage in the hands-on applications in the last two sections of the book.

Although these programs are available in different operating systems (OSs), many advise to use GNU/Linux-based computer resources. For those wishing to make the jump, various excellent online tutorials exist, such as http://www.ee.surrey.ac.uk/Teaching /Unix/. When running the different commands, keep in mind that commands can differ by OS (see box 8.1). You can also refer to the online webpage to this book (http://www .intro-statistical-genetics.com) that will have hyperlinks (and if applicable updated links) and updated programs to those discussed here.

A1.2 RStudio and R

RStudio can be downloaded at https://www.rstudio.com/products/rstudio/download/. For this book we have used the freely available RStudio Desktop Open Source License.

The version that is used in this book is RStudio Desktop 1.1.463. You will need to select the appropriate OS and follow the instructions on the website. This means downloading it and saving it in the appropriate directory. Note that RStudio also requires that you have the base program R, which you can download at http://cran.rstudio.com/. In this book we use the version R-3.5.2. Since there are many online tutorials and books on using RStudio and R, we do not replicate that material here.

A1.3 PLINK

Perhaps the most popular open-source free software program for QC (quality control) and GWAS analyses is PLINK [1, 2]. It was developed by Shaun Purcell and colleagues and facilitates multiple types of data handling and usage.

The home PLINK site is http://zzz.bwh.harvard.edu/plink/.

PLINK can be downloaded from http://zzz.bwh.harvard.edu/plink/download.shtml.

The PLINK 1.90 beta version can be downloaded from https://www.cog-genomics.org /plink2/.

It is important that you download the version that matches your operating system (i.e., Linux (x86_64), Apple Mac). Versions are constantly being updated, so readers should check this site regularly. In this book we use PLINK 1.90 and 2.0. Specifically, we use PLINK v1.90b6.7 64-bit, which was updated on December 2, 2018.

The basics of working with PLINK in the command line environment are described in detail in chapter 8 (section 8.2). Since there are many adequate resources, such as an excellent PLINK tutorial and active and helpful FAQs, mailing lists, and others, we do not repeat that information here. At the time of writing this book, PLINK 1.90 was in beta version and contains the same options. A clear advantage is that the new version is much faster. We recommend that neophytes follow the tutorial: http://zzz.bwh.harvard.edu/plink/tutorial.shtml. Once you have downloaded the zip archive with PLINK, no further installation is necessary. For usage, you need to type "plink" or "./plink" (depending on your OS) from the command line, followed by the options that you will run (see chapter 8).

A1.4 GCTA

In chapter 9 we use GCTA, which stands for Genome-wide Complex Trait Analysis and is available at https://cnsgenomics.com/software/gcta/#Overview [3]. It is developed by Jian Yang together with Peter Visscher, Mike Goddard, and Hong Lee. GCTA has multiple uses that can aid in the understanding of the genetic architecture of complex traits and supports many additional analyses not described in this introductory book. In this book we use the release v1.92.0beta1, released on February 1, 2019. To download GCTA, go to https://cnsgenomics.com/software/gcta/#Download.

The executable files only support a 64-bit OS on the x86_64 CPU platform. Please note that the macOS and Windows versions include an additional file (`libomp.dylib` for macOS and `libiomp5md.dll` for Windows) in the `/bin` directory that needs to be located in the same folder as the executable file.

A1.5 PRSice

PRSice is polygenic score (PGS) software for calculating, applying, evaluating, and plotting the results of PGS, which we use in chapter 10 [4]. In this book, we used the version 2.1.4 of

PRSice-2 [5]. PRSice can be downloaded from https://choishingwan.github.io/PRSice/. Note that depending on your OS, the downloaded files might have different extensions (i.e., `PRSice_linux`, `PRSice_mac`, `PRSice_win32.exe`, or `PRSice_win64.exe`). Please adapt the syntax accordingly or rename the respective file deleting the OS extension to execute the PRSice syntaxes in this book. See also box 8.1 for OS-specific commands.

In order to plot graphs you will likely need an updated version of R, which is described on their website. To install all of the required R packages that you will need for PRSice, type the following commands:

```
Rscript PRSice.R --dir
```

A1.6   Python

In chapter 10 we also use the statistical package LDpred, which is a software written in Python 3 [6]. To be able to run the examples you need to be sure that you have a Python version greater or equal to 3.5 installed in your computer. We recommend the installation of the Anaconda distribution of Python, because it contains the package management system `pip` that can be used to install and manage software packages written in Python. You can find Anaconda at https://www.anaconda.com/.

There are two versions of Python Anaconda. One is based on Python 2 and the other one in Python 3. Some statistical packages presented in this book are based on Python 3 (LDpred), while others are currently written in Python 2 (LDSC and MTAG). However, you *only need 1 version of Anaconda*. Anaconda is also an environment manager and makes it very easy to go back and forth between Python 2 and 3 with a single installation of Anaconda.

Once you have installed a version of Python in your system, you can verify which version is the default version in your system by following these commands:

- Windows: Open the Anaconda Prompt
- macOS: Open Launchpad, then open Terminal
- Linux-CentOS: Open Applications - System Tools - Terminal.
- Linux-Ubuntu: Open the Dash by clicking the upper left Ubuntu icon, then type "terminal."
- Enter the command `python`. This command runs the Python shell. If Anaconda is installed and working, the version information it displays when it starts up will include "Anaconda." To exit the Python shell, enter the command `quit()`.

A1.6.1   How to switch from Python 3 to Python 2

You can navigate from different version of python using the Anaconda system. This is done by creating different Python *environments*.

If you want to create a new environment with Python 2 installed, type:

```
conda create -n python2 python=2.7 anaconda
```

This will install the version of Python 2.7 and create two environments to work with. Beware that you need an Internet connection to run the previous command, and it will take a few minutes.

Once you created the new environment, you can switch to Python 2 by doing the following:

In Windows:

```
activate python2
```

In Linux, macOS:

```
source activate python2
```

To see the list of installed environments, you can type:

```
conda env list
```

This will show all the Python environments in your system. The * symbol shows which environment is active.

```
# conda environments:
#
base                     /Users/nb17719/anaconda2
python2                  /Users/nb17719/anaconda2/envs/python2
python3              *   /Users/nb17719/anaconda2/envs/python3
```

To deactivate a particular environment (and return to base), you can use the following command:

In Windows:

```
deactivate python2
```

In Linux, macOS:

```
source deactivate python2
```

In case Python 2 is installed in your system, you can use the following command to create a new environment called python3:

```
conda create -n python3 python=3.6 anaconda
```

You then can use the previous commands to switch between different Python environments.

A1.6.2   Installing packages in Python

The Python software used in this book requires additional packages that can be installed in your environment.

To run LDpred, for instance, you will need the following packages installed in your system:

1.   h5py http://www.h5py.org/
2.   scipy http://www.scipy.org/
3.   libplinkio https://github.com/mfranberg/libplinkio.

The first two packages, h5py and scipy, are commonly used Python packages and are pre-installed on many computer systems. The last package, libplinkio, can be installed using pip (https://pip.pypa.io/en/latest/quickstart.html), which is also preinstalled in many systems.

With pip, one can install libplinkio by typing the following command:

```
pip install plinkio
```

A1.7   Git

Git is a the most used version control system—a category of software tools that help a software team manage changes to source code over time. We recommend the use of Git to download the latest version of LDpred, LDSC, and MTAG in your system. Git is available for Windows, Linux, and macOS from https://www.atlassian.com/git/tutorials/install-git. You will find a very complete guide on how to install Git in different systems.

A1.8   LDpred

Once you have all the Python and pip packages installed, you can install LDpred in your system. This can be done by downloading the source files from https://github.com/bvilhjal /ldpred/archive/master.zip or by typing the following if Git is installed in your system:

```
git clone https://github.com/bvilhjal/ldpred.git
```

To run the code presented in the last part of chapter 10, we assume that the LDpred software is installed in a folder called ldpred and that Python 3 is installed in your system and can be invoked by typing python3 in your terminal.

To be certain that all the packages are installed, you can type:

```
python3 ldpred/setup.py
```

To get the best start possible, run the following command that runs a series of tests with LDpred:

```
python3 ldpred/test.py
```

test.py uses the test files contained in the /test _ data directory downloaded with the software to calculate a series of PGS using LDscore.

A1.9   LDSC

LDSC is a software written in Python 2 that performs LD score regression and can be used to estimate heritability and genetic correlations of multiple traits.

To install LDSC using Git, type:

```
git clone https://github.com/bulik/ldsc.git
```

To run LDSC, you need to switch to Python 2. You can activate a Python 2 environment following the instructions discussed previously in this appendix.

To install all of the packages needed to run LDSC, it is possible to create a new Python environment using the following command:

```
conda env create --file ldsc/environment.yml
```

In Windows:

```
activate ldsc
```

In Linux, macOS:

```
source activate ldsc
```

To check that LDSC is correctly installed, type:

```
python2 ldsc/ldsc.py -h
```

A1.10   MTAG

MTAG is a software written in Python 2. You can install the latest version of MTAG with Git using the following command:

```
git clone https://github.com/omeed-maghzian/mtag.git
```

To work properly, MTAG needs the following Python packages installed in the system:

- numpy (>=1.13.1)
- scipy
- pandas (>=0.18.1)
- argparse
- bitarray (for ldsc)
- joblib

To install all of the packages needed to run MTAG, it is possible to create a new Python environment using the following command:

```
conda env create --file mtag/environment.yml
```

In Windows:

```
activate mtag
```

In Linux, macOS:

```
source activate mtag
```

Further documentation is available at https://github.com/omeed-maghzian/mtag/wiki.

A1.11   Using Windows for this book

Windows users may experience some differences, depending on the operating system. It is possible to install Ubuntu on Windows 10, which will work in a virtually identical manner to Linux. For a tutorial on how to install Unbuntu, refer to https://tutorials.ubuntu .com/tutorial/tutorial-ubuntu-on-windows#0. Using the terminal on Ubuntu, it is possible to use most packages but also Bash, Z-shell, Korn, and other shell environments without virtual machines or dual-bootin. You will also be able to run tools such as SSH, git, apt, and dpkg directly from your Windows computer.

References

1.   S. Purcell et al., PLINK: A tool set for whole-genome association and population-based linkage analyses. *Am. J. Hum. Genet.* **81**, 559–575 (2007).

2.   C. C. Chang et al., Second-generation PLINK: Rising to the challenge of larger and richer datasets. *Gigascience* **4** (2015) (available at https://doi.org/10.1186/s13742-015-0047-8).

3.   J. Yang, S. H. Lee, M. E. Goddard, and P. M. Visscher, GCTA: A tool for genome-wide complex trait analysis. *Am. J. Hum. Genet.* **88**, 76–82 (2011).

4.   J. Euesden, C. M. Lewis, and P. F. O'Reilly, PRSice: Polygenic risk score software. *Bioinformatics* **31**, 1466–1468 (2014).

5.   S. W. Choi and P. F. O'Reilly, PRSice 2: Polygenic risk score software (updated) and its application to cross-trait analysis. *Eur. Neuro.* **29**, S832.

6.   B. J. Vilhjálmsson et al., Modeling linkage disequilibrium increases accuracy of polygenic risk scores. *Am. J. Hum. Genet.* **97**, 576–592 (2015).

Appendix 2: Data Used in this Book

A2.1 Introduction

At the start of each chapter, we describe the data that will be used so that readers can ensure they are able to actively follow all exercises. For the practical exercises in part II of this book in chapters 8, 9, and 10, we use a combination of publicly available data that you can download and additional data that we have simulated for an individual phenotype for body mass index (BMI). To the download data used in this book, refer to the dedicated companion website for this book: http://www.intro-statistical-genetics.com.

A2.2 Description of simulated data

For part II of the book, we use publicly available molecular genetic data from the 1000 Genome Project and simulated phenotypes based on publicly available genome-wide association study (GWAS) results for BMI [1]. We use PLINK [2] for the data preparation and GCTA [3] software to simulate the data. For the theory behind the simulations, see also [4].

First, we downloaded the data, extracted HapMap3 SNPs, and ran some basic QC:

```
### Download 1000G data here: https://www.dropbox.com/s
/k9ptc4kep9hmvz5/1kg _ phase1 _ all.tar.gz#In Linux wget
https://www.dropbox.com/s/k9ptc4kep9hmvz5/1kg _ phase1 _ all
.tar.gz

### Unzip

tar -xzf 1kg _ phase1 _ all.tar.gz

### Extract hm3 snps

./plink \
```

```
--bfile 1kg_phase1_all \
--out 1kg_hm3 \
--extract w_hm3.snplist \
--make-bed

### Run some basic QC

.\plink \
--bfile 1kg_hm3 \
--geno 0.02 \
--mind 0.1 \
--hwe 0.001 \
--out 1kg_hm3_qc \
--make-bed
```

Next, we downloaded the meta results from the BMI GWAS at http://portals.broadinstitute .org/collaboration/giant/images/c/c8/Meta-analysis_Locke_et_al%2BUKBiobank_2018 _UPDATED.txt.gz. We subsequently aligned the column names for SNPs and reference alleles and switched the sign of the effect in case risk and reference alleles are coded in the opposite direction between GWAS results and genetic data. We saved the rs number and the effect size in a tab-separated text file named "causal_loci_BMI.txt" Finally, we simulated a BMI phenotype using GCTA software based on the effect sizes of the GWAS results per SNP and a heritability of 20%.

```
### Simulate BMI phenotype

./gcta64 \
--bfile 1kg_hm3_qc \
--simu-qt \
--simu-causal-loci causal_loci_BMI.txt \
--simu-hsq 0.2 \
--simu-rep 1 \
--out pheno_BMI_sim \
--thread-num 5
```

A file named pheno_BMI_sim.phen will contain the simulated phenotypes for all indi-viduals together with the family id and an individual id.

A2.3   Health and Retirement Study

In part III of this book, and specifically in chapter 11, we turn to more advanced applications such as causal modeling and regression analysis using polygenic scores. Here we provide several practical examples based on real findings from the research literature using the publicly available data from the Health and Retirement Study (HRS), a longitudinal large-scale study of the U.S. population age 50 and above and household members and spouses (see box 7.1).

To obtain access to the data, you first need to register on the HRS website (see box A.1 for full instructions). You need to provide your e-mail address, username, and some additional information to receive a password that, in tandem with your username, allows you to access the data download platform. Eventually, you can enter the download portal with your username and password and most likely, we anticipate that you will get very nervous. It seems like a jungle of data, because HRS is a long-time, ongoing study with excellent (and multiple) resources. Luckily, the different data waves in HRS have been merged and harmonized into the so-called RAND datasets. Nonetheless, do read the introduction and documentation information that is provided before you start using the data.

As you know, throughout this book, we use R since it is not only an excellent working environment but also open source and freely available. On the download page of HRS, data are mainly provided in SPSS, Stata, and SAS format. It is possible to download these files and follow the instructions in the accompanied document to combine the ASCII files containing the data with the respective software information and subsequently read these files into R, for example, with the `foreign` package for SPSS and Stata format. However, ideally, try to use the following script to directly download the data via RStudio. Again, all you need is your username and password. But first it is necessary to install the required software packages:

Box A.1
Obtaining access to HRS data

There are two types of HRS data products when applying for access—public data and restricted/sensitive data. Many phenotypes are found in the public data, including but not limited to demographic, socioeconomic, health information, and genetic PGS. To access the HRS public data, go to: https://hrs.isr.umich.edu/data-products/access-to-public-data to register and obtain a username and password. Once registration has been confirmed, use the username/password to download data files. Restricted/sensitive data include biomarker and health data, Medicare and Social Security, and genotypic data. The application process varies from signing supplemental agreement, to access only through remote virtual desktop or encrypted physical media. To apply for genetic data other than PGSs, follow the steps on https://hrs.isr.umich.edu/data-products/genetic-data/products.

```
#### Install the Health and Retirement Study on your
computer
# First we need to install and activate a couple of
helper packages:
# devtools works in the background for us, allowing us
to download
# The packages (lodown, etc.) to download the data
# Important: for the R code, all quotation marks need to
be in this style: ""

install.packages("devtools")
library(devtools)

# We download and activate the helper package, called
lodown

install _ github("ajdamico/lodown")
library(lodown)

#### Let's download data

# If you would like to download all of the HRS data that
you see on
# the website, this would be the command
# Attention! This might take a while

lodown("hrs", output _ dir=getwd(),
        your _ username= "your username",
        your _ password= "your password")
```

For this exercise, we do not need all of the data available from HRS; rather, you can focus
on specific versions and subsets, which we can download like this:

```
#### Focus on specific datasets
# First, create a list of available datasets and define
the download
# destination using output _ dir. In our case we park it
in the R
# project folder

hrs _ cat <-
  get _ catalog("hrs",
```

```
            output _ dir=getwd(),
            your _ username="your username",
            your _ password="your password")

# If you execute the command:

View(hrs _ cat)

# in RStudio, you can see the available datasets in
# parallel to the website

# First, define the first dataset we wish to download,
namely the
# PGSs: PGENSCORE3.zip

PGS <- subset(hrs _ cat, grepl("PGENSCORE3.zip",
file _ name))

# Download the data

PGS <- lodown("hrs", PGS,
                  your _ username="your username",
                  your _ password="your password")

# Next define the second dataset we wish to download,
# namely the RAND longitudinal data:
randhrs1992 _ 2016v1 _ STATA.zip.
# Please note that the version of the dataset can change
-- check this viewing the
# hrs _ cat file and adapt the script accordingly.

Pheno <- subset(hrs _ cat, grepl("randhrs1992 _ 2016v1 _
STATA.zip", file _ name))

#### Download the data

Pheno <- lodown("hrs", Pheno,
                  your _ username="your username",
                  your _ password="your password")
```

The next step is to read the data into RStudio. Since we want to analyze the rich environmental (socioeconomic and demographic) information in the RAND data together with the information from the PGSs, we need to merge both data frames. In HRS, there is a household identifier (hhid) and a person identifier (pn) and together they uniquely identify individuals in the data, and therefore can be used to merge the PGSs with the phenotypes.

```
# Read in the data frame for the PGS

hrs _ PGS <- readRDS("PGENSCORE3/pgenscore3e _ r.rds")

# Read in the data frame for the Phenos

hrs _ Phenos <- readRDS("randhrs1992 _ 2016v1 _ STATA/
randhrs1992 _ 2016v1.rds")

# Merge data frames
hrs _ GePhen <- merge(hrs _ PGS,hrs _
Phenos,by=c("hhid","pn"))
```

Note that the HRS PGS data are available for African ancestry (PGENSCORE3A_R) and European ancestry (PGENSCORE3E_R), which we will focus on the latter in our examples. While the RAND data contains 37,495 individuals, genotype information is available for 12,090 individuals of European ancestry and only those who were in both original data frames end up in the merged one.

The detailed description of the genetic data collection, QC, and score construction are accompanied in your zipped download folder under "docs" and can be found online [5]. Briefly, genotyping was conducted by the Center for Inherited Disease Research (CIDR) in 2011, 2012, and 2015. Genotype data on over 19,000 HRS participants was obtained using the llumina HumanOmni2.5 BeadChips (HumanOmni2.5-4v1, HumanOmni2.5-8v1, HumanOmni2.5-8v1.1), which measures approximately 2.4 million SNPs. Individuals with missing call rates > 2%, SNPs with call rates < 98%, HWE p-value < 0.0001, chromosomal anomalies, and first-degree relatives in the HRS were removed. PGSs have been constructed with PRSice (see chapter 10), and all available information has been included in the scores. All of this type of should make sense to you and if it does not we strongly encourage that you read the introductory chapters before embarking upon any kind of data analysis.

For the PGS-phenotype file, we end up with 12,090 unrelated individuals. However, there are still multiple members of the household, and we want to focus on individuals who do not share their micro environment, because this might bias estimates due to factors such as gene-environment correlation (see also chapter 6). We therefore select only one individual per household.

```
#### Select one individual per household
# First, we install and load the helper package "dplyr"

install.packages("dplyr")
library(dplyr)
```

```
# Then we create a variable (whh _ count), which numbers
individuals
# within housholds

hrs _ GePhen _ ci<- hrs _ GePhen %>% group _ by(hhid) %>%
dplyr::mutate(wf _ count=row _ number())

# We keep arbitrarily always the first person in the
# household

hrs _ GePhen _ uni<- hrs _ GePhen _ ci[which(hrs _ GePhen _
ci$wf _ count=="1"),]

# Finally, we clean the working space in RStudio
rm(hrs _ PGS,hrs _ GePhen _ ci,hrs _ GePhen,hrs _ Phenos)
```

The data file we are using from here is named hrs_GePhen_uni and contains 8,451 individuals and more than 10,000 variables. Be sure before embarking on the analyses in chapter 11 that you also have the same number.

A2.4   Data used by chapter

Chapter 8

All chapter 8 data can be downloaded from this book's companion website, http://www
.intro-statistical-genetics.com:

- ALL.chr21.vcf.gz
- BMI_pheno.txt
- 1kg_EU_qc.bim, 1kg_EU_qc.bed, 1kg_EU_qc.fam
- hapmap-ceu.bim, hapmap-ceu.bed, hapmap-ceu.fam
- list.txt
- 1kg_hm3.bim, 1kg_hm3.bed, 1kg_hm3.fam
- individuals_failQC.txt
- hello_world.sh

The hapmap-ceu file can be found at http://zzz.bwh.harvard.edu/plink/dist/hapmap-ceu
.zip. This file contains the genotypes, map files, and two extra phenotype files that we
describe shortly and is called hapmap-ceu.zip.

All of the populations can be found at http://zzz.bwh.harvard.edu/plink/res.shtml.

Chapter 9

- 1kg_EU_BMI
- 1kg_EU_Overweight
- 1kg_ hm3_ qc
- 1kg_hm3_pruned
- 1kg_samples.txt
- BMI_pheno.txt
- 1kg_samples_EUR.txt
- hapmap-ceu

Chapter 10

- 1kg_hm3_qc
- score_rs9930506.txt
- BMI.txt
- BMI_pheno.txt
- Obesity_pheno.txt
- 1kg_EU_qc
- pca.eigenvec
- 1kg_samples.txt
- BMI_score_MULTIANCESTRY.best
- BMI_LDpred.txt

Chapter 11
See the previous section.

Chapter 12

- EA2_results.txt.gz
- Giant_Height2018.txt.gz
- eur_w_ld_chr.tar.bz2
- w_hm3.snplist.bz2
- GWAS_EA_example.txt
- GWAS_CP_example.txt
- LD-Hub_genetic_correlation_example.txt

References

1.   A. E. Locke et al., Genetic studies of body mass index yield new insights for obesity biology. *Nature* **518**, 197–206 (2015).

2.   S. M. Purcell et al., PLINK: A tool set for whole-genome association and population-based linkage analyses. *Am. J. Hum. Genet.* **81**, 559–575 (2007).

3.   J. Yang, S. H. Lee, M. E. Goddard, and P. M. Visscher, GCTA: A tool for genome-wide complex trait analysis. *Am. J. Hum. Genet.* **88**, 76–82 (2011).

4.   M. R. Robinson et al., Genotype-covariate interaction effects and the heritability of adult body mass index. *Nat. Genet.* **49**, 1174 (2017).

5.   E. B. Ware, L. L. Schmitz, A. M. Gard, and J. D. Faul, HRS Polygenic Scores—Release 3 (2018) (available at http://hrsonline.isr.umich.edu/modules/meta/xyear/pgs/desc/PGENSCORES3DD.pdf).

Glossary

Active gene-environment correlation (rGE). Also known as niche creation, this is when individuals actively select or create environments that are associated with their own genetic predispositions.

Additive genetic effects. When two or more genes contribute to a phenotype or when alleles of a single gene combine such that their combined effects on the phenotype equal the sum of their individual effects.

Admixture. See **Genetic admixture**.

Allele. Refers to each of the two or more alternative forms of a gene found at the same place on a chromosome that arise by mutation.

Ancestry components. A predefined number of subgroups with distinctive allele frequencies, inferred from genome-wide data, which are then used to assign the ancestry of each individual without specifying the population to which the individual belongs.

APOE gene. This gene provides the instructions to make a protein called apolipoprotein E, which combines with fats to form molecules called lipoproteins, which are responsible for packaging cholesterol and other fats and carrying them through the bloodstream. Normal levels of cholesterol are important to prevent cardiovascular diseases, including heart attack and stroke. There are different alleles of the APOE gene (e2, e3, e4), with e3 found in around half of the population. The e4 is well known for increasing the risk of late-onset Alzheimer's disease.

Assortative mating. In genetic research refers to a mating structure in which pairs of individuals that are (genetically) similar to each other mate with a higher probability than expected under random mating.

Autosomal chromosomes. The 22 numbered non-sex chromosomes.

Bayesian approach. A statistical approach that uses methods that combine a likelihood function with a prior probability to calculate a posterior probability. The posterior probability can be used to estimate parameters and to quantify available knowledge regarding the parameter. Given a set of assumptions about the underlying model, it can provide a rigorous assessment of uncertainty.

Bonferroni correction. A correction used to solve the multiple testing problem in genome-wide association studies (GWASs) because millions of regression tests are performed in parallel and would be likely to reap many false positives if a standard significance threshold was adopted. In a GWAS, the Bonferroni-corrected p-value of significance is $p < 5 \times 10^{-8}$.

Bottleneck effect. A temporary reduction in population size that causes the loss of genetic variation.

Broad-sense heritability (H^2). The ratio of the total genetic variance (V_G) to the total phenotypic variance (V_P) where $V_P = V_G + V_E$, and where V_E is environmental variance.

Candidate gene studies. These refer to work that focused on predefined genetic loci of interest based on what was thought to be a priori knowledge of the loci's biological function or impact on the trait being examined. Particularly, many early studies could not be replicated.

Chromosome. A single molecule of DNA that comprises part of the genome.

Clumping. A procedure in which only the most significant SNP (i.e., lowest p-value) in each LD block is identified and selected for further analyses, clumping reduces the correlation between the remaining SNPs, while retaining SNPs with the strongest statistical evidence.

Common ancestors. An ancestor shared by two or more individuals.

Cryptic relatedness. This refers to the presence of close relatives (second cousin and closer) in data sources. Such observations cannot be considered as independent. For analyses, typically, one of the individuals in a pair is removed or the dependence structure is considered in the statistical model, (e.g., including a genetic-relatedness matrix in a mixed-linear model).

Deep phenotyping. A precise and comprehensive analysis of the phenotype by providing more specificity, linkage of new types of data and measurements, and new connections between trait subtypes and genetic variations. It is phenotyping in a much more fine-grained and multifaceted way, often using advanced algorithms and data linkage.

Denisovans. Denisovans are an extinct subspecies of archaic hominins represented by fossil remains from a Denisova Cave in southern Siberia; genome sequence data indicate that Denisovans are a sister group to Neanderthals.

Differential susceptibility model. This model is sometimes also referred to as the level of plasticity or the orchid–dandelion hypothesis, which hypothesizes that there are genotype groups that are sensitive to both negative and positive environments.

DNA. Deoxyribonucleic acid is a molecule carrying the genetic instructions to build and reproduce living organisms in the form of a double helix.

Environmental variance (V_E). Environmental quantitative variance among individuals with the same genotype.

Evocative (also known as reactive) gene-environment correlation (rGE). When an individual's partially genetically inherited traits evoke reactions from others in the environment (e.g., being shy, agressive).

Evolution. Refers to the change in heritable characteristics of populations over successive generations. It forms the basis of our understanding not only about the origin of the human species but also the underlying genetic architecture and disease mutations.

Exogeneous variable. Often used in G×E studies to refer to a variable whose value is determined by factors outside of the causal system under study.

Experimental factor ontology (EFO). EFO is an ontology of variables used in molecular biology including aspects of disease, anatomy, cell type, cell lines, chemical compounds, and assay information. In this book we discuss EFO in relation to groupings in the NHGRI-EBI GWAS Catalog.

Fitness. Also sometimes referred to as evolutionary fitness. This is how well a species adapts to its environment. It can be defined differently in different models. In humans it is generally defined as the expected number of offspring of an individual of that genotype left to reproduce in the next generation.

Founder effect. The effect of genetic variation as a new population, or species, is founded. It is a type of genetic drift, when a small group splits from the main population to found a colony, where the founders may be selective and not represent the full genetic diversity of the original group.

Gene. A gene is a sequence of nucleotides in the DNA that codes for a molecule (e.g., a protein).

Gene-environment correlation (rGE). This is the process by which an individual's genotype influences or is associated with exposure to the environment.

Gene-environment interaction (G×E). G×E defines an interplay between a gene and an environmental factor in which the effect of the gene on a phenotype is modifiable by the environment, and vice versa.

Genetic admixture. A population is admixed if it received gene-flow from another population. This happens when two or more previously isolated and genetically differentiated populations interbreed. The results are new genetic lineages, and it is usually only recent flow. The concept is sometimes also used to describe individuals that have ancestors from several different populations.

Genetic drift. A change in allele frequencies over time in a population of finite size due to random transmission of parental alleles from parents to offspring and due to the fact that some individuals randomly produce more offspring than others, irrespective of their genotype.

Genetic recombination. A process during **meiosis** that produces new combinations of alleles, by breaking pieces of DNA and recombining them. This process leads to diversity of DNA sequences.

Genome. The genome is the collection of DNA of an individual.

Genome-wide association study (GWAS). A GWAS is designed to adopt an unbiased hypothesis-free approach to discover genetic loci that are associated with a phenotype. They often combine data from multiple studies to gather the largest sample possible. First introduced in 2005, over 4,000 GWASs have been published by 2020 identifying thousands of genetic associations and their biological function.

Genotype. Describes part of an individual's DNA that influences their phenotype.

Germ line mutation. A mutation that will be inherited by the offspring of the organism.

GWAS Catalog. Includes data from all published GWASs (https://www.ebi.ac.uk/gwas/).

GWAS-heritability. The fraction of phenotypic variance of a trait explained by genome-wide significant genetic variants—sometimes also by polygenic scores based on GWAS findings.

Haplotype. This is a multilocus genotype on a single chromosome. If three successive genotypes at polymorphic sites are AT, CG, and CA, the haplotypes may be ACA and TGC (or ACC and TGA, AGC and TCA, or AGA and TCC).

HapMap. HapMap is an international project to develop a haplotype map, with the goal of identifying genetic similarities and differences among human populations and providing a reference panel for human genetic variation. The project has made large amounts of data publicly available.

Hardy–Weinberg equilibrium (HWE). The HWE states that allele and genotype frequencies are constant over generations in the absence of evolutionary forces such as natural selection, mutation, or migration. Violation of the HWE law indicates that genotype frequencies are significantly different from expectations. In a GWAS, it is generally assumed that deviations from HWE are the result of genotyping errors.

Heritability. A population measure defining the proportion of variance in a phenotype explained by genetic variance within a population. We can differentiate between broad-sense heritability, including both additive and non-additive genetic effects such as epistasis and dominance, and narrow-sense heritability focusing on additive genetic effects only.

Heterozygous. The carrying of two different alleles of a specific SNP. The heterozygosity rate of an individual is the proportion of heterozygous genotypes. High levels of heterozygosity within an individual might be an indication of low sample quality whereas low levels of heterozygosity may be due to inbreeding.

Hominidae. Modern humans, their fossil ancestors, and extinct relatives thereof, up to (but not including) chimpanzees.

Homozygous. The proportion of individuals in a population that are homozygous at a particular locus. In other words, when an individual has two of the same allele, regardless of whether it is dominant or recessive.

Inbreeding. When individuals who are related to one another produce offspring. The inbreeding coefficient measures the excess of homozygous individuals in a population relative to the expectation under the Hardy–Weinberg equilibrium.

Linkage disequilibrium (LD). A measure of nonrandom association between alleles at different loci at the same chromosome in a given population. SNPs are in LD when the frequency of association of their alleles is higher than expected under random assortment. LD concerns patterns of correlations between SNPs.

Major allele. It is the allele of a SNP with the highest frequency.

Manhattan plot. This is a type of scatterplot of location of a genetic variant and the transformed p-value that is one of the main graphical techniques in a GWAS used to display the distribution of genetic hits.

Meiosis. The process by which haploid gametes are generated, during which genetic recombination occurs.

Mendelian traits. In contrast to polygenic or complex traits, Mendelian traits are only related to a single locus (e.g., Huntington's disease).

Minor allele. The allele of a SNP with the lowest frequency.

Minor allele frequency (MAF). The frequency of the least often occurring allele at a specific location. Most studies are underpowered to detect associations with SNPs with a low MAF and therefore exclude these SNPs.

Mutation. A permanent change in the sequence that makes up a gene (see box 1.1 for explanation and definition of various types of mutations).

Narrow-sense heritability (h^2). The ratio of the additive genetic variance to the total variance of a quantitative character or V_A / V_P. It is the proportion of variance in a phenotype within a population that is associated with additive genetic variance.

Natural selection. The increase or decrease of particular genetic traits as a function of the differential fitness and the reproductive success of individuals.

NHGRI-EBI GWAS Catalog. See **GWAS Catalog**.

Null hypothesis. Refers to the testable hypothesis that there is no difference between a hypothesized value and an estimated statistic (for example, a value in a population or the correlation between two variables).

Observational study. Refers to a sample of the population where the researcher—in contrast to an experiment—does not create variation in the independent or exposure variable.

Pairwise linkage disequilibrium (pairwise LD). The strength of association or co-occurrence between alleles at two different genetic markers.

Passive gene-environment correlation (rGE). The association between the genotype a child inherits from her or his parents and the environment in which the child is raised.

Phenotype. The observable trait of an individual, ranging from physical traits (hair color, height) to disease status (diabetic) to behavior (risk-taker, age at first sexual intercourse, educational attainment).

Population stratification. The presence of multiple subpopulations (e.g., individuals with different ancestral background) in a study. Because allele frequencies can differ between subpopulations, population stratification can lead to false positive associations and/or mask true associations. An excellent example is the chopstick gene, where a SNP, due to population stratification, would be wrongly assumed to be a true association due to differences in allele frequencies of those of Asian and European ancestry who have a different usage of chopsticks for purely cultural rather than biological reasons.

Population structure. There is population structure when mating is more likely to occur between some subsets of the population than between others, typically due to geographical structure. Individuals located in geographical proximity to each other are more likely to mate. Population structure is also used to describe a population in which allele frequencies differ between different geographic regions.

Power. The power of a statistical test is the probability of correctly rejecting a false null hypothesis.

Principal component analysis (PCA). A statistical technique used to emphasize variation and bring out strong patterns underlying data, with the aim of minimal loss of information. It is a visualization technique and way to reduce the dimensionality of data from high to lower dimensional data stepwise, extracting independent dimensions with the highest variance explanation in the data.

Pruning. A method to select a subset of markers that are in approximate linkage equilibrium. In PLINK, this method uses the strength of LD between SNPs within a specific window (region) of the chromosome and selects only SNPs that are approximately uncorrelated, based on a user-specified threshold of LD. In contrast to clumping, pruning does not take the p-value of a SNP into account.

Random variable. A variable that takes on different values (for example, possible outcomes of an experiment) and for which each value can be associated with a probability.

Relatedness. Indicates how strongly a pair of individuals are genetically related. A conventional GWAS assumes that all subjects are unrelated (i.e., no **cryptic relatedness**). Without appropriate correction, the inclusion of relatives could lead to biased estimations of standard errors of SNP effect sizes. Note that specific tools for analyzing family data have been developed.

Significance level (α). A concept used in statistical hypothesis testing to determine when to reject a null hypothesis. If the probability of observing an outcome is extreme or more extreme than the observed outcome under the null hypothesis is less than the significance level, then the null hypothesis is rejected. It is chosen to be the greatest probability of type I error or false positive findings tolerated for a statistical test.

Single-nucleotide polymorphism (SNP). A variation in a single nucleotide (i.e., A, C, G, or T) that occurs at a specific position in the genome. A SNP exists as two different forms (e.g., A vs. T). These different forms are called alleles. A SNP with two alleles has three different genotypes (e.g., AA, AT, and TT).

SNP arrays. Microarrays that are used to simultaneously genotype several thousand to several hundred thousand SNPs for a single sample.

SNP-heritability. The fraction of phenotypic variance of a trait explained by all SNPs in the analysis.

SNP-level missingness. The number of individuals in the sample for whom information on a specific SNP is missing. SNPs with a high level of missingness can potentially lead to bias and are typically excluded from analysis.

Social control model. Also referred to as the social push model, this is a G×E model that hypothesizes that genetic associations are attenuated or dampened in the presence of socially restrictive environmental contexts.

Spurious relationship. When two variables are correlated and it is mistakenly believed that this association is causal.

Summary statistics (in a GWAS). The results obtained after conducting a GWAS, including information on chromosome number, position of the SNP, SNP(rs)-identifier, MAF, effect size (odds ratio/beta), standard error, and p-value. These statistics are used, for example, to create polygenic scores.

Test statistic. A numerical summary of the data used to measure support for the null hypothesis. A test statistic may have a known probability distribution (e.g., such as x^2) under the null hypothesis or its null distribution is estimated.

Type 1 error. The rejection of a true null hypothesis.

Type 2 error. Accepting a false null hypothesis.

Variance. Signified by σ^2 and is the average of the squared deviation of the observations of an outcome of variable (often *random variable*) from its mean, divided by the number of individuals (sample size).

Variance-covariance matrix. Used to describe the variances and covariances between variables. It is sometimes also referred to as the dispersion matrix or simply the covariance matrix.

Variant Call Format (VCF). A type of data format that can store genomic information for genotyped, imputed data, and sequencing data. It is very flexible because various types of information can be stored (see chapter 7).

Weak instrument assumption. One of the statistical assumptions underlying the instrumental variable (IV) and Mendelian Randomization approach to evaluate whether the polygenic score is a good instrument. The IV needs to be correlated with the treatment and generate sufficient variation in the independent variable (see chapter 13).

Within-family analysis (sometimes also called family fixed-effect regression). An analysis that examines the effect of different genotypes among family members. Used to establish a true genetic effect that is not confounded by population stratification or genetic nurture, such as the differences in genotype across siblings as a random experiment (i.e., because alleles are transmitted randomly through meiosis).

Notes

Chapter 1

1. *Somatic cells* refer to any cell of a living organism other than reproductive cells (i.e., other than a gamete, germ cell, gametocyte, or undifferentiated stem cell). There are around 220 types of somatic cells in humans.

2. It is worth pointing out here that the independent assortment is violated by genetic linkage. That is, genes on a chromosome will be inherited together with a frequency in inverse proportion to the physical distance between them.

3. An insertion or deletion of bases in the genome is referred to as indel in molecular biology.

4. Although we note that an RNA gene could have alleles. RNA is defined elsewhere in this chapter.

5. *Amino acids* are the building blocks of proteins that catalyse most chemical reactions that occur in the cell. They provide the structural elements of cells and help to bind cells together into tissues.

6. *Polymers* are large molecules that are made when many small molecules join together.

7. *Diploid cells* contain two complete sets of chromosomes. This is opposed to haploid cells that have half the number and only contain one complete set of chromosomes.

8. We realize that this is an oversimplification because of introns and alternative splicing, which we do not cover in this introductory book.

9. *Exons* are any gene that encodes a part of the final mature RNA that is produced by that gene (after introns have been removed).

10. The *exome* is the part of the genome that consists of exons (i.e., the coding regions).

11. *Introns* are the noncoding sections of an RNA transcript that are spliced out before the RNA molecule is translated into a protein. Introns are also in the gene and do not remain in the transcript for long before they are spliced out.

12. A *conserved gene* is one that has remained essentially unchanged throughout evolution and indicates that it is unique and essential. Note that conservation is always relative to a phylogenetic group.

13. A *peptide bond* is a chemical bond that is formed between two molecules. Amino acids are joined by a series of peptide bonds which in turn form a peptide.

14. A *eukaryote* is an organism that has complex cells where the genetic material is organized into a membrane-bound nucleus, such as animals, plants, and fungi. In contrast, prokaryotes are organisms that lack nuclei and most other complex cell structures, such as bacteria.

15. *Cytoplasm* is the material within a cell excluding the nucleus.

16. A *polypeptide* is a sequence of amino acids that are linked. A single polypeptide chain may comprise the entire primary structure of a protein, but more complex proteins are formed when two or more polypeptides link together.

Chapter 2

1. A random variable is one in where the possible values are outcomes of a random phenomenon.

2. Latent variables (from hidden in Latin) refer to those that are not directly observed; rather they are inferred from other variables are directly observed and measured. One common example is using measures of the Big 5 personality traits to infer personality or others such as well-being, quality of life, or social belonging.

3. Some of this basic terminology is used differently across the disciplines. In the social sciences, for instance, sample often refers to the dataset that is used (i.e., a sample of individuals from a larger population). In genetics, sample often refers to data from one individual. Another term that is used very differently is cohort. In genetic research the term cohort is often used to refer to what social scientists would call a sample or dataset and demographers would define as a birth cohort. Geneticists often use the term cohort to refer to the data regardless of whether the research design is a cohort design, which may be confusing for outsiders.

Chapter 3

1. Hominidae refers to the taxonomic family of primates that includes the living (extant) and extinct humans, chimpanzees, gorillas and orangutans.

2. Recall from chapter 1 that a germ line is a series of germ cells that have developed or descended from earlier cells in a particular species. They continue through each successive generation.

3. We note that these classic theories are often framed in purely heterosexual terms, since they are based on biological reproduction. We recognize that many gay and lesbians reproduce, most often through previous heterosexual relationships or increasingly via assisted reproductive technology [40].

Chapter 4

1. http://depts.washington.edu/chargeco/wiki/ResultsSharingFormat.

Chapter 5

1. A recent study tested Cheverud's Conjecture, which states that phenotypic correlations are reasonable proxies for genetic correlation among humans with evidence for an association between both phenomena [77].

2. Cis-regulatory modules (CRMs) refers to the part of the DNA, often 100–1,000 DNA base pairs in length, where transcription factors bind and regulate the expression of nearby genes.

3. https://rdrr.io/github/DudbridgeLab/avengeme/man/sampleSizeForGeneScore.html.

Chapter 7

1. We note that probe design is a very complex topic, and we simplify the discussion here. If a site is biallelic, only one variant may be targeted but you might have probes complementary to each strand to allow for cross-checking.

2. Within the FTO gene, rs9930506 showed the strongest association with BMI ($p = 8.6 * 10^{-7}$), hip circumference ($p = 3.4 * 10^{-8}$), and weight ($p = 9.1 * 10^{-7}$). Homozygotes for the rare "G" allele were 1.3 BMI units heavier than homozygotes for the common "A" allele. Source: https://www.snpedia.com/index.php/FTO.

3. https://www.ncbi.nlm.nih.gov/assembly?term=GRCh38&cmd=DetailsSearch.

Chapter 8

1. The prefix of a file is the first part of the name of the file before the dot (.). The suffix refers to the file type that is listed after the dot. For example, in the file `text.txt`, `text` is the prefix and `.txt` is the suffix.

2. Output is printed to `stdout`, which means standard output in Unix-like operating systems such as Linux and macOSx, where `stdout` is defined by the POSIX standard. POSIX stands for Portable Operating

System Interface for Unix, which is a set of standards that define how Unix operating systems operate with one another and the command structure. In the terminal, `stdout` defaults to the user's screen.

3. Alternatively, we can also remove one random individual from pairs that have a high degree of relatedness.

4. Strand orientation refers to the direction in which genes are read (transcribed). In different genotyping platforms as well SNP databases, the same SNP can be defined as being on the forward (plus) or as being on the reverse (minus) strand.

Chapter 9

1. PLINK requires three parameters for pruning: a window size in variant count or kilobase (if the "kb" modifier is present) units, a variant count to shift the window at the end of each step, and a variance inflation factor (VIF) threshold. At each step, all variants in the current window with VIF exceeding the threshold are removed.

2. Note that we are using the terminology from the 1000 Genomes Project, where American, for instance, refers to the CEU group of Utah residents with Northern and Western European ancestry. For more information, see http://www.internationalgenome.org/category/population/.

Chapter 10

1. https://github.com/crahal/GWASReview/blob/master/tables/Manually_Curated_Cohorts.csv.

2. Bootstrapping is a technique that allows us to assign measures of accuracy to sample estimates.

3. A point normal mixture statistical distribution (also called spike and slab) is a probability distribution derived by the combination of two probability distributions. Part of the SNPs are distributed according to a point mass distribution while the others are distributed according to a continuous normal distribution.

4. MCMC methods refer to a class of algorithms that sample from a particular probability distribution. You are able to obtain a sample of the desired distribution by constructing and then observing a Markov chain after a number of steps. A larger number of steps indicates a closer match to the desired distribution. A Markov chain refers to a stochastic model of a sequence of possible events that is "memoryless" or, in other words, where the probability of each event only depends on the state of the previous event.

5. Remember that an infinitesimal model is the polygenic model that assumes that a quantitative (continuous) trait is controlled by an infinite number of loci, all of which has an infinitely small effect.

Index

Note: Page numbers in *italics* indicate figures.

Adaptation, 63, 117, 138
Additive model
  applied example, 300–301
  in association analysis, 218, 220
  graphical depiction of, 43–44
  upper bound of, 112
Admixture
  genetic, 58, 61–63, 73, 93
  and GWAS results, 142
  human dispersal, *57*, 58
Africa. *See also* African
  distribution population, 60
  East Africa, 60
  in GWAS, 96
  human dispersal out of, 55–58, 226
  PCAs, 61, 229, *230*
  population structure, 264
  sub-Saharan, 56, 58
  West African, 60–61
African. *See also* Africa
  ancestry groups, 61, 264–265, *266*
  difference in allele frequencies, 60–61, 226
  genetic diversity, 58, 72
  in GWAS, *95*, 96
  inability to apply European ancestry polygenic
    scores, 264–266
  LD in, 84
  in 1000 Genomes Project, *157*
  scientists, 163
African Americans, 61, 74, 103, 366
Amino acids, 15–16, 18–20
  definition of, 405n5
Anaconda, 267. *See also* Python
  installation of, 383–385
  LDSC and MTAG, 333
Ancestors, 56, *57*
  of Australian Aborigines, 56
  colonizing, 61
  common, 70

Anthropometric traits, 78, 115, 155, 315
  and GIANT consortium, 16
Area under the curve (AUC), 120, 365
Asia
  calculating LD scores, 324
  data, 96, 161
  East, 229, *230*, 266
  in GWAS, *95*, 96
  migration out of, 56, *57*
  PCAs, 61, *230*
  polygenic score distribution in, 266
  population stratification, 226, *230*
  rise of genetic research, 96
  Southeast, 226
Asian. *See* Asia
Assortative mating
  in Hardy–Weinberg equation assumption, 70–71
  in MR assumptions, 347–349
  by traits, 68, 144
Autism, 6, 115
  genetic overlap with, 162
  heritability of, 26
  incorrect controlling for population stratification, 110
  polygenic score applications, 103
Autosomal chromosomes
  definition of, 9
  heterozygosity across, 90, 203
  to identify duplicated or related individuals, 206
  in PLINK, 239

Behavioral traits
  complex, 65, 78, 113, 115, 130, 315, 366
  and polygenicity, 250
  problem in predicting complex ones,
    360–363, 365
Belsky, Daniel W., 137, 139
Bias
  ascertainment, 367
  differential, 104, 110

Bias (cont.)
  due to assortative mating, 349
  due to duplicated or related individuals, 218, 236
  due to gene-environment correlation, 394
  due to heterogeneity in sample or selection, 104, 341
  due to pleiotropy, 351
  due to population stratification, 206, 231
  by genetic confounding, 27, 295
  by genetic drift, 60
  healthy volunteer, 143, 310
  mortality, 65, 367
  in MR due to canalization, 348
  in MR due to weak instrument, 347
  by not considering genetic inheritance, 121
  omitted variable, 308, 339–340
  publication, 135, 351
  reduction by controlling for family fixed effects, 134
  in results, 85, 110
  simultaneity, 346
  systematic by chip differences, study design errors, or low-quality variants, batch effects, 90, 202–203, 206–207, 211, 249
  in use of polygenic score, 110, 244, 273
Bioecological model. See Social compensation model
Bipolar disorder, 44, 104, 113–114, 116
  heritability of, 26
Birth weight, 132
Boardman, Jason D., 131, 133, 136, 137, 141, 309–310
Body mass index (BMI), 4, 14, 23, 27, 113, 115, 155, 209, 352, 354
  association analysis, 218–222
  by birth cohort, 46, 277, 299–307
  and FTO gene, 138, 195–196, 245–246
  in GCTA relatedness matrix, 238–240
  gene-environment correlation, 277, 289, 299–307
  heritability of, 26
  and mortality in MR, 353
  and pleiotropy, 118
  polygenic score applications and calculations, 252–253, 257–272, 289–295
  in sample data in this book, 175–176, 184, 218, 239, 245
  simulation in example data, 389
Bonferroni correction, 84–85
Bottleneck effect, 55, 56, 68–69, 73
Broad Institute, 166, 327

Candidate-gene research, 134–135
Cardiovascular disease, 4, 91, 92, 130, 161, 353
Cell, 3, 5, 11, 15–16, 18, 20, 91, 112, 115, 378–379
  cycles, 11
  diploid, 12, 16
  division and mitosis, 10–12
  egg and sperm, 9, 12, 72
  eukaryotic, 19
  sickle cell disease, 117
  somatic, 8, 11

Centimorgan, 169, 172
Central dogma of molecular biology, 3, 8, 15–20
Cholesterol, 115, 116, 120, 354
Clumping, 226, 248–249, 252–255, 259, 267
Complex traits, 8, 15, 23, 28, 42, 78, 105, 111, 130, 131, 310. See also Genome-wide Complex Trait Analysis
Conley, Dalton C., 142
Consent, 80, 155, 359, 367–369, 370–371, 373
Contextual triggering model. See Diathesis-stress model
Coronary heart disease, 28, 60–61, 91, 353
Correlation, 33–34
  bivariate, 25
  versus causation, 40–50, 87, 112
  estimating genetic, 328–333
  gene-environment (rGE), 7, 129, 136, 143–146, 295, 308, 315–316, 320
  genetic, 7, 27, 101, 114–117, 288, 294, 348, 350
  between genotypes and phenotypes, 339
  LD and SNPs, 72, 85, 156, 217, 223–225, 267
  phenotypic, 24, 114–115, 134
  positive or negative, 41, 42, 93
  in twin models, 24–25, 79
  zero-order, 131
Covariance, 36–40, 59, 330, 335. See also Variance
Covariance matrix. See Variance-covariance matrix
Culture and cultural, 23, 59, 65, 93, 131–132, 141, 143

Daw, J., 309
dbGaP—Database of Genotypes and Phenotypes, 163
Deep phenotyping, 124, 179, 371, 379
Dementia, 119, 352, 354–355
Demographic
  diversity, 77, 79, 96–97, 162
  history, 62, 364
  science, 66, 309
  variables, 155, 300, 360, 363
Denisovans, 56, 57
Diabetes, 130, 353, 364
  heritability of, 26
  type 1, 26, 113
  type 2, 4, 6, 14, 26, 60–61, 78, 112–113, 115, 116, 159, 168, 362
Diathesis-stress model, 122, 129, 136–140, 138, 307
Differential susceptibility model, 129, 137, 138, 139–140, 305, 310
Diploid cells, 10, 12, 16, 405n7
Direct pleiotropy, 117–119, 120–121, 294, 348, 351–353. See also Indirect pleiotropy
Discordant sex information, 90, 203
  identification of individuals with, 205–206
Dominance, 20, 22
Domingue, Benjamin W., 122
Dudbridge, Frank, 103, 120, 406n3 (ch. 5)

Educational attainment and education, 4, 14, 43, 78, 354
 and causal relationship with fertility, 340, 346
 conducting GWAS of, 317–320
 estimation intergenerational transmission of, 296–299
 in G×E models, 139, 176, 250
 genetic scores compared to other typical variables, *287*
 GWAS of, 112, 279, 348
 and health, 123, 136
 heritability of, 26
 and MTAG, 119, 333–336
 and natural selection, 65
 and number of children ever born, 117
 polygenic score applications, 279–295
 in rGE models, 144
 and social policy applications, 359–360, 362, 365
Endogeneity, 45, 308, 339, 340, 345–346, 352
Epistasis, 22, 105, 244, 349
Erlich, Yaniv, 359, 360, 368
Ethics, 120, 160, 341
 and application of polygenic scores, 365
 and data governance and management, 163, 180, 369
 and data security, 82
 and data sharing, 177
 and equity, 361
 in genomics research, 359–373
 and informed consent, 368–369
 and personal information risks, 366
 and privacy, 367–368
 review, 79, 164, 370–372
Ethnicity, 132–133, 367. *See also* Race
 as confounder, 142
 as not interchangeable with ancestry, 59
Exome, 18, 405n10
 sequencing, 158, 179

Family and families
 environment, 140, 146
 heritability, 24–26, 104–105, 111, 239
 history, 11, 63, 359, 361, 363–365
 identifier (ID) in data, *167*, 168, 193–194, 200–201, 246, 254
 interaction with, 4, 132
 members, 16, 23–25, 63, 370
 relatedness, 166, 168
 secrets and genetics, 366–369
 studies, 24–25, 48, 81, 120, 133–134, 231
 and within-family analysis (within-family fixed-effect model), 47, 223, 231, 355
Fisher, R. A., 342
 infinitesimal model, 107
Fitness, 55–56, 63, 66, 70–71, 73, 111, 117, 378
Founder effect, 55–56, 61, 68–69, 73, 190–191, 202, 236
Freese, Jeremy, 309

GEDmatch, 367–368
Gene-environment correlation (rGE)
 active, 145
 definition of, 129, 143–144
 evocative, 145
 passive, 144
 research designs, 146
Gene-environment interaction (G×E)
 applied models of, 299–307
 bioecological theory of, 137, *138*
 and causal modeling, 119
 challenges and solutions, 141–142, 308–310
 and defining environment, 130–133
 definition of, 129
 diathesis-stress theory of, 136–137
 differential susceptibility theory of, 139–140
 and history of research, 133–136
 research designs, 140
 social control theory of, 140
 theories of, 136–140
Gene-environment interplay, 129–147
 correlation (rGE), 129
 interaction (G×E), 118, 129
 interaction and heterogeneity, 122–123
Genetic instrumental variable regression (GIV), 352, 355
Genome-wide association study (GWAS), 77–98
 correction for multiple testing, 84–85
 definition of, 6, 78
 heritability, 26, 111, 324–328
 heterogeneity, 89–90, 130
 history of, 91–93
 and lack of diversity, 93–97, 159
 Manhattan plots, 85–87, 316–319
 meta-analysis, 82, 209–210
 NHGRI-EBI GWAS Catalog, 77, 78–79, 91, 93, 97, 164
 and polygenic scores, 105
 QC, 90–91, 209–210
 and Q-Q plot, 36, 320, *321*
 research design, 79–80
 and sample selectivity, 161–162
 and sample size, *94*, 108
 significant SNPs, 108
 summary statistics, 114, 164–165, 336
 and transferability of results across ancestry groups, 59–61, 110
Genome-wide Complex Trait Analysis (GCTA), 25, 48, 175, 178, 185, 214, 217, 218, 241
 calculating genetic relatedness with, 237–238
 calculating heritability with, 238–240
 installing, 218
Genomic Relationship Matrix (GRM), 48, 238–239
Genotype
 calls, 171, 202, 204
 correlation between, 85
 definition of, 5–6

Genotype (cont.)
  formats and files in PLINK, 166–170, 172–173, 196
  frequencies in a population, 59, 69–71, 171, 208
  heterozygous or homozygous, 20, 69
  imputation of, 156, 158, 171
  and interaction with environment, 130–147
  in MR, 342–343
  proxy, 217, 224–225
  quality, 174, 203
  rate, 81, 190–192
  and relationship to phenotype, 5, 39, 43, 48, 85,
    117–118, 218, 220, 367
Genotyping and sequencing arrays, 154–155, 158–159
  limitations of arrays, 158
Geography, 55–56, 61–63, 73, 142, 226, 241, 285, 368
  ggplot and ggplot2, 229–231, 292–293, 331–332
Goddard, Mike, 382
Guo, Guang, 137

HapMap
  data, 184, 188–191, 194–196, 199–202, 205, 207,
    218, 224
  project, 73, 155–156, 177, 180, 226, 324
Health and Retirement Survey (HRS), 158, 176, 180
  access to, 163
  cross-trait prediction and genetic covariation using,
    288–295
  description of, 155
  gene-environment interaction using, 300–307
  genetic confounding using, 295–299
  genomic data, 166
  polygenic score applications using, 278–288
Height, 4, 6, 9, 11, 23, 27, 34, 35–39, 43, 60, 66–67, 78,
    87–89, 93, 110–111, *116*, 155, 250, 323–324, 348
  and NBA player Shawn Bradley, 16, *17*
  and polygenic score application, 289–295, 317–319,
    326–331
Heritability
  broad-sense, 22
  common misconceptions of, 23
  definition of, 22
  estimated from summary statistics, 316, 324–328, 330
  estimated with GCTA, 238–240
  estimates for selected phenotypes, *26*
  GWAS-based, 25, 286
  hidden, 27, 111, 288, 309, 362
  missing, 27, 111, 288, 362, 352, 362
  narrow-sense, 22
  SNP-based, 25, 79, 178, 299, 324
  twin or family, 24–25, 79, 111
Heterogeneity, 27, 46–47, 77, 83–87, 89–90, 110,
    112, 122, 164, 204–205, 211, 288, 309
Homo sapiens, 56–58, 174

Illumina, 25, 90, 154, 155, 158
Imputation, 72, 77, 81, 89, 153, 158, 171
  detailed description of, 156

  and LD, 155, 179
  quality, 82, 91, 156, 158, 171, 191, 207, 209–210, 268
  reference panels, 159, 180
  software, 180
  uncertainty, 223, 254
Indirect pleiotropy, 117, 124, 294, 352. *See also*
    Direct pleiotropy
Inheritance, 3, 8, 16, 30, 72, 107, 121, 139, 166, 295,
    299, 342, 348
  distinction from heritability, 23
  rules of, 5
Instrumental Variable (IV) approach, 43, 120,
    339–340
  in an MR framework, 343–346
  statistical assumptions, 347–349
Insurance, 359–360, 367, 369–370
Intelligence, 60, 139, 266, 359–362
Intergenerational transmission, 295–297

Keller, Matthew, 104, 141, 142, 308

Lambda, 315, 320, 323, 327–328, 330
LD. *See* Linkage disequilibrium
LDpred, 243–245, 267–273, 278, 350
Lee, Hong, 382
Linear regression, 39, 82–83, 87, 218–220, 251,
    261–265, 283, 324
Linkage disequilibrium (LD), 55–56, 59, 60, 68, 72,
    78, 84, 111, 114, 115, 146, 153, 171, 207, 217, 241,
    244, 348
  accounting for in calculating polygenic scores
    using LDPred, 267–272
  definition of, 21
  and haplotype blocks, 71–73
  identifying independent SNPs through, 223–226
  and imputation, 155–158
  score regression, 114–115, *116*, 316, 324, 327–328,
    331, 348
Linux, 74, 184–186, 251, 325, 334
LocusZoom, 315, 320–321, *322*
Longevity, 90, 379

macOS, 184–187, 325, 334
Manhattan plot, 7, 79, 336
  description of, 85–87
  plotting a, 316–319
Martin, Alicia, 60, 103, 110, 266, 367
Mediation model, 34, 43, 45–47, 117
Meiosis, 10–12, *10*, 72, 223, 231, 342
Mendelian Randomization (MR), 7, 12, 42–43, 45,
    119–120, 244, 299, 339–356
  applications, 352–355
  bidirectional, 352
  in the IV model, 332–347
  and randomized control trails, 341–342
  using polygenic scores in, 351–352
  violation of statistical assumptions of, 347–351

Mendelian traits, 13–14, 23, 30, 363
Meta-analysis. *See* Genome-wide association study: meta-analysis
Mills, Melinda C., 93, 96, 142, 161
Minor allele frequency (MAF), 12, 13, 28, 48, 81, 84, 88, 89, 159, 199, 207–208, 210, 224, 225
Missing heritability. *See* Heritability, missing
Mitosis, 10–12, *10*. *See also* Meiosis
Moderation model and moderators, 34, 43, 45–47, 123, 131, 133, 136, 139, 141, 146, 307
Molecules, 8, 16, 18–20, 169
Monogenic traits, 3, 11, 13–15, 78, 101, 105, 243, 245, 359, 361, 372
Mortality, 65, 118, 162, 352–353, 363, 367, 379
Multi-Trait Analysis of Genome-wide association summary statistics (MTAG), 7, 119, 316, 333–336
Mutation, 10–12, 58, 63, 66, 69–73, 115, 117, 208, 223, 361, 363
  de novo, 11
  detailed description of, 11, 66
  germ line, 11, 66
  hereditary, 11
  rare, 158
  somatic or acquired, 10

Neale, Benjamin M., 79, 98, 109, 165, 315
Neanderthals, 56, *57*, 74
Neuroticism, 4, 6, 93, 115, *116*, 119
Nonadditive genetic effects, 22–23, 27
Novembre, John, 60, 61, 226
Null hypothesis, 39, 83, 85, 240, *256*, 265, 320, 331

Obesity, 23, 46, 96, 104, 113, 123, 132, 134, 138, 141, 195, 245, 250, 259, 300, 307, 352–353, 362–363, 365
Okbay, Aysu, 279, 280, 317
Omnigenic traits, 3, 15

Paternal effects, 296, 299
Personality, 6, 28, 115, 144–145, 340, 346, 406n2 (ch. 2)
Phenotype
  binary or dichotomous, 83, 87
  complex, 15 (*see also* Complex traits)
  definition of, 6
  and genetic correlation, 114
  heritability of, 22–26, 48, 111–112
  influence of environment on, 22
  intermediate, 45
  measurement of, 34, 80, 83
  polygenic (*see* Polygenic)
  quantitative or continuous, 83, 87
  and relationship to genotype, 5, 39, 43, 48, 85, 117–118, 218, 220, 367
  shared genetic architecture of, 113–114
  sources of heterogeneity of, 89
  types of, 13–14
Pleiotropy, 101, 333, 347–348, 362
  antagonistic, 117

  definition of 15, 115–119
  developmental, 117
  direct, 117, 120–121, 294, 330, 348, 350–353
  directional, 104
  indirect, 117, 294
  molecular-gene, 117
  selectional, 117
PLINK software, 166–169
  association analysis in, 218–223
  attaching a phenotype in, 197
  binary files, 170, 244, 254, 268
  calculating LD scores, 272
  checking for LD between two markers in, 223–226
  command line environment, 184, 186–187
  data formats, 166–170, 172–175
  data management, 193–198
  descriptive statistics in, 199–200
  genetic formats for imputed data, 171–172
  genetic relatedness in, 236–238
  importing data, 191–192
  installation, 184, 382
  introduction to, 184
  merging genetic files in, 196
  missing values in, 200–201
  monogenic score construction in, 245–247
  notes, warnings, and error messages, 191
  opening files, 189
  Oxford file formats, 172–173
  population stratification in, 226–231, 232–235
  QC of genetic data in, 202–211
  recoding binary files, 189
  running, 186
  running scripts in, 188
  selecting individuals and markers, 193–196
  10 commandments for new users, 187
  variant call format (VCF) files, 174–175
  in various operating systems, 185
  version 2.0, 171
Plomin, Robert, 144, 359, 360
Polygenic. *See also* Polygenic scores
  versus monogenic and omnigenic, 13–15, 245, 359, 361
  traits, 8, 28
Polygenic scores, 27, 45, 65. *See also* Polygenic
  accounting for LD in, 267–272
  and application across ancestry groups, 59–61, 110, 264–266
  and AUC, 365
  Bayesian approach to, 267–272
  calculating using LDPred, 267–272
  calculating using PRSice, 247–259
  challenges and solutions for, 103–104
  and common variants, 111
  confidence intervals of, 263–264
  construction of, 107–108, 243–273
  for continuous phenotype, 252
  controlling for confounders with, 120–122

Polygenic scores (cont.)
  definition of, 101, 105
  and differential bias, 110–111
  for education, 65
  and gene-environment interaction, 122–123, 130,
    135, 299–310
  and genetic confounding, 119–120
  for height, 16, 38
  in HRS, 155, 176
  and independent target sample, 109–110
  and intelligence, 360–361
  as IV (instrumental variable), 346, 351–352
  and missing and hidden heritability, 111–112
  monogenic, 245–246, 361
  and multitrait analysis, 119, 333
  normal distribution of, 106
  origins of, 105–106
  out-of-sample prediction, 278–288
  overview of working with, *102*
  and pleiotropy, 115–119, 362
  and policy, 359
  and population stratification, 110–111
  and predicting other phenotypes, 113–114
  pruning and thresholding method in creating,
    247–251
  R^2 of, 261–262, 285–286, *287*
  in regression model with covariates, 261
  and relatedness, 110
  selection of SNPs for, 108
  standardization, 260, 291
  summary statistics to create, 164–165
  using different p-value thresholds, 255–257, 316
  validation and prediction, of 108–109, 260
Polymorphic sites, 3, 8, 20, 210
Population stratification. *See also* Population
    structure
  and ancestry, 93, 103–104, 110, 206, *230*,
    231–235
  and assumption of independence in MR, 349
  controlling for, 82, 87, 221, 228
  and geography, 89, 133, 285
  and inflation of polygenic score, 110
  and inflation of R^2, 110, 262–265
  and number of principal components to control for,
    251
  and PCA, 58, 109, 226–235
  Q-Q plots to reveal for unaccounted, 211
  and use of lambda to detect inflation, 323
  use of linear mixed models instead of PCA to
    control for, 231
Population structure. *See also* Population
    stratification
  assumptions of, 70
  common misnomers of, 59
  different LD in, 60
  and geography, *62*, 285
  and stratification, 58–59, 103, 226, 264

Prediction. *See also* Predictive power
  accuracy and precision, 104, 120, 273,
    360
  across ancestry groups, 59
  cross-trait, 288–295
  and environmental G×E influence, 309,
    360
  and missing and hidden heritability, 111
  out-of-sample or independent target sample, 49,
    109, 277–279
  for personalized medicine, 97, 363–366
  polygenic, 16, *102*, 104, 107–108, 250
  power required for, 119, 251, *256*
  public understanding of genetic, 366–367
  and R^2, 103–104, 109–110, 251, *256*
  and understanding biological mechanisms
    trade-off, 112–113, 348, 355
  and weak instrument assumption in MR, 347
Predictive power, 4, 107, 111, 113, 279, 348–349, 351,
    355, 360. *See also* Prediction
Principal components. *See also* Principal
    Component Analysis (PCA)
  and ancestry, 110
  graphical representation of, *62*
  inspection of, 229–235
  number to control for, 61, 109, 165, 251, 262,
    285
  and PCA, 58–59, 226
  ranking of, 228
Principal Components Analysis (PCA). *See also*
    Principal components
  calculating, 214
  definition of, 58–59, 226
  example in applied model, 262–265
  and geography, 61–62
  number to control for, 61, 109, 165, 251, 262,
    285
  and population stratification, 226–235
Privacy
  and consent, 80, 359, 361
  protection, 369, 373
  and public genetics to solve crimes and find people,
    367–368
PRSice
  commands, 185, 252, 254–255
  to construct polygenic scores, 251–259
  for different operating systems, 185, 251–252
  how to install, 251, 382–383
  input files for, 254–255
  types of output in, 253–254
Python
  Anaconda distribution, 333
  how to install, 267, 383–386
  for LDpred software, 267
  for LDSC, 324–329
  for MTAG, 333–336
  versions of, 267

Quality control (QC)
  of genetic data, 90, 202–210
  genome-wide association meta-analysis, *80*, 82, 165, 171, 209–210
  per-individual, 203–206
  per-marker, 206–209
  protocol, 202
Quantile-Quantile (Q-Q) plot, 36–37, *37*, 51, 87, 211, 315, 320–322, *323*

R^2, 72, 103–104, 109, 110–112, 120, 123, 223–225, 249–257, 261–265, 273, 279, 285–286, *287*, 311, 320, 348
Race. *See also* Ethnicity
  is not a biological construct, 56, 59
  is not ancestry, 59
  is a social construct, 56, 59
  and misconceptions in genetics, 60, 133
Rahal, Charles, 93, 96, 142, 161
Rare variants, 13–14, 104, 147, 158, 159, 207, 210, 363
  common versus, 16, 97
  and imputation, 156
  and missing and hidden heritability, 27, 111
  in QC, 90
Replication of research, 48–49, 78, *95*, 134–135, 139, 143, 161, 315
RStudio, 204, 381

Sample selectivity, 143, 162, 309–310, 367
Sample size
  in candidate-gene studies, 134–135
  in GWAS, 26–27, 79–80, 89, 91, 93–94, *95*, 103–104, 107–108, 110, 119, 159, 279, 288, 333
  MAF threshold dependent upon, 208, 210
  p-values influenced by, 87–88
  required to detect G×E, 142–143, 146, 309
  and statistical tests, 34–36, 87–88, 114
  target to apply a polygenic score, 120
Sanger sequencing method, 155, 180
Schizophrenia, 14, 26, *116*
Sex chromosomes, 9, 196, 199, 205
Sexual dimorphism, 56, 68, 162
Sexual reproduction, 3, 8, 9–10, 28, 144, 342
Shanahan, Michael, 136, 137, 138, 140
Single-nucleotide polymorphisms (SNPs)
  ambiguous, 249
  chips, 111, 154, 158
  definition of, 6, 12–13
  diagnostic checks for, 210–211
  and GWAS, 78, 159, 164
  heritability, 24–26, 48, 111–112, 115, 178, 238–239, 299, 324
  in high LD, 224–225
  and imputation, 156
  to include in polygenic scores, 106, 108, 112, 141, 250, 252, 279
  LD pruning to remove redundant, 224–225

  low-call rate of, 207
  with a low MAF, 207
  missingness by, 200
  online browsers, 29, 79, 98
  and polygenic scores, 105
  select independent, 247–248, 267
  selecting one, 195
  and SNPedia, 79, 98
  visualization of, *8*
Smith, George Davey, 343
Smoking, 11, 14, 42, 45, 115, 122, 131–132, 136, 140–141, 147, 155, 310, 341, 354, 362–363
Social compensation model, 129, 137–139, 140
Social control or social push model, 129, 137, *138*, 140
Somatic cells, 8, 10–11, 405n1

Tropf, Felix C., 104, 142
Turkheimer, Eric, 119–120, 132–133, 137, 139
Turley, Patrick, 119, 333
23andme, 97
  ancestral diversity, 63
  data format, 180
  health tests from, 360
  importing personal data from, 192–193
  large data, 161, 211
  privacy and public genetics, 368
  selecting markers, 195
  selective release summary statistics, 288
  selling of data, 369
Twin studies, 24–25, 48, 78–79, 112–113, 206, 231, 237, 288, 339
  heritability, 24, 299, 328
  and missing heritability, 27, 111, 288

UK Biobank, 49, 79, 93, 97, 98, 109, 122, 158, 161, 163, 165, 178, 231, 309–310, 315, 336, 353, 367
Unix, 74, 184–187, 381

Variance, 34–36, 48, 67, 107, 109, 111, 119, 134, 240, 250, 344. *See also* Covariance
  dominance, 22
  environmental, 22
  genetic, 22–23, 26–27, 58, 114, 117, 377
  interaction, 22
  inverse weighting, 88–89
  partitioning or decomposition of, 22, 240, 250, 251, 262–265, 285, 288, 36
  phenotypic, 22–25, 120, 130, 133
  sources of, 22, 377
Variance-covariance matrix, 36–37, 335–336, 403
Visscher, Peter, 23, 111, 118, 250, 343, 382
Vulnerability model. *See* Diathesis-stress model

Windows, 388
Wray, Naomi, 103, 110